Innovations and Emerging Technologies in Wound Care

Innovations and Emerging Technologies in Wound Care

Edited by

Amit Gefen

ELSEVIER

ACADEMIC PRESS

An imprint of Elsevier

Academic Press is an imprint of Elsevier
125 London Wall, London EC2Y 5AS, United Kingdom
525 B Street, Suite 1650, San Diego, CA 92101, United States
50 Hampshire Street, 5th Floor, Cambridge, MA 02139, United States
The Boulevard, Langford Lane, Kidlington, Oxford OX5 1GB, United Kingdom

Library of Congress Cataloging-in-Publication Data
A catalog record for this book is available from the Library of Congress

British Library Cataloguing-in-Publication Data
A catalogue record for this book is available from the British Library

ISBN: 978-0-12-815028-3

For information on all Academic Press publications visit our website at https://www.elsevier.com/books-and-journals

Publisher: Mara Conner
Acquisition Editor: Fiona Geraghty
Editorial Project Manager: Joshua Mearns
Production Project Manager: R.Vijay Bharath
Cover Designer: Matthew Limbert

Typeset by TNQ Technologies

Contents

Contributors

Pierre Badel Mines Saint-Etienne, Univ Lyon, Univ Jean Monnet, INSERM, U 1059 Sainbiose, Centre CIS, Saint-Etienne, France

Dimitri Beeckman Ghent University, Faculty of Medicine and Health Sciences, Department of Public Health and Primary Care, University Centre for Nursing and Midwifery, Skin Integrity Research Group, Gent, Belgium

Christine Thies Berke Wound Ostomy Services, Nebraska Medicine, Omaha, NE, United States

Joyce Marie Black College of Nursing, University of Nebraska Medical Center, Omaha, NE, United States

Ulrike Blume-Peytavi Charité-Universitätsmedizin Berlin, Department of Dermatology and Allergy, Clinical Research Centre for Hair and Skin Science, Berlin, Germany

Kath M. Bogie Department of Biomedical Engineering, Case Western Reserve University, Cleveland, Ohio, United States; Louis Stokes Cleveland Veterans Affairs Medical Center (LSCVAMC), Cleveland, Ohio, United States

Christelle Bonod Laboratory of Tissue Biologie and Therapeutic Engineering, UMR 5305, CNRS, University of Lyon 1, Lyon, France

David Brienza Rehabilitation Science and Technology, University of Pittsburgh, Pittsburgh, PA, United States

M. Bucki TexiSense, Montceau-les-Mines, France

Evan Call Adjunct Faculty, Microbiology, Weber State University, Ogden, UT, United States

Fanette Chassagne Mines Saint-Etienne, Univ Lyon, Univ Jean Monnet, INSERM, U 1059 Sainbiose, Centre CIS, Saint-Etienne, France; Thuasne, Saint-Etienne, France

Adam Cheney Adjunct Faculty, Rhetoric, University of North Texas, Denton, TX, United States

Alberto Corrias Department of Biomedical Engineering, National University of Singapore, Singapore

Alex Demers Department of Chemical, Paper and Biomedical Engineering, Miami University, Oxford, OH, United States

Y. Dror Department of Molecular Microbiology and Biotechnology, Faculty of Life Sciences, Tel Aviv University, Tel Aviv, Israel

Leigh Toni Fleming Centre for Precision Technologies, University of Huddersfield, Huddersfield, United Kingdom

A. Freeman Department of Molecular Microbiology and Biotechnology, Faculty of Life Sciences, Tel Aviv University, Tel Aviv, Israel

Rinat Friedman Department of Biomedical Engineering, Faculty of Engineering, Tel Aviv University, Tel Aviv, Israel

Bérengère Fromy Laboratory of Tissue Biologie and Therapeutic Engineering, UMR 5305, CNRS, University of Lyon 1, Lyon, France

Daniel Garcia-Martinez National Centre for Advanced Tribology at Southampton (nCATS), Department of Mechanical Engineering, Faculty of Engineering and Physical Sciences, University of Southampton, Southampton, United Kingdom

Amit Gefen Department of Biomedical Engineering, Faculty of Engineering, Tel Aviv University, Tel Aviv, Israel

Alexander Golberg Porter School of Environment and Earth Sciences, Tel Aviv University, Tel Aviv, Israel

Priyanka Gupta Department of Textile Technology, IIT Delhi, Hauz Khas, Delhi, India

Sandra Guzman SHRS-Rehab Sci and Tech, University of Pittsburgh, Pittsburgh, PA, United States

Jan Kottner Charité-Universitätsmedizin Berlin, Department of Dermatology and Allergy, Clinical Research Centre for Hair and Skin Science, Berlin, Germany; Ghent University, Faculty of Medicine and Health Sciences, Department of Public Health and Primary Care, University Centre for Nursing and Midwifery, Skin Integrity Research Group, Gent, Belgium

Louis W.M. M. Krieger Department of Chemical, Paper and Biomedical Engineering, Miami University, Oxford, OH, United States

Bipin Kumar Department of Textile Technology, IIT Delhi, Hauz Khas, Delhi, India

Pak Hung Leung Centre for Precision Technologies, University of Huddersfield, Huddersfield, United Kingdom

Maria Fabiola Leyva-Mendivil National Centre for Advanced Tribology at Southampton (nCATS), Department of Mechanical Engineering, Faculty of Engineering and Physical Sciences, University of Southampton, Southampton, United Kingdom; Bioengineering Science Research Group, Department of Mechanical Engineering, Faculty of Engineering and Physical Sciences, University of Southampton, Southampton, United Kingdom

Georges Limbert National Centre for Advanced Tribology at Southampton (nCATS), Department of Mechanical Engineering, Faculty of Engineering and Physical Sciences, University of Southampton, Southampton, United Kingdom; Bioengineering Science Research Group, Department of Mechanical Engineering, Faculty of Engineering and Physical Sciences, University of Southampton, Southampton, United Kingdom; Biomechanics and Mechanobiology Laboratory, Biomedical Engineering Division, Department of Human Biology, Faculty of Health Sciences, University of Cape Town, South Africa

Jérôme Molimard Mines Saint-Etienne, Univ Lyon, Univ Jean Monnet, INSERM, U 1059 Sainbiose, Centre CIS, Saint-Etienne, France

Hairshkumar Narayana Department of Materials Engineering, University of British Columbia, Vancouver, British Columbia, Canada

S. Niroomandi Univ. Grenoble Alpes, CNRS, Grenoble INP, TIMC-IMAG, Grenoble, France

C. Ophir Department of Molecular Microbiology and Biotechnology, Faculty of Life Sciences, Tel Aviv University, Tel Aviv, Israel

Ning Pan Biological & Agricultural Engineering, UC Davis, Davis, CA, United States

Yohan Payan Laboratoire TIMC-IMAG, CNRS & University Grenoble Alpes, France

A. Perrier Univ. Grenoble Alpes, CNRS, Grenoble INP, TIMC-IMAG, Grenoble, France; TexiSense, Montceau-les-Mines, France

Noelle Remoué Laboratory of Tissue Biologie and Therapeutic Engineering, UMR 5305, CNRS, University of Lyon 1, Lyon, France

Tom Ridge Department of Chemical, Paper and Biomedical Engineering, Miami University, Oxford, OH, United States

Ana Scalamandré Department of Biomedical Engineering, Case Western Reserve University, Cleveland, Ohio, United States

Noga Shabshin HaEmek Medical Center, Israel; Hospital of the University of Pennsylvania, Penn Medicine, United States

Pramod Shankr Department of Textile Technology, IIT Delhi, Hauz Khas, Delhi, India

Dominique Sigaudo-Roussel Laboratory of Tissue Biologie and Therapeutic Engineering, UMR 5305, CNRS, University of Lyon 1, Lyon, France

Bryan Wee Siang Soh Department of Biomedical Engineering, National University of Singapore, Singapore

Jessica L. Sparks Department of Chemical, Paper and Biomedical Engineering, Miami University, Oxford, OH, United States

Lisa Tucker-Kellogg Centre for Computational Biology, Duke-NUS Medical School, Singapore; Cancer and Stem Cell Biology, Duke-NUS Medical School, Singapore

F.J. Vermolen Delft Institute of Applied Mathematics, Delft University of Technology, Delft, The Netherlands

Karthik Vishwanath Department of Physics, Miami University, Oxford, OH, United States

Annika Vogt Charité-Universitätsmedizin Berlin, Department of Dermatology and Allergy, Clinical Research Centre for Hair and Skin Science, Berlin, Germany

Daphne Weihs Faculty of Biomedical Engineering, Technion-Israel Institute of Technology, Haifa, Israel

Foreword

Chronic wounds and burns that do not normally progress to healing for periods exceeding several weeks are one of the most important global medical concerns, currently affecting many millions of patients and accumulating billions in annual health-care costs, in each country. Pressure ulcers (now often termed pressure injuries) and diabetic foot ulcers account for the majority of these chronic wounds and occur when soft tissues are excessively deformed and distorted, either in static or cyclic loading, and typically between bony prominences that transfer bodyweight forces and a supporting surface (e.g., a mattress, seat, or shoe insole). Chronic wounds are caused by a combination and interaction of etiological factors, but impaired sensation such as due to a central nervous system injury or disease, or peripheral neuropathy, is almost always involved. Similarly, a common injury risk is the inability to respond to discomfort or pain by changing body postures or otherwise responding to the applied tissue distortions (e.g., as in neuromuscular diseases where there is not sufficient muscle power). A difficulty or inability to communicate the discomfort or pain information to a person who provides care (e.g., concerning pediatric, psychiatric, or dementic patients who are not verbal) is a similar injury risk factor. The insensate symptoms can be caused by an acute injury or a chronic illness or by anesthesia or analgesia, e.g., in operation rooms and in childbirth. Accordingly, the risk for chronic wounds can accompany a person over a lifetime or be temporary, e.g., during surgery or recovery. Pressure ulcers/injuries can also be caused by medical devices or other objects that contact the skin and purposely or unintentionally apply sustained (localized) tissue deformations and mechanical stresses in tissues. Other wounds that may become chronic, difficult, and costly to treat are burns, surgical wounds, large traumatic wounds, and wounds resulting from removal of large tumors. All wound types bear the risk of infections, including by antibiotic-resistant bacteria as well as by fungi, and some may deteriorate to organ system failure, renal failure, osteomyelitis, and sepsis, which are all life-endangering conditions. The variety of patient ages, populations, medical conditions, and health-care facilities and settings including home care to which chronic wounds are relevant is enormous. In that aspect, the position of chronic wounds is unique in medicine. The fact that chronic wounds impact such a variety of patient groups partially explains why these wounds are already one of the greatest challenges to global health and will become more so with the aging of populations and epidemic spread of diabetes.

This book provides the most complete and up-to-date, and, in several cases, the only quantitative explanation of the mechanobiology and biomechanics, which causes the chronicity of wounds, including the contributions of extrinsic factors to the halted healing. The book further covers all the dimensional scales of the phenomenon, starting from whole-body pathophysiology and biomechanics to the study of tissues and their interactions internally in the body and with their environment. The book addresses tissue changes that occur in aging and chronicity that increase the fragility of skin and subdermal tissues and their risk of damage, which eventually evolves to chronic wounds. At the microscopic levels, the book describes experiments and modeling of cell-scale phenomena that occur during injury and healing, including the mechanical function of living cells and how they are recruited and collectively migrate to repair tissue damage. The book spans from basic science perspectives, as in the aforementioned examples, to practical aspects such as laboratory testing methods and computational (finite element) modeling for evaluating the efficacy of preventative medical devices and interventions. Accordingly, this book provides a complete picture of the present state of knowledge of the tolerance of human cells, tissues, and organs to sustained mechanical and thermal loads, as well as compiled understanding of the risk for formation and development of chronic wounds.

This cross-disciplinary book bridges the well-recognized gaps between clinicians and biomedical researchers or bioengineers, between basic scientists and practitioners, and between academia and industry—all of which are known to hinder progress in the field. The book is valuable for all professionals involved in the prevention, diagnosis, and treatment of chronic wounds and burns, concerning both clinical work and research and development, in academia, medical settings, and industry. The book contains contributed chapters of the leading experts on the subject matter and from all around the world. It is the best synthesis of wound research knowledge with regards to the tissue damage cascades and healing processes relevant to chronic wounds. The book further describes the state-of-the-art technology in wound prevention, diagnosis, prognosis, and treatment. In addition, it depicts the current and emerging research directions and future technology trends in the field of wound prevention and care, at the early stages of research and development in academia and industry. The book will hence be pivotal as a source of scientific knowledge for all professionals and graduate students focusing on the diagnosis, prevention, early detection, management, treatment, and care of chronic and nonhealing wounds.

Professor Amit Gefen

The Editor

CHAPTER 1

Mathematical models of healing of burns

F.J. Vermolen

Delft Institute of Applied Mathematics, Delft University of Technology, Delft, The Netherlands

Chapter Outline

1. Introduction

Serious burn injuries and other deep tissue injuries often accompany with problematic complications on the patient's skin. These complications can be the formation of hypertrophic scar tissue, which worsens the appearance of the skin, and can leave itchy sometimes painful sensations. Another major complication is the development of a severe contraction that in the worse cases leads to disability of the patient. If this is the case, then the contraction is usually referred to as a *contracture*. Combinations of development of excessive scar tissue and contractures are very frequent. Contractures are often treated by splinting, applying flaps, or skin grafting (skin transplantation) (see Refs. [1,2] and references therein).

To optimize therapy that is applied to treat burn injuries, a thorough understanding of the physiological processes that are occurring in the skin after trauma is crucially important. Developing such understanding requires experimental studies, where these studies contain patient (clinical) data, as well as animal experiments and in vitro experimental studies. Clinical data involve patient data, which are very useful, but highly uncontrollable because each patient has his or her own genetic pattern, gender, pigmentation, lifestyle, and age.

Innovations and Emerging Technologies in Wound Care. https://doi.org/10.1016/B978-0-12-815028-3.00001-8

Another complicating factor arises from the fact that burn traumas are often the result of accidents, which makes the wound architecture more unpredictable. In vitro studies, although very useful and very well controllable, focus only on few aspects, and it is questionable whether the observations from these in vitro experiments are representative for the wound behavior in human skin. All these experimental observations should be seen as models for the evolution of posttrauma human skin. Animal-based studies for therapy are receiving less and less support because of ethical concerns, and it is not sure whether experimental results observed on animals are transferrable to humans.

A recent development is the construction of in silico (computer-based) models for (human) skin evolution after trauma. Although in silico models cannot entirely replace experimental studies, we do believe that these models can be used to reduce the number of (animal-based) experiments and that these models can contribute to the understanding of various phenomena in biological medicine. Furthermore, these models can be of great value for the prediction of different scenarios that patients may be faced with. The computer models can be used to retrieve, quantify, and analyze scenarios that have not been measured. The collaboration between sciences such as mathematics and physics has existed for centuries. This differs largely from the collaboration between biologists, physicians, and mathematics. The last-mentioned collaboration only exists during the latest past decades. A complicating factor that arises in biological sciences is the large amount of uncertainty. The success rate of a therapy often varies from patient to patient. Possible reasons for the variability of the success rate are not only gender, lifestyle, and age but also genetic constitution or the presence of (unknown) other diseases. All these aspects render many biological processes in the human body even more unpredictable. These uncertainties imply that the various input parameters that are needed for the computer simulations contain uncertainties as well. Appropriate mathematical formalisms will have to deal with this degree of uncertainty.

This chapter reports on a review about various modeling studies and approaches in the context of skin evolution after serious burn trauma. The work that has been done in our group will be paid most attention. The models that have been studied will entail *agent-based* modeling, in which each cell is treated individually, as well as *continuum-scale* models. Some new results from *cellular automata* models will be presented as well. This modeling type is applied to wound closure, but it can be extended very easily to other phenomena. We will end up with some conclusions and discussion regarding modeling and the current models.

2. Biological mechanisms

In the introduction, it has been mentioned that various mathematical modeling approaches exist for the simulation of posttrauma skin evolution. Before the description of the various mathematical methodologies, we first review the sequence of the most important biological processes that occur during skin evolution after serious trauma, such as wounding or burn injury. A more detailed description of the physiology of wound healing processes, such as contraction, angiogenesis, wound closure, and chains of other biological subprocesses, can be found in Ref. [3].

Wound healing is a very complicated process, in which many biological processes interact, and therefore we can only briefly summarize the sequence of processes. In very superficial wound, only the epidermis (upper layer of skin) is damaged and repair amounts to restoring the epidermis. The epidermis consists of several layers, but most importantly keratinocytes constitute this layer and this layer is restored by a combination of keratinocyte migration and proliferation. To this extent, this repair mechanism is relatively straightforward compared to the restoration of deeper wounds. If it comes to deep tissue injury, then the dermis is damaged. The dermis consists of tiny blood vessels (capillaries), immune cells (macrophages), collagen (extracellular matrix), and fibroblasts. The first step in wounds that have been caused by cuts or incisions is the flow and loss of blood from the capillaries. The blood cells, platelets, make the blood coagulate and this (temporarily) closes the wound such that further loss of blood and ingress of contaminants and pathogens is prevented. This phase is referred to as *hemostasis*. The formed connective tissue (fibrin), which is there as a temporary repair barrier, is subsequently replaced with collagen. This process is sequenced as follows: the platelets, which were responsible for the coagulation of blood, secrete several chemical agents that attract the immune cells. The immune cells phagocytize debris, pathogens, and damaged tissue. This phase is referred to as the *inflammatory phase*. Furthermore, the immune cells secrete growth factors, such as TG-beta, to attract the fibroblasts. The fibroblasts, in turn, migrate toward the wound site and replace the fibrin network with collagen. This is done via a sequence of disassembling the fibrin by the use of matrix metallic proteases and the regeneration of collagen. This phase is referred to as the *proliferative phase*. The aforementioned phases typically take place within weeks, depending on the size, geometry, and location of the injury. The final phase restores the oriented (nonisotropic) collagen to randomly oriented (isotropic) collagen. This slow process takes several months or even years. This process is referred to as the *remodeling phase*.

In normal wound healing, although the fibroblasts secrete collagen as long polymeric molecules that are oriented according to their migration pathway, the patient does not encounter any complications that limit his or her ability to move any joints. However, in the case of severe burn trauma over large surfaces of the patient's skin and in a large depth into the skin, several complications can occur. The most important complications are excessive scar formation and excessive contraction. We will describe the fundamental biological mechanisms behind these complications. Owing to the presence of TG-beta and local dilatations as result of fibroblastic forces exerted on the environment just around the circumference of the wound, some of the fibroblasts in the wound region differentiate to myofibroblasts. The myofibroblasts secrete more collagen than the fibroblasts do, as well as they exert larger contractile forces to their immediate environment. This makes the tissue in, and around, the wound region contract and the wound edge is displaced toward the wound center.

If this contraction happens in such an extent that the patient faces a reduction of joint mobility as its result, then the contraction is called a *contracture*. Furthermore, the excess on production of collagen is responsible for the formation of excessive (hypertrophic) scar tissue. Our research aims at finding the likelihood whether contraction takes place in such a large extent that the patient faces loss of joint mobility. This is crucially important in predicting whether the patient will face disability or any other mobility limitation.

3. Methodologies

This manuscript focuses on the modeling of the proliferative phase and a little on the inflammatory phase. To model the migration of the cells, as well as the proliferation, cell death, and differentiation, several modeling approaches exist to simulate at various scales. Small-scale (micro) models incorporate individual cells as individual entities that are moving around in the wound region. Each cell is modeled in interaction with the other cells and its immediate surroundings. These small-scale models are commonly referred to as agent-based or particle methods or cell-based models. On the other hand, large-scale (macro) models do not consider individual cells but merely incorporate cell densities. Cell densities can be interpreted as a number of cells per unit of volume (or area in 2D cases). These macro-scale models are characterized by partial differential equations, which model mass balances. Cellular automata models can be considered as a modeling framework in both the macro- and the micro scales. We will consider these three modeling frameworks.

3.1 Agent-based models

This model class explicitly deals with the migration of each cell, as well as its differentiation, proliferation, apoptosis (programmed cell death), and possible mutation. The reader who is interested in mathematical implementation and in various applications of agent-based models is referred to Refs. [4–14]. The construction of our agent-based models was much inspired on the experimental study in Ref. [15]. We will explain the common procedures that are followed in setting up these models. Initially the cells are positioned over a computational domain. The cells migrate as a result of various cues, such as haptotaxis, chemotaxis, or any form of taxis. Haptotaxis and chemotaxis, respectively, reflect the migration of cells as a result of the gradient of the concentration of a chemical through an extracellular matrix or through a liquid. Other migratory modes are random walk (diffusion), durotaxis, tensotaxis, and hard impingement through physical contact as a result of cell collision or drag as a result of local displacement of the extracellular matrix. Random walk is incorporated by the use of a vector Wiener process for the displacement vector, in which all the components of this vector receive an independent (uncorrelated) stochastic contribution as samples from the normal distribution with zero mean and the product of the time step and the macroscopic diffusivity as the variance, see for instance Ref. [16]. In general, the location $\mathbf{X}_j$ of a cell j is described by the following stochastic differential equation

$$d\mathbf{X}_i(t) = \mathbf{v}(t, \mathbf{X}_1, \ldots, \mathbf{X}_N)dt + \sigma(t, \mathbf{X}_i)d\mathbf{W}(t), \tag{1.1}$$

where $\mathbf{v}$ represents the deterministic component of the migratory velocity, and σ represents the standard deviation of cell migration per unit of time as a result of random walk. Finally, $\mathbf{W}$ represents the vector Wiener process. All these migratory modes are summed to obtain the total displacement vector of each cell over a time step. This displacement vector is added to the current location of the cell, so that the location of the cell at the next discrete time level follows. In the case of chemotactic or haptotactic modeling, the concentration gradient needs to be evaluated on the position of the cell. The concentration gradient can be evaluated from analytic expressions (by the use of Green's functions, for instance, see Ref. [17]) or from finite element solutions for the concentrations. In the latter case, one needs a postprocessing step from the finite-element solution by the use of an approximation of the derivatives of the concentration. Furthermore, one needs to map the gradient field onto the current position of the cell, which does not always coincide with a finite element mesh point. The numerical procedures have been described in Ref. [18]. Next to cell migration, one needs to determine whether or not the cell changes state, such as by differentiation, proliferation, or cell death. As the initial state of all cells is not known apriori, one commonly models this transition by random processes. From a probability point of view, it is realistic to use an exponential distribution to sample from in order to determine whether the cell changes state.

The use of the exponential distribution is also common in modeling the occurrence of subsequent telephone calls at a reception as a random (stochastic) process. Such a process is commonly referred to as a Poisson process. The same is done for the simulation of the appearance of cells on the rim of the computational domain, where commonly a Poisson process is used to model the number of cells that appear there. One can use for instance a uniform distribution to model the position of the newly appearing cell on the rim of the domain. Next to the modeling of cell migration and state of change, it is of crucial importance to model contractive behavior. Therefore, the forces that are exerted by the cells (predominantly the myofibroblasts and to a lesser extent fibroblasts) are incorporated through a mechanical balance over the extracellular matrix. In our studies, we have used the immersed boundary approach, in which the boundary of the cell has been simplified into a low-degree polygon (such as a square or triangle) on which we assign delta forces on each midpoint of the cell boundary segment. Combined with the mechanical balance in the extracellular matrix, this gives a displacement field at each nodal point of the finite element mesh. After mapping from the finite element mesh onto the cellular positions, the mechanical displacement over each time step is also added to the total displacement vector of each cell over each time step. In the current simulations, we only experimented with standard Hooke's Law combined with viscoelasticity, which linearly relates the local strain tensor to the local stress tensors. Furthermore, we have assumed isotropy of the extracellular matrix so far. These simulations enable the computation of the strain energy density over the domain, which might be useful in the quantification of the amount of pain that the patient experiences. Furthermore, the boundary of the wound displaces as a result of the forces that the cells exert on their immediate environment, which quantifies the reduction of wound area due to the cellular forces. This decrease of area is a measure of the contraction. If the reduction is large, then the contraction is large. As the myofibroblasts pull firmly on their environment and because these cells tend to shorten the polymeric chains in the collagen matrix, we have incorporated a permanent stress state by means of a phenomenological modeling technique in which a permanent force is incorporated on each element (that is a linear triangle in our case) face (boundary segment). This force grows with the presence of myofibroblasts. This models the permanent contraction of the wound. More details regarding the modeling features can be found in Refs. [4,14]. In Ref. [4], the model incorporates the attraction of white blood cells toward the wound site as a result of the secretion of chemokines by the platelets. Subsequently, the white blood cells release the TG-beta growth factor that attracts the fibroblasts. The pulling forces by the fibroblasts, as well as the plastic forces that contract the wound tissue have been incorporated. The model is able to predict the evolution of the contraction of the wound. The model also simulates the evolution of the isotropy of the collagen and fibrin network and the replacement of the fibrin network with collagen. Of course, one should bear in mind that the model is applicable to very small regions and that in principle, it is able to model only part of a larger wound or to model very small wounds, for which there is actually no wound contraction. An important characteristic of such agent-based formalisms is the presence of random processes. These stochastic components make that each simulation should be considered as a sample from an unknown statistical distribution. Therefore, the results should always be evaluated by the use of statistical methods, which enable the prediction of the likelihood that certain processes, such as excessive contractures, take place. This statistical nature of the results will be described later in more detail. We should also bear in mind that a large number of cells will make the numerical framework expensive in the sense of computation time because of all the cell–cell interactions. In this chapter, we consider a case study in which fibroblasts and myofibroblasts are considered. Fig. 1.1 shows the wound area as a function of time. It can be seen that initially the wound area increases as a result of the pulling

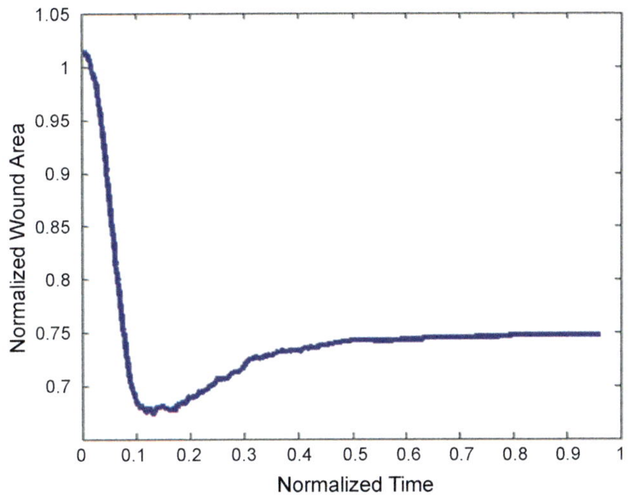

Figure 1.1
One sample run of the normalized wound area as a function of time.

forces of fibroblasts that are still around the wound site. Subsequently, the wound contracts as a result of fibroblasts and myofibroblasts that are entering the wound site. Next to exerting pulling forces, the myofibroblasts shorten the polymeric chains by which permanent displacements and deformations result. It can be seen that in the course of time, the wound is retracting toward the initial state. However, the initial state is not reached because of the permanent displacements resulting from the chain-bond shortening mechanism by the myofibroblasts. An example of a run has been given in Fig. 1.1. Clinical experiments, for instance Ref. [19], confirm this behavior.

As we have done sampling from the statistical distributions, we computed the cumulative probability distribution. The resulting histogram and cumulative probability distribution are shown in Fig. 1.2. It can be seen that about 20% of the wounds develop a final contraction to 60% of the initial wound area. These contractions are classified as serious contractions. Furthermore, in Fig. 1.3, we show the following scatter plots: the final wound area versus the minimal wound area (with sample correlation of about 0.57), the wound area after five time steps versus the final wound area (with sample correlation of about 0.01), the final wound area after 10 time steps versus the final wound area (with sample correlation of about 0.18), the final wound area versus the plastic forces (with sample correlation of about −0.66), the final wound area versus the temporary fibroblastic forces (with sample correlation of about −0.05), and the minimal wound area versus the temporary fibroblastic forces (with sample correlation of about −0.24). As for the practitioners it is crucially important to design their therapies based on the observations on their patients, we have quantified these correlations. For the doctors, it is important to

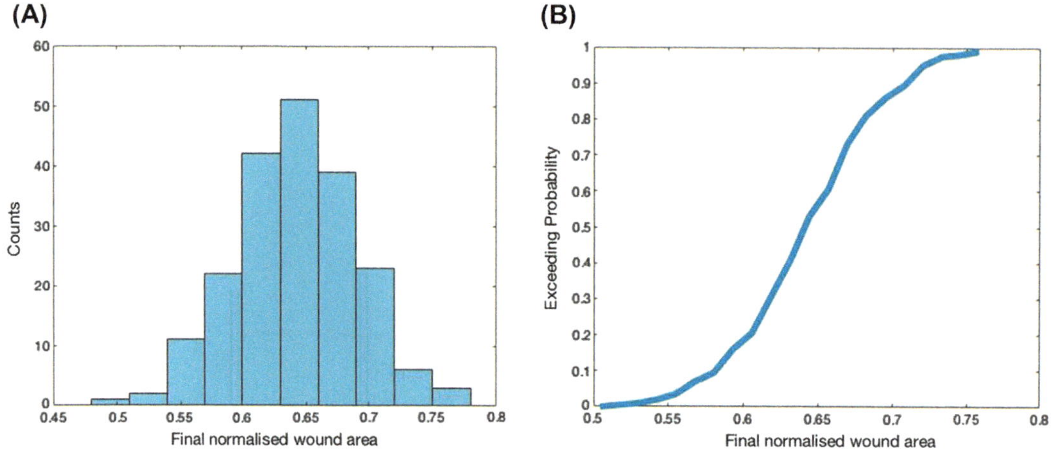

Figure 1.2
(A) Histogram and (B) cumulative probability distribution based on 200 runs with the agent-based model.

know in advance whether the wound will develop a dramatic contraction. From the results in Fig. 1.3, it can be seen that the final wound area correlates slightly positively with the wound area after 10 time steps (which is near the time at which the minimal area is reached), whereas the correlation with the wound area after 5 time steps (area at the early stages) is not significant. The correlation between the minimal area and the final wound area is significantly positive. Furthermore, the correlation between the final wound area and the plastic force is significantly negative, whereas the correlation between the temporary cell forces and the final normalized wound area is not significant. The correlation between the minimal normalized wound area and the temporary cell forces is moderately negative. From these simulations, it follows that the model predicts that the final contraction becomes more severe with a lower minimal wound area and with a lower wound area in the intermediate stages, whereas the change of area in the early stages hardly impacts the final contraction, which implies that the earliest stages of skin evolution is unfortunately not useful to the doctors for the sake of prediction of the permanent contraction. Furthermore, the plastic (morphoelastic) forces have a large, positive impact on the severity of the contraction. Temporary forces do positively impact the severity of the minimal wound area, but hardly impact the final contraction.

The scatterplots of the elasticity of the skin versus the final wound area and minimal wound area are shown in Fig. 1.4. It can be seen that the stiffness, in the form of the Young's modulus, significantly impacts the minimal wound area and final wound area with correlation coefficients of about 0.57 and 0.63, respectively. Lower stiffness gives a larger contraction, but the ultimate strain energy, which is proportional to the Young's modulus, probably is not influenced that much.

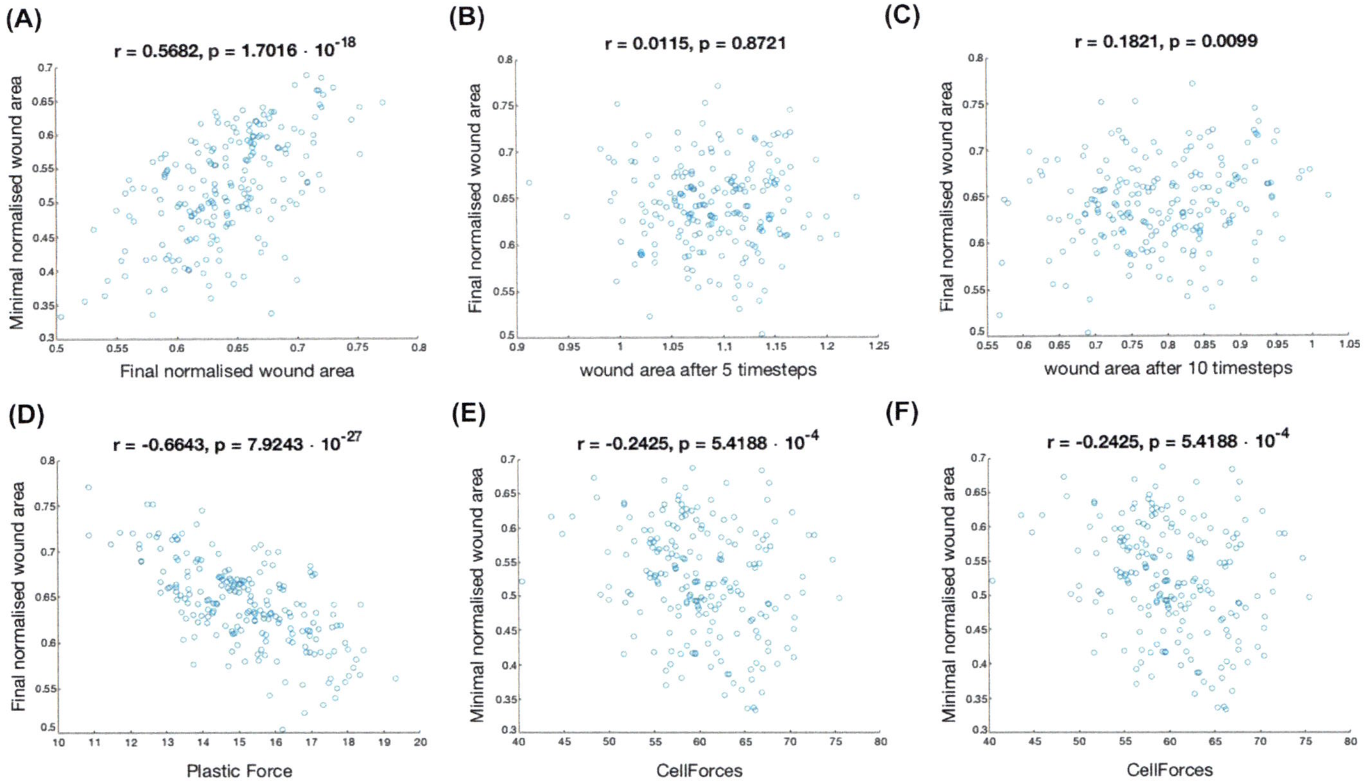

Figure 1.3

Scatterplots and correlations for the (A) final wound area versus the minimal wound area (with sample correlation of about 0.57), (B) the wound area after 5 time steps versus the final wound area (with sample correlation of about 0.01), (C) the final wound area after 10 time steps versus the final wound area (with sample correlation of about 0.18), (D) the final wound area versus the plastic forces (with sample correlation of about −0.66), (E) the final wound area versus the temporary fibroblastic forces (with sample correlation of about −0.05), (F) the minimal wound area versus the temporary fibroblastic forces (with sample correlation of about −0.24).

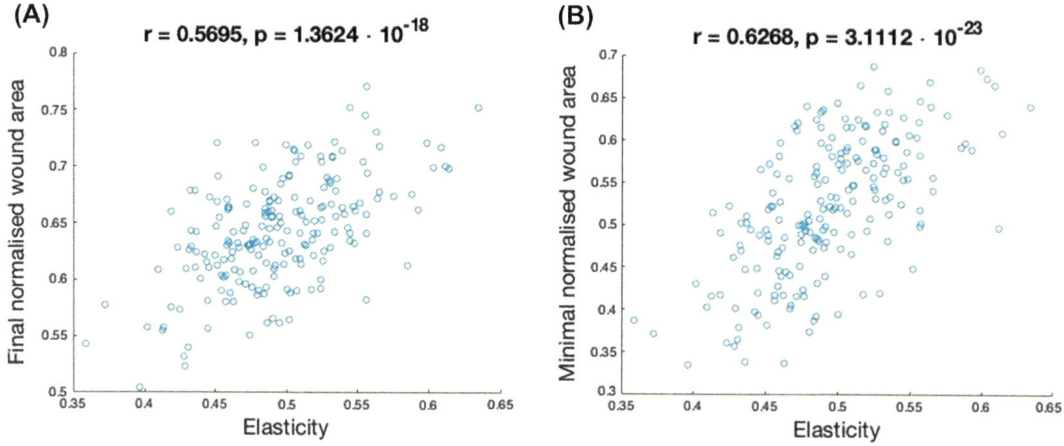

Figure 1.4
Scatterplots and correlations for the (A) final wound area and (B) minimal wound area versus the nondimensional elasticity of skin.

3.2 Continuum-based models

Continuum-based models are based on partial differential equations and are suitable for larger scale simulations. This modeling class is applicable on tissue scale. The models consist of partial differential equations that are based on conservation principles, such as mass conservation in the case of chemokines, and cell densities. In general, the partial differential equations for the various cell densities and chemokine concentrations read as

$$\frac{\partial u}{\partial t} + \nabla \cdot \mathbf{J}(t, \mathbf{x}, u) = F(t, \mathbf{x}, u), \tag{1.2}$$

where u denotes the unknown, being a cell density or a concentration, $\mathbf{J}$ denotes the flux, and F denotes the reaction term. Phenomena such as random walk and chemotaxis are modeled by diffusion terms and convective terms, in which the convective velocity is proportional to the gradient of the concentration that triggers chemotaxis of the cell type involved. These terms are incorporated into the flux expression $\mathbf{J}$. Cell division, differentiation, and cell death are modeled as "reaction-like" terms in the partial differential equations and hence incorporated in the F-term. Hence, contrary to the cell-based (agent-based) class of models, these models incorporate all the aforementioned biological processes by deterministic models, and hence the outcomes from the simulations from this modeling class should be interpreted as the average behavior. Some examples of continuum-based modeling studies in the context of medical biology are given in Refs. [20–29].

Regarding the mechanics, we have extended this modeling class with morphoelasticity, which entails three mechanical states of the domain of computation, which we explain now.

The first state is the initial domain configuration, which is referred to as the *initial state*, before any contraction sets in. Of course, it should be realized that this state is artificial in the sense that the computational domain is always mechanically active with fibroblasts and other cell types exerting their forces on the tissue. As a result of the forces that are exerted by the fibroblasts, the computational domain deforms, which results into the *current state* at the time that is considered. In classical Hooke's law with the infinitesimal strain approximation, the computational domain would instantaneously be set into the initial state of the domain if the forces are removed. In the morphoelastic formulation, this assumption is relaxed and the domain gradually "deforms" back toward the initial state, but it will not reach the actual initial state. The state that will be reached represents a new equilibrium. This new equilibrium is referred to as the *current equilibrium state*. In the mathematical model that has been developed for the evolution of the wound region, these three states have been taken into account and the current equilibrium state is the plastically deformed state. Mathematically, this model results into a balance of momentum, in which the inertia has been incorporated, as well as viscoelasticity. We believe that from a biological point of view, this modeling framework is very accurate. Although from a biological point of view, it may not be necessary to involve all these terms (inertia and the viscous part in viscoelasticity). For numerical reasons, it is more attractive to incorporate these physical complications despite its counterintuitive unnecessary complications. The reason is that otherwise the system would exhibit a very hyperbolic nature from a mathematical point, which may give conflicts as a result of the boundary conditions and give complications regarding monotonicity of the solution. The addition of the viscous term gives an elliptic term, which renders the problem parabolic, which is easier to solve and copes better with the boundary conditions. This state mimics the *permanent contraction* of the wound.
The simulations in Ref. [25] incorporate the migration of fibroblasts toward the wound site, as well as the differentiation of fibroblasts toward myofibroblasts. The collagen density is modeled by a reaction equation. The fibroblast and myofibroblast densities are modeled by reaction transport equations that contain reactive terms from the proliferation of fibroblasts, differentiation from fibroblasts to myofibroblasts, and death of cells. All these terms are under the interference with the concentration of TG-beta. Further terms in the partial differential equations for the (myo)fibroblasts are transport terms by the use of diffusion and chemotaxis. Next to these entities, the concentration of a generic growth factor that acts like TG-beta has been incorporated. All partial differential equations have been enriched with a term that models *passive convection* because of the displacement of the finite element mesh as a result of the mechanical forces. Furthermore, the morphoelastic formulation has been implemented. For more details, we refer to Ref. [25].

As a very simple example of a continuum-based model, we consider wound closure, which is modeled by the Fisher–Kolmogorov equation

$$\frac{\partial u}{\partial t} - D \operatorname{div} \operatorname{grad} u = \alpha u(1-u), \tag{1.3}$$

with appropriate boundary and initial condition. Here u denotes the cell density, which can be interpreted as the number of cells per unit area (volume). The results from the Fisher–Kolmogorov equation are shown in Fig. 1.5. This very simple partial differential equation models the cell density, u, over time and space. The *div grad* term represents diffusion, which corresponds to random walk. The right-hand side represents cell proliferation up to an equilibrium $u = 1$, which can be considered as a normalized cell saturation. The Fisher–Kolmogorov equation admits self-similar solution in terms of $u(t,x) = u(x - ct)$ in the one-dimensional case. Furthermore, it can be shown that the equilibrium $u = 1$ is stable, see Ref. [30].

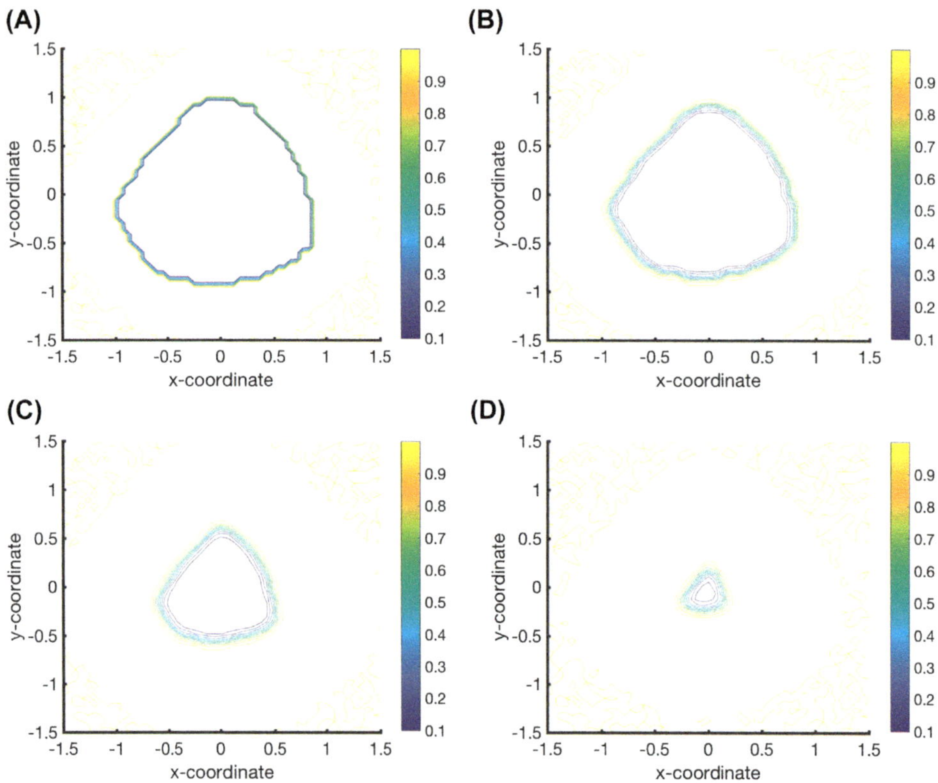

Figure 1.5
Several snapshots at consecutive times during wound closure using the Fisher–Kolmogorov model, which models a cell density. Dimensionless times are at $t = 0$ (A), 10 (B), 30 (C), and 50 (D).

3.3 Cellular automata models

The last class of models that is considered in this manuscript is the class of cellular automata models, where in particular we consider a spatial Markov chain model. The main principles are the following: (1) the computational domain is divided into mesh points, which are referred to as lattice points; (2) each lattice point can be a number of discrete states, which mimic for instance that the point falls within the wounded region or in the undamaged (healed) region. Multiple states are also possible. (3) The transition from one state to another state is determined by some (phenomenological) law. This law can be based on (virtual) energy, which applies to cellular Potts model [31–33], or just by chance [34]. Vermolen and Pölönen [34] have used this modeling framework to simulate the progression of skin cancer under the influence of long burden exposure to UV radiation. We note that these models can also be used to simulate microscopic processes such as cell migration or deformation.

In Ref. [34], the state from healthy tissue to cancer is irreversible. The model is based on the following principle. First, at each mesh point that is not in cancer state, the stochastic parameter ξ is sampled from the standard uniform distribution, which is $\xi \sim U(0,1)$. Subsequently, the likelihood that the mesh point will change state from healthy state (H) to cancer state (C) is computed by inserting the time step Δt into the exponential distribution

$$P(H \rightarrow C) = \int_{t}^{t+\Delta t} \lambda \exp(-\lambda(s-t))ds. \tag{1.4}$$

Here λ represents the effective transition probability rate that the current node changes its state from the healthy state to the cancer state. Change of state on a mesh point occurs if

$$\begin{cases} \text{If } \xi < P(H \rightarrow C), \text{ then change state;} \\ \text{else stay in healthy state.} \end{cases} \tag{1.5}$$

This principle is adjusted to wound healing by reversing the wounded state to the healthy state and by changing the cancer state to the wounded state. Furthermore, it is proved rigorously there, using probabilistic arguments, that the effective transition probability rate, λ, in the exponential distribution is linearly proportional to the number of neighbors that are already in cancer state. Furthermore, it has been proved that the probability rate is inversely proportional to the distance between adjacent lattice points. These principles can be adopted for wound closure simulations. From a biological point of view, the transition results from the combination of cell proliferation and migration of cells to their neighborhood.

A major merit of cellular automata modeling is its simplicity regarding its implementation. Furthermore, the model allows to be programmed to be very efficient in its computational time (for instance, using parallel computing techniques such as on a Graphical Processing Unit (GPU) or any parallel computing environment). Another advantage is that the model contains a small parameter space, which amounts to having just few input parameters, which may be hard to measure or to determine by alternative procedures such as inverse modeling (regression). We could easily obtain a good fit between experimental studies and our cellular automata model in Ref. [34]. A small parameter space allows a straightforward quantification of uncertainty. We display some wound closing results in Fig. 1.6, where we show several snapshots during the healing of a superficial wound. It can be seen that the shape evolution of the wound can be very complex because of random processes taking place with respect to migration and cell division. Next to this, we also graphically illustrate a possible consistency between the cellular automata modeling

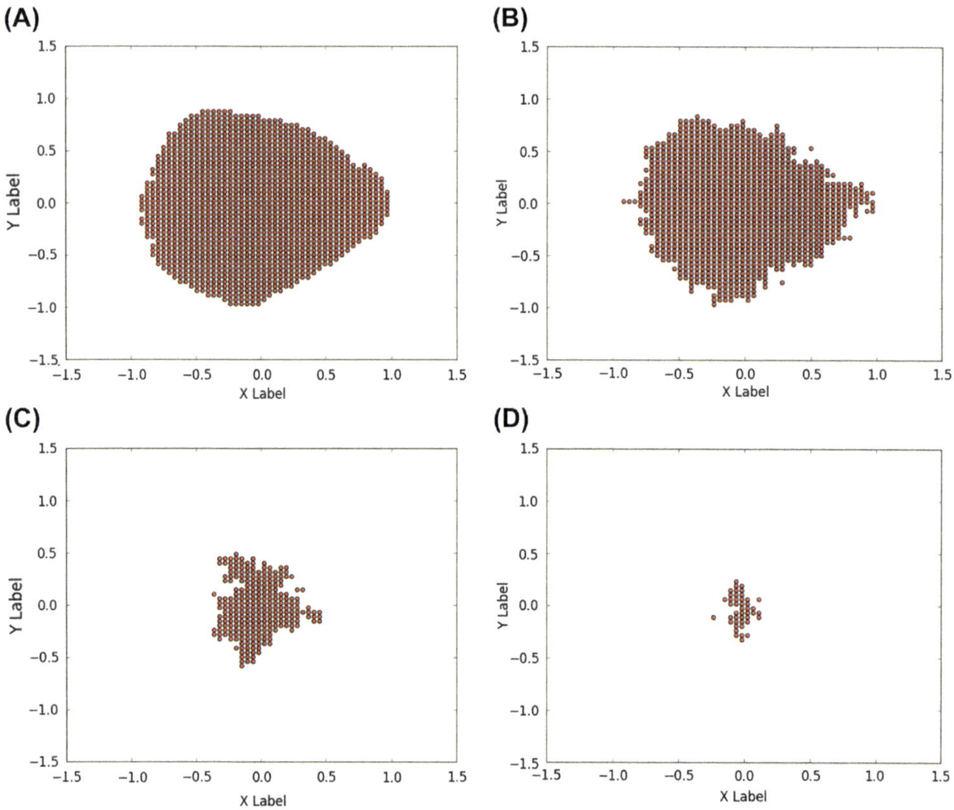

Figure 1.6
Several snapshots at consecutive times during wound closure using a spatial Markov chain cellular automata model. The red area represents the wound area. Dimensionless times at $t = 0$ (A), 20 (B), 80 (C), and 110 (D).

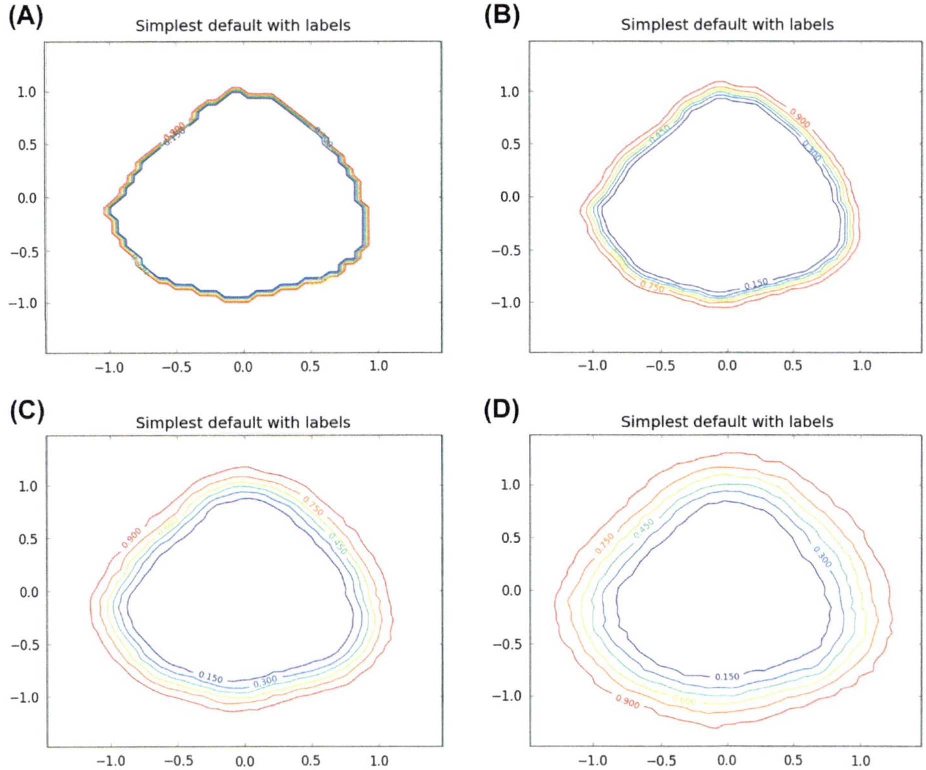

Figure 1.7
Several snapshots at consecutive times during wound closure using the averaging of 1000 runs of the spatial Markov chain cellular automata model. Dimensionless times at $t = 0$ (A), 10 (B), 40 (C), and 100 (D).

framework and a simple partial differential equations framework based on the Fisher–Kolmogorov partial differential equation. To illustrate the consistency between the Fisher–Kolmogorov model and the Markov chain cellular automata model, 1000 runs of the cellular automata model have been carried out. Subsequently, for each lattice point, the state has been averages over all runs at all times. The results are shown in Fig. 1.7.

Although the results are not exactly the same in Figs. 1.5 and 1.7, it can be seen that the results are comparable from a qualitative point of view. To make the models comparable from a quantitative point of view, more parameter fitting needs to be done, which is beyond the scope of the current paper and which is meant to be illustrational. We hereby think that it is possible that the averaged cellular automata model results can be mapped onto the Fisher–Kolmogorov equation, which implies that it is very plausible that the Fisher–Kolmogorov equation merely describes the evolution of the probability as a function of time and space. We want to investigate this matter in a more rigorous mathematical setting.

4. Discussion

Of course a model is the modeler's representation of a biological phenomenon such as the posttrauma contraction of skin tissue. Some mathematical models come in as complicated formalisms, in which much of the underlying biology and physics is incorporated. Another class of mathematical models is simple, in the sense that such a model can be used to reproduce the most important features of a phenomenon, and therefore such models are referred to as *phenomenological* models. A major advantage of such phenomenological models is their low parameter space, which amounts to having relatively few input parameters. If such models have been calibrated to a series of clinical case studies, then the predictive value of such models is high, and then these models can be used to predict the likelihood whether a patient with major burn trauma will develop any complications such as contractures or hypertrophic scar tissue. On the other hand, if one wants to gain more insight into the fundamental mechanisms behind the biological process of wound contracture, then one has to involve a high degree of physics into the modeling. In most cases, this amounts to having a large parameter space, which entails a large number of input parameters. Another drawback of a complicated model is the uncertainty of the input parameters. Many input parameters have never been measured or their approximate values are known under circumstances that differ from the circumstances in the study of interest. Furthermore, if a model contains very many input parameters, then it is often also difficult to calibrate the model to experimental results.

Hence, each simulation should be considered as a sample from a statistical distribution in the same way as measurements in experimental studies. Therefore, the outcomes of a simulation as such are not of particular interest; however, a series of outcomes can be used to determine the likelihood whether complications such as hypertrophic scar formation or wound contracture will take place. This type of computational framework requires thorough sampling, where each sample takes a full model run by the use of finite element methods for instance. Furthermore, the impact of therapy can be quantified by adjusting input parameters or boundary conditions. An example is the use of splinting after skin grafting, which amounts to changing and adjusting the boundary conditions to the equations for the mechanical balance.

Such a sampling framework amounts to Monte Carlo simulations in which one takes very many samples to compute statistical measures such as averages and standard deviations, which in turn allow the computation of a cumulative probability distribution for the exceeding probability, as well as estimation of correlations between several input and output variables. An example can be shown in Fig. 1.2, where the cumulative likelihood that a wound contracture is worse than the value on the horizontal axis is shown. The input values were calibrated by the clinical experiments by El Hadidy et al. [19]. In this study, a continuum-based model was used. It can be seen that relatively few wounds develop large contractures in this setting. The parameters of uncertainty were the apoptotic rate of myofibroblasts and the initial wound area. These parameters have been sampled from statistical distributions around the mean.

We note that Monte Carlo simulations typically suffer from slow convergence, in which the error, which is commonly quantified by the standard deviation of the mean of multiple sample sets, is inversely proportional to the square root of the total number of samples, that is

$$\widehat{E}_N = \frac{s_N}{\sqrt{N}}, \tag{1.6}$$

where $\widehat{E}_N$, s_N, and N, respectively, denote the estimate of the Monte Carlo error, sample standard deviation, and the number of samples. The above expression has been motivated in Ref. [35]. Therefore, the realization of each sample should be efficient and the use of multilevel Monte Carlo techniques is highly recommended. The multilevel Monte Carlo technique is based on computing samples at various mesh resolutions, and the idea is that one can use fewer samples at high resolutions so that one computes many "computationally cheap" samples and just a few "computationally expensive" samples. An excellent paper in which this issue is dealt with mathematically has been written by Ref. [36].

The cellular automata model, which we have used in the current study, is by far mostly differing from physics; however, its very low parameter space allows a straightforward parameter sensitivity analysis and Monte Carlo simulation. Hence, a disadvantage of such modeling is its large distance from physics by which this model is not suitable for obtaining a thorough understanding in the fundamental mechanisms of wound closure or healing. On the other hand, the complicated models require many input parameters, which are not always available. The current formalisms for modeling the evolution of skin after major burn trauma entail both macroscopic and microscopic aspects from the dermis. The cellular automata formalism can be applied to the restoration of both the dermis and the epidermis. So far, the models lack the interplay between the dermis and epidermis and this relation, although of significant importance in particular for the chemical constitution, has not been dealt with extensively in the wound healing literature. This issue could be of interest in future studies. Furthermore, we are going to expand the probabilistic modeling nature of our simulation procedures in the near future.

If it comes to the simulation of the underlying mechanics in terms of contraction and (local) displacements, it is crucially important to let the models distinguish between temporary and permanent displacements and deformations. The temporary displacement and contraction results from the actual pulling mechanisms by the (myo)fibroblasts, whereas the permanent displacement and contraction results to plastic deformations as a result of long burden and large pulling forces, predominantly, by the myofibroblasts. For this reason, it is recommended to investigate the models of morphoelasticity, which allow permanent deformations. Our preliminary study where the permanent forces are incorporated via adding artificial forces to the elements in the finite element triangulation also works well from a phenomenological point of view. Both approaches decompose the mechanical displacements and deformations into a temporary and a plastic part.

5. Conclusions

Owing to the complexity of the biological processes, complicated, physics-based models are necessary. On the other hand, these models need the input of many variables, whose values are not known well. The simple models can be of great value. The easy tailoring of the simple models to (clinical) experiments is a great asset for the prediction of how skin evolves posttrauma. On the other hand, the physical insights that the complicated models provide are also valuable. Therefore, in the Author's perception, both types of models are important.

References

[1] Hop MJ, Langenberg LC, Hiddingh J, Stekelenburg CM, van der Wal MBA, Hoogewerf CJ, van Koppen MLJ, Polinder S, van Zuijlen PPM, van Baar ME, Middelkoop E. Reconstructive surgery after burns: a 10 year follow-up study. Burns 2014;40:1544–51.

[2] Stekelenburg CM, Marck RE, Tuinebreijer WE, de Vet HC, Ogawa R, van Zuijlen PP. A systematic review on burn scar contracture treatment: searching for evidence. J Burn Care Res 2015;36(3):e153–61.

[3] Young A, McNaught C-E. The physiology of wound healing. Surgery 2011;29(10):475–9.

[4] Boon WM, Koppenol DC, Vermolen FJ. A multi-agent cell-based model for wound contraction. J Biomech 2016;49(8):1388–401.

[5] Britton NF, Chaplain MAJ. A qualitative analysis of some models of tissue growth. Math Biosci 1993;113:77–89.

[6] Byrne H, Drasdo D. Individual–based and continuum models of growthing cell populations: a comparison. J Math Biol 2009;58:657–87.

[7] Groh A, Louis AK. Stochastic modeling of biased cell migration and collagen matrix modification. J Math Biol 2010;61:617–47.

[8] Madzvamuse A, George UZ. The moving grid finite element method applied to cell movement and deformation. Finite Elem Anal Des 2013;74:76–92.

[9] Neilson MP, Mackenzie JA, Webb SD, Insall RH. Modeling cell movement and chemotaxis using pseudopod–based feedback. SIAM J Sci Comput 2011;33(3):1035–57.

[10] Rey R, Garcia–Aznar JM. A phenomenological approach to modelling collective cell movement in 2D. Biomechanics Model Mechanobiol 2013;12(6):1089–100.

[11] Vermolen FJ, Gefen A. A phenomenological model for chemico–mechanically induced cell shape changes during migration and cell–cell contacts. Biomech Model Mechanobiol 2013;12(2):301–23.

[12] Vermolen FJ, Gefen A. A semi–stochastic cell–based model for in–vitro infected 'wound' healing through motility reduction. J Theor Biol 2012;318:68–80.

[13] Vermolen FJ, Mul MM, Gefen A. Semi–stochastic cell–level computational modeling of the immune system response to bacterial infections and the effects of antibiotics. Biomechanics Model Mechanobiol 2014;13(4):713–34.
[14] Vermolen FJ, Gefen A. Semi-stochastic cell-level computational modelling of cellular forces: application to contractures in burns and cyclic loading. Biomechanics Model Mechanobiol 2015;14(6):1181–95.
[15] Reinhart-King CA, Dembo M, Hammer DA. Cell–cell mechanical communication through compliant substrates. Biophy J 2008;95:6044–51.
[16] Steele JM. Stochastic calculus and financial applications. New York: Springer; 2001.
[17] Evans LC. Partial differential equations graduate studies in mathematics. 2nd ed., vol. 19. Providence, Rhode Island: Americal Mathematical Society; 1998.
[18] Vermolen FJ, Harrevelt SD, Gefen A, Weihs D. A particle finite element-based framework for differentiation paths of stem cells to myocytes and adipocytes: hybrid cell-based and finite-element modelling. In: Cerrolaza M, Shefelbine SJ, Garzon-Alvarado D, editors. Numerical meth ods and advanced simulation in biomechanics and biological processes. Elsevier Academic Press; 2018. p. 171–85.
[19] El Hadidy M, Tesauro P, Cavallini M, Colonna M, Rizzo F, Signorini M. Contraction and growth of deep burn wounds covered by non-meshed and meshed split thickness skin grafts in humans. Burns 1994;20:226–8.
[20] Adam JA. A simplified model of wound healing, with particular reference to the critical size defect: one–dimensional model. Math Comput Model 1999;30:23–32.
[21] Prokharau PA, Vermolen FJ, Garcia–Aznar JM. A mathematical model for cell differentiation as an evolutionary and regulated process. Comput Methods Biomech Biomed Eng 2014;17:1051–570.
[22] Gaffney EA, Pugh K, Maini PK. Investigating a simple model for cutaneous wound healing angiogenesis. J Math Biol 2002;45(4):337–74.
[23] Javierre E, Moreo P, Doblare M, Garcia–Aznar JM. Numerical modeling of a mechano–chemical theory for wound contraction analysis. Int J Solids Struct 2009;46(20):3597–606.
[24] Javierre E, Vermolen FJ, Vuik C, van der Zwaag S. A mathematical analysis of physiological and morphological aspects of wound closure. J Math Biol 2009;59(5):605–30.
[25] Koppenol DC, Vermolen FJ. Biomedical implications from a morphoelastic continuum model for the simulation of contracture formation in skin grafts that cover excised burns. Biomechanics Model Mechanobiol 2017;16(4):1187–206.
[26] Maggelakis SA. A mathematical model for tissue replacement during epidermal wound healing. Appl Math Model 2003;27(3):189–96.
[27] Maggelakis SA. Modeling the role of angiogenesis in epidermal wound healing. Discrete Cont Sys 2004;4:267–73.
[28] Olsen L, Sherratt JA, Maini PK. A mechanochemical model for adult dermal wound closure and the permanence of the contracted tissue displacement role. J Theor Biol 1995;177:113–28.
[29] Sherratt JA, Murray JD. Mathematical analysis of a basic model for epidermal wound healing. J Math Biol 1991;29:389–404.
[30] Kaliappan P. An exact solution for travelling waves of ut = Duxx + u – UK. Physica D 1984;11:368–74.
[31] Merks MH, Koolwijk P. Modeling morphogenesis in silico and in vitro: towards quantitative, predictive, cell-based modeling. Math Model Nat Phenom 2009;4(4):149–71.
[32] Alarcon T, Byrne H, Maini P, Panovska J. Mathematical modeling of angiogenesis and vascular adaptation. In: Paton R, McNamara L, editors. Studies in multidisciplinary, vol. 3; 2006. p. 369–87.
[33] Graner F, Glazier J. Simulation of biological cell sorting using a two- dimensional extended Potts model. Phys Rev Lett 1992;69:2013–6.
[34] Vermolen FJ, Pölönen I. Uncertainty quantification on a spatial Markov- Chain model for the progression of skin cancer. J Math Biol 2019 [accepted].

[35] Koehler E, Brown E, Haneuse SJ-PA. On the assessment of Monte Carlo error in simulation-based statistical analyses. Am Stat 2009;63(2):155–62.
[36] Cliffe KA, Giles MB, Scheichl R, Teckentrupp AL. Multilevel Monte Carlo methods and applications to elliptic PDEs with random coefficients. Comp Visual Sci 2011;14:3–15.

Further reading

[1] Vermolen FJ, Gefen A. A semi–stochastic cell–based formalism to model the dynamics of migration of cells in colonies. Biomechanics Model Mechanobiol 2012;11(1–2):183–95.

CHAPTER 2

Ultrasound and magnetic resonance imaging methods for determining pressure injury risk

Sandra Guzman[1], David Brienza[2]
[1]SHRS-Rehab Sci and Tech, University of Pittsburgh, Pittsburgh, PA, United States; [2]Rehabilitation Science and Technology, University of Pittsburgh, Pittsburgh, PA, United States

Chapter Outline

1. Introduction

Understanding that the most severe pressure injuries originate below the surface of the skin [1] where their presence is hidden from view has motivated the development of methods that use medical imaging technologies to better understand the risk factors that contribute to pressure injury development and to assess an individual's risk of developing a pressure injury. Imaging of the tissues beneath the skin has enabled researchers to both observe the pathogenesis of pressure injuries and to reveal associations between risk and anatomical characteristics. Ultimately, clinical implementation of these methods could lead

Innovations and Emerging Technologies in Wound Care. https://doi.org/10.1016/B978-0-12-815028-3.00002-X

to early detection of tissue damage so that treatment can be initiated sooner with better outcomes and to the identification of instances of elevated risk where preventive interventions can be employed. Diagnostic ultrasound and magnetic resonance imaging (MRI) have been the two technologies most commonly deployed for these purposes. This chapter is a review of the progress made toward using ultrasound and MRI technologies to achieve a better understanding of pressure injury risk factors and toward the clinical use of these technologies to assess wounds at an early stage in their development and an individual's risk of developing a pressure injury.

Traditional clinical techniques for detecting and grading the severity of pressure injuries rely on an assessor's visual assessment of the condition of the skin. For example, the international staging system [2] describes severity according to skin color, the appearance of tissue type, temperature, and mechanical characteristics that can be observed from the outside of the body. The limitations of such a system are clear when one considers that the injured tissue may lie beneath the skin hidden from view and the depth of the injury is unknown. The obvious advantage of detecting the pressure injury at an early stage in its evolution is that interventions can be started to prevent progression to a more severe wound. The use of imaging technologies, particularly ultrasound, has shown some potential for early detection of pressure injuries. The same techniques have also been used to track progression and healing.

Researchers have studied tissue damage that develops under intact skin because this pathology typically leads to the most severe pressure injuries. The pathophysiology of these deep tissue pressure injuries is inherently different than the injuries developing in superficial tissue skin tissue due to the various other types of tissues involved (e.g., fat, muscle, and connective tissue). Studies have demonstrated that the composition of the tissues under load affects the tissue resilience and tolerance to load, specifically during sitting conditions. An individual's risk of developing a pressure injury may be dependent on anatomical features such as tissue thicknesses of skin, fat, and muscle tissue layers, the mechanical properties of those tissues and their ability to sustain viability, and the nature of bony prominences.

Both MRI and diagnostic ultrasound have advantages and disadvantages with respect to their utility for assessing pressure injury risk and studying risk factors. Ultrasound requires less financial investment in instrumentation, facilities, and maintenance compared with MRI, making ultrasound the preferred modality for the clinical applications such as early detection of tissue damage and for assessing risk. However, typical ultrasound systems require a skilled operator for both acquiring and interpreting the images. The process in which ultrasound scans are obtained includes operator interaction with the patient by moving and adjusting the probes' position and orientation until a desired feature of the tissue is displayed. The motion of the probe and resulting dynamic image helps the operator to identify tissue structures that may be more difficult to recognize in a static image. MRI, on the other hand, requires no interaction between the operator and the patient. Results from MRI scans are a series of static images in which tissue structures are displayed with relatively high resolution and definition. These distinct characteristics have led to different applications of the two imaging modalities. Ultrasonographic technology in clinical practice and research is becoming popular for the assessment and diagnosis of tissue injuries, while MRI, on the other hand, is being used for research aimed at characterizing the composition, deformation, and displacement of the soft tissues.

Diagnostic ultrasound has been used to compare the measurements of the features of soft tissues with MRI, given that the measurements in MRI are very well standardized. It has been demonstrated that ultrasound can be used for characterizing tissues in a comparable way with MRI. Both techniques have the potential to be used for pressure injury management and prevention, but the use for pressure injury assessment is not yet standardized. Identifying the most representative characteristics of tissues for pressure injury development is one of the biggest challenges. In this chapter, a literature review of the characterization of the tissues and the findings in the pathogenesis of the developing of tissue injuries is presented. Interpretation of images for deriving the most representative features of the tissues could lead to the development of tools for identifying personalized risk of pressure injuries. Identifying those persons with more potential to develop pressure injuries could inform health-care decisions to target preventive measurements. For example, preventive interventions such as cushioning surfaces under load-bearing tissue can be tailored to individual needs, or recommendations for limits on sitting time and pressure-relief protocols can be altered in response to the defined risk.

2. Overview and comparison of diagnostic ultrasound and magnetic resonance imaging technologies

2.1 Ultrasound technology

Ultrasound uses the reflections of sound waves from within the body to create tissue images. A transducer transmits longitudinal sound waves into the tissues of the body. When these propagating waves hit a boundary between tissues with different acoustic properties, such as between the soft tissue and bone, a portion of the energy is reflected back toward the transducer as an echo, where the received signal can be converted to an image representing structures within the tissue represented by the reflections. Stronger reflections are assigned a proportionally brighter image signal. The depth of the boundary is determined by the time of flight between the transmission of the wave and the reception of the reflection. The magnitude of the reflection and hence the brightness of the images depends on the difference in acoustic properties between the two tissue types [1,3].

Spatial resolution and imaging range for diagnostic ultrasounds are dependent on the frequency of the transmitted sound waves. Higher frequencies with shorter wavelengths generate higher resolution images, and lower frequencies penetrate further into the body to achieve greater depth range. The so-called high-frequency ultrasound (HFUS) at 20 MHz provides a viewing depth of 8–34 mm with strong signals to a depth of approximately 18 mm beneath the contact point of the cone tip [4]. HFUS is adequate for visualizing the epidermis, dermis, hypodermis, deep fascia and, in some regions of the body, tendons, and ligaments. The high frequency provides a high resolution, allowing the visualization of a greater number of details but with less penetration and shorter focal lengths [3,5]. A conventional musculoskeletal diagnostic ultrasound device operating at a frequency of 3.5 MHz typically has a resolution of 1 mm with a depth range of 10–20 cm, but when the frequency increases to 50 MHz, the resolution rises to 75 μm with a loss of penetration to less than 1 cm [6].

The use of HFUS has been shown to be effective in measuring skin thickness and has been employed for the examination of healthy skin, the assessment of pathological lesions, and the evaluation of wounds and scars [5,7]. In the past decade, ultrasonography has been used for research in hospitals for characterization, detection, and monitoring of pressure injuries in patients at risk before clinical signs become visible. Studies have reported that the detection of tissue injuries is better with the use of ultrasound rather than with common methods for examination before visible signs begin [8]. Ultrasonography has the potential of becoming a convenient modality in the management of pressure injuries in people at high risk, and there are several efforts for the interpretation of the images, though the use and techniques are not standardized yet for this purpose.

2.2 Magnetic resonance imaging

MRI generates images by sensing energy discharged from hydrogen nuclei in water molecules after they have been stimulated by radio frequency (RF) electromagnetic signals. In an MRI imaging sequence, the protons in the hydrogen nuclei are aligned by a strong magnetic field in the MRI scanner. An RF pulse is generated at a right angle (90 degrees RF pulse) to the magnetic field causing the protons to rotate, absorb energy, and become magnetized in the transverse direction. After the 90 degrees, RF pulse ceases and a 180 degrees RF pulse is applied, the protons rotate back into alignment with the magnetic field and release energy that is detected by the scanner to form the MRI image. The time needed for this realignment varies according to tissue type. Molecules with faster recoveries, that is, those molecules that give up their energy more quickly, can be detected by weighting the energy measurement toward a time relatively soon after the 90 degrees RF pulse. These are referred to as T1-weighted images. As T1 images measure energy relatively soon after the 90 degrees RF pulse ceases, these images show high signal intensity for tissues that give up energy rapidly, such as fat.

During the recovery process after the 90 degrees RF pulse ceases, the protons also lose their transverse magnetization. The slower the transition from transverse magnetization back to longitudinal magnetization, the higher the signal detected toward the end of the imaging cycle. Images generated from energy measurement weighted toward the end of the cycle are referred to T2-weighted images. Late in the cycle, molecules such as water, which do not give up energy quickly, produce a high signal intensity. T2-derived images are therefore good at showing inflammation. Fat would appear dark in the T2 image because it would have given up nearly all of its energy by the time the energy is recorded for the T2-weighted image. Because T1 images are derived from energy measurement early in the cycle, when molecules still have relatively high energy, T1 images offer better resolution compared with T2-derived images. Pixel intensity in the images also depends on the density of protons.

In pressure injury applications of MRI, T1 images are generally used for measurements of tissue dimensions where resolution and soft tissue differentiation are most important, and T2 images are used for assessing tissue damage where detecting inflammation is critical.

3. *Imaging method's use in risk factor identification*

The clinical challenges in the prevention and treatment of pressure injuries have led to different approaches to understanding their development, including identifying and assessing risk factors [2,9]. Reported studies [1,2] suggest the shape of the bone, the composition of the tissue, and the tissue's biomechanical properties all impact the concentration of mechanical forces and deformation under load, the two necessary conditions for development of a pressure injury. But the current tools for assessing risk do not consider the detailed assessment of anatomy of the person [10].

With the aim of understanding the development of tissue injuries, some studies have evaluated the characteristics of soft tissue around a bony prominence (e.g., ischial tuberosity), focusing on the role of tissue deformation [11,12]. In practice, evaluating these deformations is not an easy task. Ultrasound and MRI have been used recently to investigate the concentrations of forces that exist near a bony prominence [1–3].

MRI played an essential role in the elucidation of the deformation pathway to tissue damage. In the early 2000's, researchers hypothesized that tissue damage in load-bearing soft tissues was the direct result of excessive stain to tissues' cellular structure. Direct deformation damage was hypothesized to be independent of the longstanding understanding that a reduction in blood supply and impaired lymphatic function supply could initiate a cascade of events that ended in a pressure injury. MRI was essential to the studies investigating direct damage from deformation for several reasons but primarily stemming from its ability to generate highly detailed images of the tissue structure.

The high detail in the images was used to generate accurate geometric models of tissue structures in various load-bearing scenarios to quantify tissue strain [3]. The strain information was combined with tissue mechanical properties in finite element models to estimate stress and strain distribution throughout the entire area or volume of interest [4,5]. MRI technology was also used to assess tissue damage. T2-weighted MR images are able to differentiate healthy from damaged muscle tissue because inflammation results in increased signal intensity in damaged muscle compared with the intermediate signal intensity for healthy muscle [6]. Stekelenburg et al. exploited these phenomena to study direct tissue damage in a rat model and was able to show damage from deformation alone [7,8].

Tissue deformation on the buttocks of able-bodied subjects compared with individuals with spinal cord injuries has been measured using MRI. Able-bodied subjects showed a reduction of tissue thickness under the ischial tuberosity resulting from a combination of displacement and muscle distortion. On the other hand, the tissue beneath the ischial tuberosity(IT) of spinal cord injury(SCI) individuals is predominantly composed of fat and connective tissue. This means that the person did not sit directly on muscle causing more tissue strain, which suggests that these tissues might be more vulnerable to developing an injury over the bony prominence [13,14].

MRI has also been used to compare the effects of support surfaces on tissue deformation [15,16]. The tissue around the ischial tuberosities during sitting on a variety of wheelchair cushions on people with and without SCI has demonstrated that both the anatomy of the person and the type of cushion affect the deformation response of the tissues [15]. Similarly, differences in tissue composition in both density and size have been found among individuals with and without spinal cord injury [17]. Although MRI can generate highly detailed images of tissue structure, it is not a feasible tool for assessing the characteristics of tissue for determining the risk of developing pressure injuries. The characteristics of MRI limiting its practical application for routine screening include long acquisition time, high cost, and poor accessibility. These obstacles to translation to practice appear to be insurmountable with the current state of MRI technology.

Ultrasound has been used as an alternative to MRI protocols for the measurement of soft tissue geometry, tissue deformation, and bone characteristics. Ultrasound provides a potentially practical alternative solution as it is a noninvasive, inexpensive, and portable tool. It has the potential to be used to measure the bone and the surrounding tissues with the aim of identifying the individual risk factors for developing pressure injuries. Also, ultrasound has been used to measure the skin over the sacrum and ischial tuberosities of people with and without SCI. It has been shown that the skin in individuals with SCI is thinner than similar people without SCI. It has been suggested that thin skin is a risk factor for soft tissue damage, leading to pressure injuries [18]. Other studies used B-mode diagnostic ultrasound for also measuring muscle thickness, adipose thickness, total thickness, and the radius of curvature of the ischial tuberosities.

Studies have shown that the measurements using diagnostic ultrasound of the thickness of muscle, fat, and total tissue of the surrounding area of the ischial tuberosity are highly correlated with measurements using MRI, but the radius of curvature of the ischial tuberosity is poorly correlated when comparing the measurements using ultrasound and MRI [19]. The poor correlation between the two measurements of radius of curvature may be attributable to the fact that MRI does not allow differentiation between cortical bone and portions of the musculotendinous junction. On the other hand, cortical bone is easily discernible using ultrasound, making ultrasound a potentially better choice to measure bone shape.

Characterizing the tissue using ultrasound can be a challenge as there are no standard criteria for identifying its most representative features, and the measurements of the tissue could vary from one operator to another. To address the latter issue, a study evaluated the reliability of measurements when different operators assessed the thickness of the tissues overlying the ischial tuberosities and the bony dimensions by means of the radius of curvature of the lowest point of the IT in both the short and long axes of able-bodied subjects and people with SCI [20]. The study showed a good within-operator and between-operator reliability for measuring three metrics (total thickness, skin/fat, and tendon/muscle) but poor reliability when identifying the five soft tissue layers separately. On the other hand, the between-operator reliability of the ischial tuberosities measurements was zero or not existent. This could be attributable to the lack of registration points for this particular anatomical site, and the operators might consider different points for measuring the radius of curvature.

Recently, ultrasound imaging has been used for the quantification of localized soft tissue strains of buttock tissues in the frontal and sagittal planes under the ischium during sitting (see Fig. 2.1). With the goal of avoiding subjective interpretations, postprocessing of the images was done with a digital image correlation method that enables to measure the displacement field. The authors used Green–Lagrange strain tensors for measuring the deformation [9]. The strain tensor consists of measuring the change of position of pixels in a region of interest (ROI) selected manually. With this method, the authors found that soft tissue in the fat layer can be tracked. To the contrary, the displacement in the muscle region beneath the ischial tuberosity could not always be assessed reliably. The gluteus maximus and hamstring muscles did not always remain underneath the IT but slid out of the plane, leaving the IT adjacent to the fat tissue. This observation is close to previously reported results from studies using MRI for measuring the distortion of tissue [13]. The decrease of the thickness of soft tissue muscle under the ischial tuberosity is being suggested as a factor of risk of pressure injury.

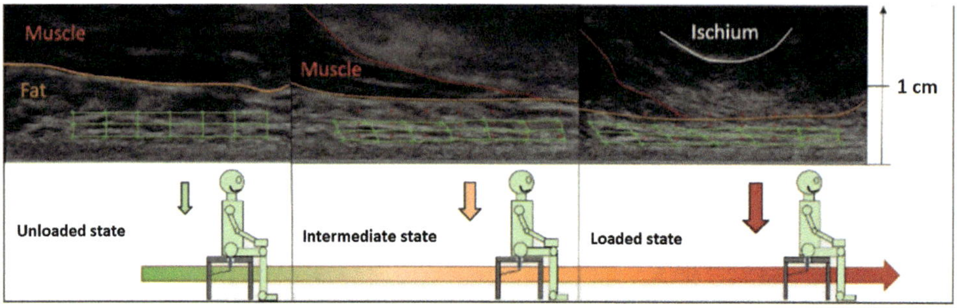

Figure 2.1
Illustration of tissue deformation during siting using 8 MHz ultrasound [9]. *Permission requested.*

The studies acknowledge the influence of the bone under the loaded tissue and that its anatomical shape affects the pressure injury development [10]. Despite the efforts to characterize the ischial tuberosities, currently, there is no standard procedure for measuring it. The reliability of the characterization of the bony prominences could be improved using postprocessing of images for avoiding variability between operators. The current efforts for assessing the shape of the ischial tuberosities characterize it as a sphere, and it has been demonstrated that there is a change in the bone after an SCI [21].

The importance of the characterization of the bone and the surrounding tissues has the final goal of a better understanding of the pressure injury development and the development of methodologies of the personal risk factors. With a more sensitive tool to identify the risk, associated preventive measures can be personalized. For example, preventive interventions such as cushioning surfaces under load-bearing tissue can be tailored to individual needs, or recommendations for limiting sitting time and specifying pressure-relief protocols can be altered in response to the defined risk. Neither of these preventive measures can be achieved with sufficient precision with the available risk assessment tools [22]. Evidence suggests that both ultrasound and MRI have the potential to be used for this purpose, but more work is needed to develop and demonstrate efficacy of practical methodologies.

4. Early detection of pressure injuries

4.1 Ultrasound imaging and magnetic resonance imaging of healthy tissue

Some efforts to characterize healthy soft tissue to better differentiate from compromised tissue during the evaluation of people at risk for developing tissue injuries have been made. See Table 2.1 and Fig. 2.2 for examples of the characteristics of healthy tissues in ultrasound images. Generally, when using ultrasound, the reflected signals come from different levels of brightness depending on the echogenicity and the reflective

Table 2.1: Identification of parameters in ultrasound and magnetic resonance imaging (MRI) images.

Tissue	Ultrasound image characteristic	T1-weighted MRI image signal intensity	T2-weighted MRI image
Cortical bone	Hyperechoic, smooth continuous	Low	Very low
Muscle	Hypoechoic, with parallel fibrous, hyperechoic bands	Intermediate	Intermediate
Nerve	Hyperechoic, relative to muscle		
Fat	Hypoechoic, a series of clearly defined layers	High	Intermediate

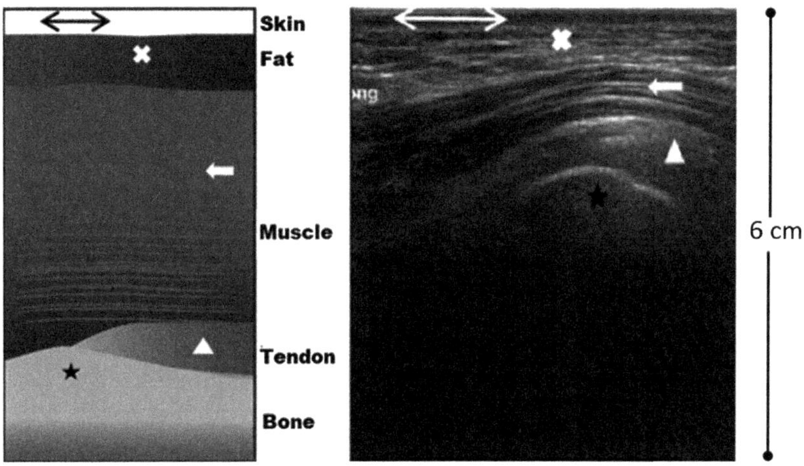

Figure 2.2
Illustration of soft tissue layers on ischial tuberosity using 12 MHz ultrasound [20]. *Permission requested, image modified from original.*

characteristics of the tissues and interfaces. For example, when the tissue contains more fluid, the waves penetrate freely, and the echo images back are low. On the other hand, ultrasound waves do not penetrate bones, so when the ultrasound waves hit the bone, the returning echo images are higher [6]. High echogenicity of the tissues provides a high brightness in the image. Some of the HFUS equipments allow color imaging in green, blue, or red, depending on the settings of the apparatus; red/black are classed as low echogenic and blue/green as higher and more homogeneous echoes [20,23]. See Fig. 2.3 for an example of healthy soft tissue layers using 20 MHz ultrasound.

MRI produces gray-scale images of the tissues; these images are presented as slices. Image pixel intensity is derived from the rate and magnitude of energy return from hydrogen nuclei after magnetic and electromagnetic pulses. Different tissues with different

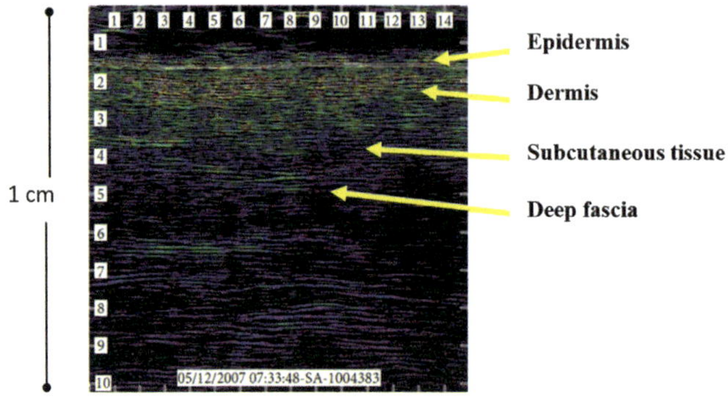

Figure 2.3

Illustration of soft tissue layers of normal tissue using 20 MHz ultrasound [24]. *Open access, small changes from original figure.*

concentrations of hydrogen can be identified in energy measurements soon after reversing electromagnetic polarity (T1) and later in the cycle (T2). For example, water exhibits very low signal on T1 images and rise to high signal on T2 [6]. On the other hand, cortical bone displays very low signals on all sequences. Muscle gives intermediate signal intensity on both T1 and T2 images though the signal intensity is slightly lower on T2. See Table 2.1 for examples of characteristics of tissues using T1- and T2-weighted MRI images.

4.2 Ultrasound assessment of pressure injury under intact skin

The ability to identify tissue changes before visual signals begin would be beneficial to prevent further damage to skin integrity. Commonly, the areas of suspected tissue injury are evaluated by visual inspection and palpation. Several efforts have been made to identify lesions under intact skin and to identify the pathogenesis of pressure injuries. HFUS at 20 MHz has been used attempting to characterize pressure injuries. Studies have identified that the lesions caused by pressure are shown as edema from the bone extending upward, and the lesions most likely caused by friction are in the superficial layers [1]. Quintavalle et al. identified three patterns of lesions: pockets of edema in subcutaneous tissue with no dermal involvement, edema extending from the subcutaneous tissue into the dermis, and edema extending from the subcutaneous tissue via the dermal/epidermal junction where it has pooled. These patterns have also been identified in other clinical studies for identifying the suspected tissue damage [24] (see Fig. 2.4 for these examples).

Diagnostic ultrasound at 10 MHZZ has also been used for detecting deep tissue injury before clinical signs begin. The difference against the HFUS is that diagnostic ultrasound is able to visualize deeper tissues than HFUS. Images with low-echoic areas under the intact skin are potentially a deep pressure injury. Studies have shown low-echoic lesions

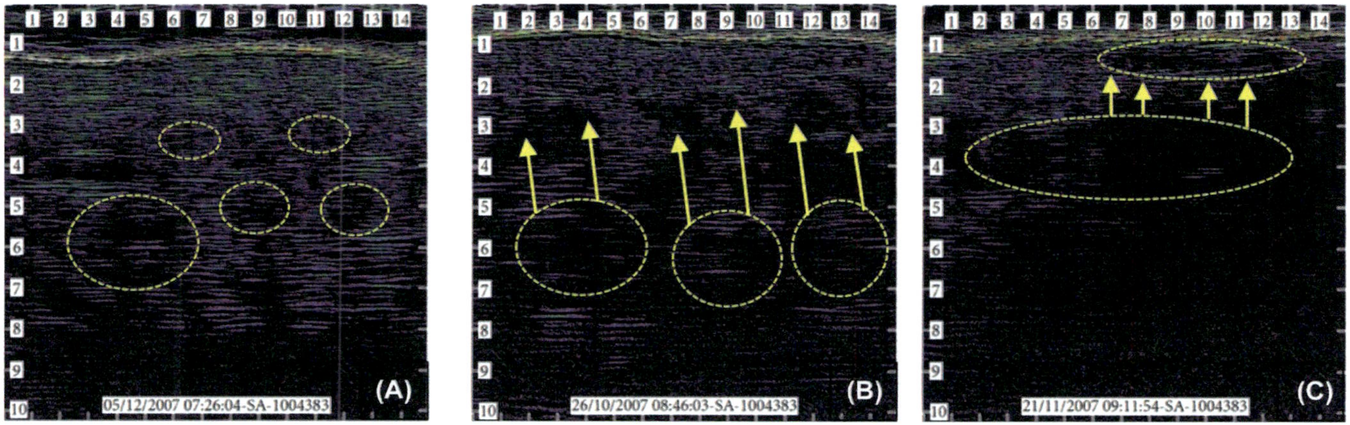

Figure 2.4

Patterns of pressure injuries under intact skin using 20 MHz ultrasound. (A) Pockets of edema in subcutaneous tissue with no dermal involvement, (B) edema extending from the subcutaneous tissue into the dermis, and (C) edema extending from the subcutaneous tissue via the dermal/epidermal junction where it has pooled [24]. *Open access, small changes from original figure.*

on ultrasound imaging significantly deeper than those detected by palpation, and all such lesions were directly adjacent to the bone. Palpation with a free-floating sensation has ultrasound imaging with low-echoic signal from near the bone to the skin [25]. Fig. 2.5 shows examples of patterns of lesions under intact skin using 10 MHz ultrasound.

Ultrasonography has also been used to assess potentially damaged tissue and to monitor the healing in follow-up studies [11,26]. Aoi et al. used 10 MHz ultrasound to periodically analyze (almost every week) 12 patients under intensive nursing care who showed deep tissue injury—related abnormal findings on ultrasonography screening and followed up until the pressure injury reached a final stage (range of follow-up, 21—92 days) [11]. They recognized four patterns of abnormal signs; these are as follow: (1) unclear layered structure has a foggy-appearing area with low contrast and rough resolution, (2) hypoechoic lesion is a small lesion with a relatively clear margin that has no or little echoic signal and may correspond to a small lesion with fluid, such as a hematoma, (3) discontinuous fascia is a broken or discontinuous line with heterogeneous echo signals, which likely means damaged, disrupted, or ruptured fascia, and (4) heterogeneous hypoechoic area is an area with low echoes that sometimes has a diffuse border. On the first ultrasonic examination, all the patients showed signs of an unclear layered structure; the researchers related this pattern to the most common abnormal finding on deep tissue injuries. On the follow-up screenings, when the patterns included discontinuous fascia and heterogeneous hypoechoic area, the pressure injuries were less likely to heal and the researchers classified these patterns as predictors of future progression [11].

The early changes in the tissue begin with inflammation. Fluid accumulation or edema in dermal or hypodermal tissue layers has been considered as early indicators of tissue injury. Ultrasonography has shown to be a reliable tool for detecting lesions under intact skin if the patterns of lesions are recognized. The early detection of changes in a lesion would minimize the risk that a pressure injury worsens because preventive interventions can be applied, resulting in better care, reduced caregiver effort, and less suffering [12]. However, further longitudinal clinical assessment is required to characterize ultrasound images against the stages and the classification of pressure injuries.

5. Imaging in etiology studies

5.1 Assessment of tissue in wounds

Depth, exudate, size, inflammation, infection, granulation, necrosis, and undermining are characteristics considered during the healing process of a pressure injury. Studies report that necrotic tissue in the undermining area delays the healing process [27]. It is important to detect the presence of necrotic tissue in undermining areas to apply appropriate treatments. However, it is difficult to observe the margins of undermined wounds by visual

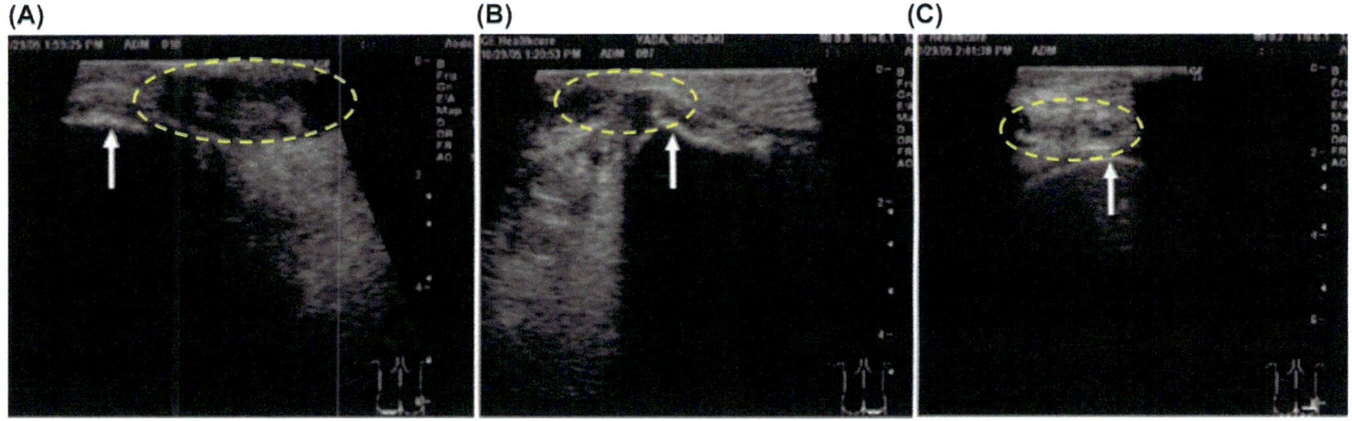

Figure 2.5

Patterns of pressure injuries under intact skin using 10 MHz ultrasound [25]. A) Low-echoic lesion containing high-echoic areas near the left ischial tuberosity (arrow). B) Low-echoic lesion near the bone (arrow) in the right ischial region. C) Area of the low-echoic lesion close to the bone (arrow) and irregular muscle layer. *Permission requested, small changes from original image.*

Table 2.2: Image characteristics of ultrasound on wounds.

Tissue	Description
Necrotic tissue	Hypoechoic, no layers, unclear borders, uneven gray
Granulated tissue	Low echoic, no layers, coarse resolution, even gray

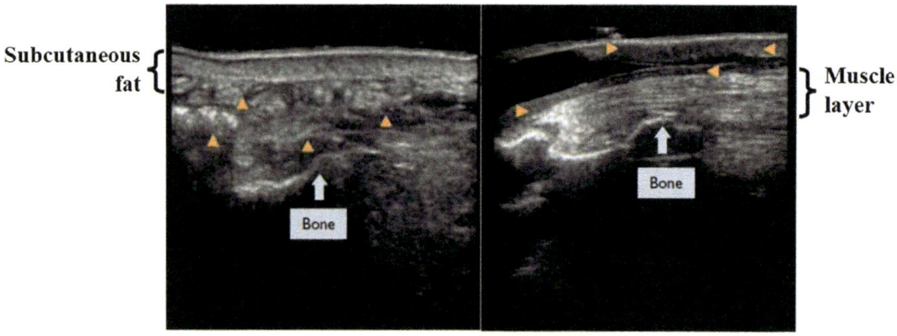

Figure 2.6
Ultrasound imaging inside an undermining area. (A) Necrotic tissue and (B) granulation tissue [28]. *Permission requested, small changes from original images.*

inspection. Diagnostic ultrasound has shown the potential to observe the characteristics of both necrotic and granulation tissue in a wound and inside an undermined region [28] and the quantity of inflammation in the wound and surrounding area [23].

When placing ultrasound (at 12 MHz) at the end of an undermining wound, by visual inspection, the necrotic tissue is shown as heterogeneous hypoechoic area. The necrotic tissue regions have been reported as a high-echoic image with no layers, unclear borders, and an uneven gray level (cloud-like image). Granulation tissue appears as a low-echoic image, which was of coarse resolution, had no layers, and even gray length of the undermining [28]. Table 2.2 and Fig. 2.6 show examples of the characteristics of tissue in wounds.

To evaluate the necrotic and granulation tissue objectively, researchers have used the ImageJ analysis software (v1.42q: NIH) and diagnostic ultrasound images for measuring the pixel uniformity by calculating both maximum and minimum values of gray in the regions of interest in injured, uninjured, and granulation tissues [29]. Findings have shown significant differences between necrotic and granulation tissue compared with not injured tissue near the wound [28].

HFUS at 20 MHz has also been used to assess the level of edema in wounds and surrounding tissues and to monitor the evolution of necrotic tissue within the wound [23]. In open wounds, measurements are taken with dressings in place as the sound waves travel

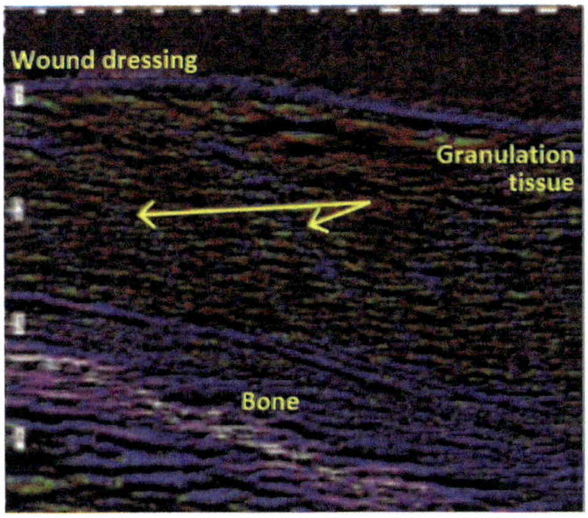

Figure 2.7
Wound bed granulation tissue using high-frequency ultrasound. The plane of the ultrasound is perpendicular to the surface of the film. *Permission requested, small changes from original image.*

well through the moist and gauze film. Studies had evaluated the healing process of a wound, comparing the scans of the wounds and surrounding peri-wound tissue as a baseline profile and also compared with scans at the tissue adjacent to the wound site to establish as a condition of the patient normal skin. By visual inspection, there are changes in the relative distribution of red to blue pixels as the wound base granulation tissue matures, see Fig. 2.7 for an example. In HFUS, blue and red are graded as low echogenic measurements, the same as level gray images using diagnostic ultrasound.

To get a quantitative assessment of the level of edema present in the tissues immediately surrounding the wound using HFUS, the scans are analyzed using a form of pixel distribution. The pixels below certain intensity (red/black) are classed as low echogenic pixels (LEPs). With this technique, the ratio of LEPs to total pixel (TP) reflects changes in dermal water content [30]. Either using diagnostic ultrasound or HFUS, the objective measurements are important to compare the evolution of wounds and to assess the healing process.

6. Injury measurement techniques using ultrasound

As a variety of clinicians may obtain and evaluate the ultrasound images, it is important that quality images are obtained and processes to evaluate them are standardized. There are no standard criteria to determine an ideal image and to measure the changes representative of tissue injury. Evaluation of density of the layers is commonly used for determining the quality when using HFUS image [31,32]. However, there are few data available using ultrasound for evaluating tissues, most of the reports assessed the images by visual inspection, others used the features of the ultrasound equipment, and some of them developed their algorithms by image postprocessing. The first characteristic to look in abnormal changes is the tissue inflammation either in the dermal or hypodermal layer [1]; this is represented by the presence of homogeneous internal echoes [33]. Therefore, for evaluating ultrasonography, the first characteristic to quantify is the echogenicity of the images. When using HFUS, commonly, the scans are analyzed using a form of pixel distribution, whereby pixels below certain intensity are classed as LEPs. To compare the echogenicity values, the most popular parameter among researchers [23,30,34,35] is the ratio of the LEPs and TPs counted.

The algorithms have been developed with the purpose to have more objective results and to avoid the requirement of knowledge to interpret. Andersen et al. developed an algorithm to postprocess cross-sectional images of pressure injuries and the surrounding skin using an HFUS at 20 MHz [36]. The algorithms quantify the measurements of the thickness of the hypoechogenic layer by using the echo intensity from the dermis versus the position by a normal distribution. Results showed that the algorithm is more objective but not more sensitive than visual inspection due to the fact that the postprocessing approach is not capable to extract thickness of a layer with noise. As for example, fat tissue is hypoechogenic, and therefore it is difficult to evaluate whether inflammation also exists due to skin exposed to pressure. In addition, static images might not be enough as the expertise of the movement of tissues during the assessment may give a more accurate result in the differentiation of low-echoic pockets. The interpretation of an injury is not only in the identification of low echoes but also in the response of the surrounding tissues. In this sense, the real-time evaluation of the movement of tissues could be interesting, but it has not been explored yet.

Other approaches with image postprocessing have measured the changes of echogenicity in an ROI by computing the root mean square of the pixel values and created a method using fractals for the distribution of echoes defined as texture [37,38]. Mlosek et al. defined the echogenicity by means of summing up the number of pixels in an ROI and used the ratio of the LEPs and the TPs for comparing the echogenicity values [34]. The higher the ratio of LEP/TP, the lower the echogenicity considered. Dermal tissue with suspected inflammation accompanied by higher water content produces overall lower echogenic values. Healthy tissue would be expected to have generally lower LEP/TP ratio values than tissue with inflammation. In the same way, other researchers adopted this latter parameter to get the quantitative assessment of the inflammation presented in the tissues [23,35].

Understanding and quantifying the echogenicity of the ultrasound images is a fundamental part in the assessment of tissue injuries, as well as the good quality images and low echoes under a high echogenic tissue could either (1) be misinterpreted as abnormal fluid secondary to injury or (2) mask true pockets of fluid secondary to injury, as, for example, the case of the thickened epidermis on the heel that is highly echogenic [39]. Other considerations to take into account when analyzing the ultrasound images are defects and artifacts resulting from bubbles in the transmission gel, improper probe placement, and improper technique related to the optimal configuration when acquiring the images that may distort or occlude image components.

7. Summary

Understanding pressure injury etiology and improving risk assessment through early detection of injury and quantification of risk factors depends on the implementation of reliable imaging modalities. Ultrasound is a good choice for clinical implementation due to its accessibility due to low cost and ability to make valid geometric measurements and to identify pockets of inflammation. MRI, although more expensive and therefore less accessible in the clinic, has been invaluable for generating precise anatomical measurements used for modeling studies that elucidated the deformation pathway to tissue injuries. In the future, technology using either ultrasound or MRI may develop into practical risk assessment methodologies that will allow better prevention strategies to be employed for improved outcomes.

References

[1] Quintavalle PR, et al. Use of high-resolution, high-frequency diagnostic ultrasound to investigate the pathogenesis of pressure ulcer development. Adv Skin Wound Care 2006;19(9):498–505.

[2] NPUAP E, PPPIA. Prevention and treatment of pressure ulcers. Quick reference guide. In: Haesler E, editor. Osborne Park, Western Australia: Cambridge Media; 2014. 2018.

[3] Todd BA, Thacker JG. Three-dimensional computer model of the human buttocks, in vivo. J Rehabil Res Dev 1994;31(2):111–9 [comment].

[4] Stekelenburg A, et al. Compression-induced deep tissue injury examined with magnetic resonance imaging and histology. J Appl Physiol 2006;100(6):1946–54.

[5] Linder-Ganz E, et al. Assessment of mechanical conditions in sub-dermal tissues during sitting: a combined experimental-MRI and finite element approach. J Biomech 2007;40(7):1443–54.

[6] McKinnis LN. Fundamentals of musculoskeletal imaging. FA Davis; 2013.

[7] Stekelenburg A, et al. Role of ischemia and deformation in the onset of compression-induced deep tissue injury: MRI-based studies in a rat model. J Appl Physiol 2007;102(5):2002–11.

[8] Loerakker S, et al. The effects of deformation, ischemia, and reperfusion on the development of muscle damage during prolonged loading. J Appl Physiol 2011;111(4):1168–77.

[9] Doridam J, et al. Feasibility of sub-dermal soft tissue deformation assessment using B-mode ultrasound for pressure ulcer prevention. J Tissue Viability 2018;27(4):238–43.

[10] Takahashi Y, et al. A new concept:'Relative position between the external force and the bony prominence'explains location-specific occurrence of superficial injury over an undermining lesion. J Tissue Viability 2017;26(1):75–8.

[11] Aoi N, et al. Ultrasound assessment of deep tissue injury in pressure ulcers: possible prediction of pressure ulcer progression. Plast Reconstr Surg 2009;124(2):540–50.

[12] Lyder C. The use of technology for improved pressure ulcer prevention. Ostomy/Wound Manag 2007;53(4):14–6.

[13] Sonenblum SE, et al. 3D anatomy and deformation of the seated buttocks. J Tissue Viability 2015;24(2):51–61.
[14] Sonenblum SE, et al. 3-dimensional buttocks response to sitting: a case report. J Tissue Viability 2013;22(1):12–8.
[15] Brienza D, et al. An MRI investigation of the effects of user anatomy and wheelchair cushion type on tissue deformation. J Tissue Viability 2018 Feb;27(1):42–53.
[16] Peko Cohen L, Gefen A. Deep tissue loads in the seated buttocks on anoff-loading wheelchair cushion versus air-cell-based and foam cushions: finiteelement studies. Int Wound J. 2017 Dec;14(6):1327–34.
[17] Wu GA, Bogie KM. Not just quantity: gluteus maximus muscle characteristics in able-bodied and SCI individuals–implications for tissue viability. J Tissue Viability 2013;22(3):74–82.
[18] Yalcin E, et al. Skin thickness on bony prominences measured by ultrasonography in patients with spinal cord injury. J Spinal Cord Med 2013;36(3):225–30.
[19] Akins JS, et al. Feasibility of freehand ultrasound to measure anatomical features associated with deep tissue injury risk. Med Eng Phys 2016;38(9):839–44.
[20] Swaine JM, et al. Adaptation of a MR imaging protocol into a real-time clinical biometric ultrasound protocol for persons with spinal cord injury at risk for deep tissue injury: a reliability study. J Tissue Viability 2018;27(1):32–41.
[21] Rittweger J, et al. Bone adaptation to altered loading after spinal cord injury: a study of bone and muscle strength. J Musculoskelet Neuronal Interact 2006;6(3):269–76.
[22] Henzel MK, et al. Pressure ulcer management and research priorities for patients with spinal cord injury: consensus opinion from SCI QUERI Expert Panel on Pressure Ulcer Research Implementation. J Rehabil Res Dev 2011;48(3):xi–xxxii.
[23] Young SR, Hampton S, Martin R. Non-invasive assessment of negative pressure wound therapy using high frequency diagnostic ultrasound: oedema reduction and new tissue accumulation. Int Wound J 2013;10(4):383–8.
[24] Porter-Armstrong AP, et al. Do high frequency ultrasound images support clinical skin assessment? ISRN Nursing 2013:2013.
[25] Kanno N, et al. Low-echoic lesions underneath the skin in subjects with spinal-cord injury. Spinal Cord 2009;47(3):225.
[26] Nagase T, et al. Ultrasonographic evaluation of an unusual peri-anal induration: a possible case of deep tissue injury. J Wound Care 2007;16(8):365–7.
[27] Matsui Y, et al. Development of the DESIGN-R with an observational study: an absolute evaluation tool for monitoring pressure ulcer wound healing. Wound Repair Regen 2011;19(3):309–15.
[28] Ueta M, et al. Use of ultrasound in assessment of necrotic tissue in pressure ulcers with adjacent undermining. J Wound Care 2011;20(11):503–10.
[29] Ferreira T, Rasband W. ImageJ user guide. ImageJ/Fiji; 2012. p. 1.
[30] Gniadecka M, Quistorff B. Assessment of dermal water by high-frequency ultrasound: comparative studies with nuclear magnetic resonance. Br J Dermatol 1996;135(2):218–24.
[31] Grap MJ, et al. Use of high frequency ultrasound to detect changes in skin integrity: an image evaluation validation procedure. Intensive Crit Care Nurs 2015;31(3):141–7.
[32] Burk RS, et al. Effects of position and operator on high-frequency ultrasound scan quality. Intensive Crit Care Nurs 2015;31(3):148–54.
[33] Yabunaka K, et al. Can ultrasonographic evaluation of subcutaneous fat predict pressure ulceration? J Wound Care 2009;18(5):192–8.
[34] Mlosek RK, et al. The use of high frequency ultrasound imaging in skin moisturization measurement. Skin Res Technol 2013;19(2):169–75.
[35] Burk RS, et al. High-frequency ultrasound: description of sacral tissue characteristics in healthy adults. J Wound, Ostomy Cont Nurs 2017;44(5):434–9.
[36] Andersen ES, Karlsmark T. Evaluation of four non-invasive methods for examination and characterization of pressure ulcers. Skin Res Technol 2008;14(3):270–6.

[37] Moghimi S, Baygi MHM, Torkaman G. Automatic evaluation of pressure sore status by combining information obtained from high-frequency ultrasound and digital photography. Comput Biol Med 2011;41(7):427−34.

[38] Moghimi S, et al. Quantitative assessment of pressure sore generation and healing through numerical analysis of high-frequency ultrasound images. J Rehabil Res Dev 2010;47.

[39] Helvig EI, Nichols LW. Use of high-frequency ultrasound to detect heel pressure injury in elders. J Wound, Ostomy Cont Nurs 2012;39(5):500−8.

CHAPTER 3

Computational modeling of the thin muscle layer, panniculus carnosus, demonstrates principles of pressure injury and prophylactic dressings

Bryan Wee Siang Soh[1], **Alberto Corrias**[1], **Lisa Tucker-Kellogg**[2,3]

[1]*Department of Biomedical Engineering, National University of Singapore, Singapore;*
[2]*Centre for Computational Biology, Duke-NUS Medical School, Singapore;*
[3]*Cancer and Stem Cell Biology, Duke-NUS Medical School, Singapore*

Chapter Outline

1. Introduction

Pressure ulcers (PUs, also called pressure injuries) are areas of soft tissue breakdown that result from sustained mechanical loading of the skin and the underlying tissues. The loading may be direct pressure or pressure with shear or friction force [1]. A PU arises when some regions of soft tissue undergoe cell death in response to the sustained pressure [2].

The etiology of PUs requires greater understanding to provide efficient prevention and therapy, and more research is necessary [3]. Possible mechanisms of injury include ischemia (disrupted blood flow), reperfusion, rupture of cellular structures, metabolic stress, and oxidative stress. Ischemic injury can result from direct sustained pressure, from repetitive moderate pressure, and/or from shear. When ischemia is followed by reperfusion

Innovations and Emerging Technologies in Wound Care. https://doi.org/10.1016/B978-0-12-815028-3.00003-1

of blood, the ischemia–reperfusion combination causes oxidative stress and tissue injury [4]. Another important mechanism of tissue injury during PU formation is increased membrane permeability or outright rupture of cell membranes [5,6]. The integrity of cell plasma membranes (the sarcolemma in muscle cells [7]) is required for health and homeostasis, and if a cell membrane experiences a large rupture, the cell undergoes lysis and death. The cellular contents released from cell lysis may include alarmins that trigger inflammation or redox active compounds such as globin proteins, which can create oxidative stress, alone or in combination with other proteins in the wound [8]. The cytoskeleton of a healthy cell has a remarkable capacity for dynamic remodeling and flexible geometry [9]. Cellular stress response signaling can help cells survive harsh conditions [10], often with transient dynamics [11], and the same pathways can make cells vulnerable to prodeath stimuli through the pathways of extrinsic apoptosis [12,13]. Extracellular matrix architecture may provide considerable protection for mechanical stability, but the extracellular matrix is in constant flux, and the mechanical capability of extracellular matrix can be decreased if there is a jump in protease activity [14]. In summary, many biological processes may contribute to injury, and more research must be focused on the PU context rather than speculated by analogy from other contexts.

As improvements in medical care allow people to live longer with comorbid conditions, immobility becomes an increasingly prevalent problem [15]. Care of immobile people requires frequent turning to prevent PU formation in vulnerable areas [16]. Despite the significant costs to stakeholders and the efforts invested in prevention/education, PU occurrence remains a serious problem. An American study reported that there are 2.5 million new cases of PUs every year, with an annual cost of USD 11.6 billion, and 60,000 patient deaths due to PUs annually [17]. Turning and off-loading can be achieved through direct repositioning and/or by use of specialized support surfaces [18], and interest has grown dramatically in the use of prophylactic dressings [19]. Preoperative preparation of surgical patients [20] has been a particularly attractive context for early stage trials of dressings, where the incremental cost is easily accommodated. Lessons learned from early efforts with prophylactic dressings may later extend to a broader variety of PU contexts.

The heels are the most common site for PUs acquired in long-term acute care facilities [21] and are the second most common site for PUs overall (23.6%), after the sacrum (28%) [22]. The heels are also the site where deep tissue injury is most prevalent [16,23,24]. As with other PUs, heel ulcers can involve serious, life-threatening complications [25]. The dorsal surface of the heel is particularly vulnerable to pressure injury. The plantar surface is well adapted to mechanical loading, including high-impact activity, but the posterior heel is not. Over the heel bone (calcaneus) is only a thin layer of skin and soft tissue [16].

The panniculus carnosus (PC) is a thin layer of muscle present in many mammals and in some areas of the human body [3]. The PC is a muscle just below the skin and attached to the skin, which allows movement of the skin independent of the skeleton or deeper muscles. The animal equivalent of the PC functions to dislodge insects or other noxious agents. The PC is considered vestigial in higher primates, where limbs are long enough to reach around the body. In the human body, specific areas have a PC muscle as a discrete structure with clear presence: the platysma of the neck, the subareolar muscle around the nipple, the dartos of the scrotum, the palmaris brevis of the hand, and the corrugator cutis ani [26]. In other areas, the PC layer may be present, absent, or sparsely present in the form of muscle strands that connect to skin but not a continuous sheet. Multiple authors concur that remnants of the PC are still found in many individuals [27]. Note that human anatomy routinely encompasses enormous variability. For example, the plantaris muscle varies in size from person to person, sometimes as large as 13 cm, but is absent in 7%–20% of individuals [28].

The function of the PC is poorly understood and frequently speculated on. The palmaris brevis is a thin PC muscle layer at the base of the hand, and it is believed to protect the nerves and vessels at the ulnar canal (beside the bone) [29]. On this basis, surgeons are advised to preserve it during hand surgery. The electromyographic function and microanatomy at the base of the hand has been more closely studied than equivalent locations on the heel. The heels of cadavers were studied by Cichowitz et al., who performed microdissections and found thin layers of PC muscle at the subcutaneous level in many individuals [3]. To the best of our knowledge, no study has yet performed a comprehensive survey of the geometric extent of the PC in human heels. They proposed that the PC may be the site of origin for PUs of the heel because muscle, as a tissue type, is comparatively vulnerable to ischemic death. From this viewpoint, people with a PC layer in the heel may be more susceptible to pressure injuries. Another viewpoint arose from studying the contractures caused by the PC after burn trauma; Greenwood hypothesized [26] that the PC may have an adaptive function to accelerate wound closure by contracting wound edges. The viewpoint inspired by the palmaris brevis instance (above) suggests the PC may serve as mechanical protection for softer structures near bone. It is this protective capability we wish to consider further in this chapter. Our approach will employ computational biomechanics, simulating finite element models of mechanical deformation and comparing two models: one with the PC and one without.

Prior research in computational biomechanics has generated many insights into the etiology and pathophysiology of heel PUs. In understanding the biomechanics of heel ulcers, mathematical models and finite element simulations have shown stress–strain behaviors of the heel under loading conditions [18,25,30,31]. This has improved understanding of tissue deformation and also the role of support surfaces or external measures to decrease the loading or to improve the load distribution. The simplest biomechanical model for studying PU risk in heels was constructed using the curved shape of the posterior calcaneus as a rigid sphere that compresses a soft and elastic tissue layer [25]. The abstract geometry of such models is extremely simple approximations of the natural phenomenon. Subsequent work from Luboz et al. performed a 3D simulation and examined internal Von Mises strains resulting from pressures applied to the foot sole to simulate a standing position [30]. The improvements of later work included more detailed models with full bone structure for a foot in addition to tendons and ligaments, including four layers: plantar skin, nonplantar skin, arch muscles, and fat between layers, all with different Neo-Hookean properties [30]. The geometry of these regions was constructed using MRI scans of subjects and can be considered a high-order approximation. This state-of-the-art work is costly to construct. For the question we address here, about the mechanical implications of having a PC layer at the heel, we seek a first-order approximation, more detailed than an abstract sphere and less costly than a clinically customized high-order model.

Another recent contribution aids our work by providing data about heel microanatomy. Cichowitz and Ashton [3] performed careful examination of foot vasculature and, in the course of that effort, came across important evidence about the PC. They found a thin PC muscle layer located within the soft tissue of the heel, just under the dermis of their subjects. Their quantification of PC geometry in the heel is our primary source of PC data. However, because their focus was the whole foot vasculature, they assessed the presence or geometry of a PC layer neither specifically at the posterior aspect of the heel nor in the transition from plantar to posterior surface. Because the dorsal (posterior) surface has very high risk of PU formation, these areas require additional study.

The high prevalence and human cost of PUs creates urgent need for improved methods of risk assessment. If the presence of a PC layer does redistribute biomechanical stress in soft tissues of the heel, as we hypothesize, then clinical imaging of the PC anatomy through means such as ultrasound might one day contribute to improved methods of risk assessment. In the meanwhile, we should not presume to know the net effect of PC anatomy on tissue viability. The firm elasticity of PC tissue might provide some protection of soft structures near bone, when loading is low enough to avoid ischemia of the muscle. On the other hand, the vulnerability of muscle tissue to ischemic injury, and the cytotoxic effects of muscle death on surrounding tissues, might cause harm during the prolonged static stress of immobility [3]. Mechanistically grounded principles for risk assessment will be important to develop, but we also need more tools for prevention, in cases where high risk has been identified. Prophylactic dressings are gaining momentum for prophylaxis of pressure injury [19], and interest may continue to grow if evidence of efficacy accumulates. The biomechanical engineering of dressings to prevent pressure injury is a relatively new field of research, and the field might benefit from biomimetic designs. If the PC has naturally evolved to play some protective function, or if our simulations show it improves load distribution, then PC anatomy may offer inspiration for the design of prophylactic dressings. This is particularly attractive option because a PC-like dressing on the outer surface of the skin might provide biomechanical protection similar to a natural PC layer under the skin, but unlike natural muscle, the dressing would not become ischemic, necrotic, or toxic after high stress.

2. Methods

2.1 Three-dimensional model of the heel

The three-dimensional (3D) geometry of our model with discrete surface layers (Fig. 3.1) was constructed in Solidworks (CAD software suite) using published values for the physical dimensions of different layers [3,27,32]. Our heel model uses simple geometrical volumes to provide a first-order approximation of the bone, tendon, soft tissue, and skin. A cylinder is used to model the lower leg, and a sphere is used to model the heel bone (calcaneus). To model the slight elevation of the lower leg from the plane, the sphere is rotated 100° instead of 90° from the coronal plane. Figs. 3.1 and 3.2 below show the 3D model of the heel constructed and used in the simulation.

The pressure applied to the calcaneus is intended to mimic the weight of one foot (2 kg), so the force was simulated to be 20 N (Fig. 3.3).

To simulate attachments between each tissue layer, the layers were merged into one solid with different mechanical properties in each region. To simulate pressure originating specifically from the calcaneus bone, we constructed the calcaneus as a separate object. The connections between bone and surrounding tissues are difficult to model because they

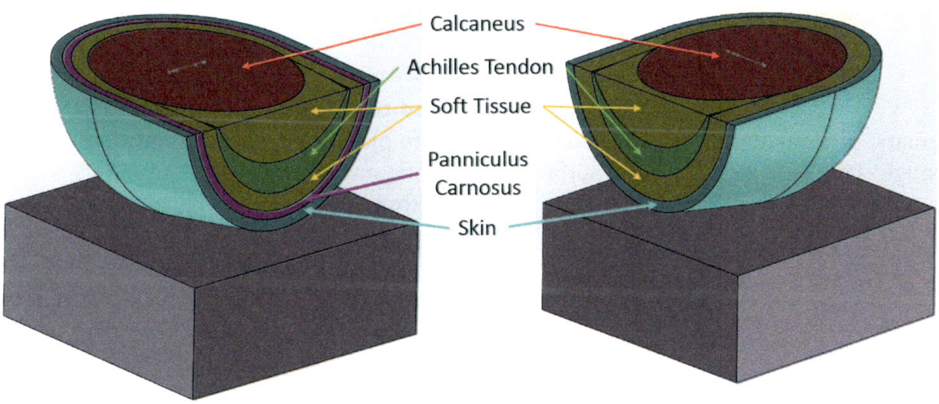

Figure 3.1: Three-dimensional geometry of the heel model, with panniculus carnosus (PC) muscle layer (left) or without (right).
The model without PC has a thicker layer of soft tissue layer, in place of the PC layer. For this and future images, red denotes bone, yellow denotes adipose-like soft tissue, green denotes tendon, purple denotes PC, and turquoise denotes skin.

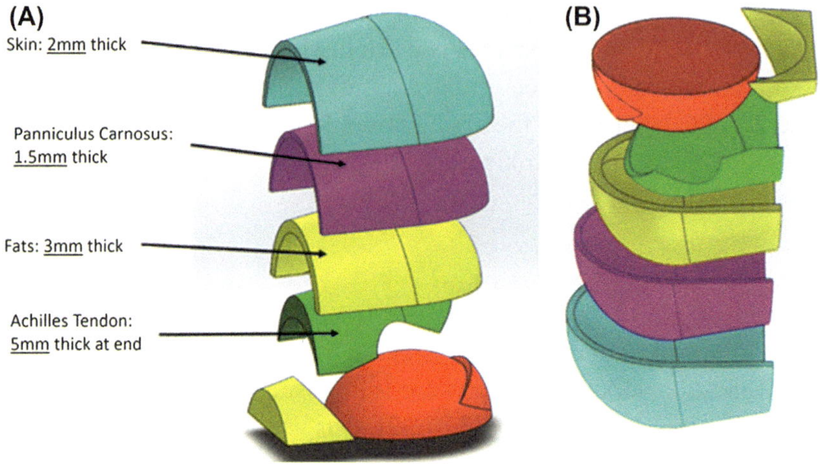

Figure 3.2: Geometry of the model layers.
(A) The thickness of each layer is annotated on an inverted orientation (skin on top and the calcaneus at the bottom) viewed from the medial–lateral perspective. (B) The same model is shown in upright orientation.

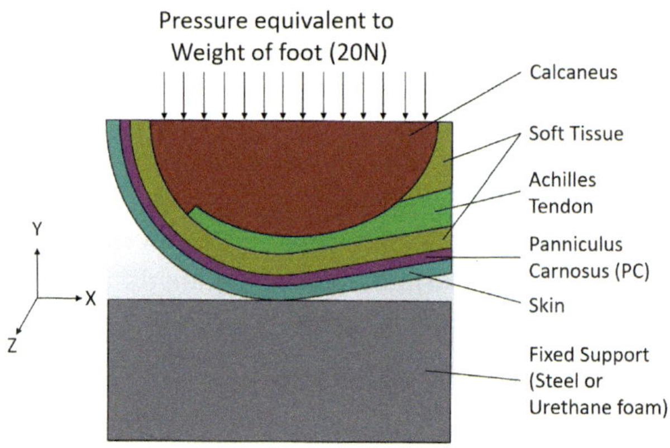

Figure 3.3: Load applied to the bone.
Schematic of simulated force, with constant pressure placed on the calcaneus and transmitted through the heel layers to a base surface. This configuration is intended to induce sustained deformation of the tissues, similar to that of a patient lying supine.

arise through many discrete microscale attachments to filamentous structures. As a macroscale approximation of the adhesion between bone and surrounding tissues, we introduced artificial friction (coefficient 1.5) that was sufficient to maintain contact along the interface between the bone and the surrounding tissues. Boundary conditions were set to prevent the model from rotating in any axis or displacing out of the frame. The top surface of the model will only displace in the Y direction, and the model will be allowed to expand in the X and Z directions to simulate later expansion of the soft tissue as it is flattened. In the superior surface of the heel, boundary conditions permitted movement only in the Y direction. This simulates resistance from the tissues present in the superior area of the heel. The bottom surface of the base will have zero rotation and displacement, to hold the model in place.

2.2 Simulations

The 3D assembly created using Solidworks was then imported into Abaqus, a software suite for finite element analysis. Mechanical properties were assigned to each layer as shown in Table 3.1. The model was meshed using tetrahedral rather than hexagonal elements due to sharp angles in the soft tissue layer. We added a base layer with mechanical properties of steel (e.g., the footrest of a standard wheelchair) or urethane foam (e.g., a mattress) [33]. The hard base is expected to accentuate the differences in stress behavior between the two models, and the soft foam is expected to minimize the differences.

Following prior work [31,38,39], we used Neo-Hookean hyperelastic properties (Table 3.2) for the more elastically deformable layers, namely the soft tissue, skin, and muscle. Including this deformability improves the model because strong deformations of the softer layers would violate a general elastic model. Here, the shear modulus of the PC is taken as that of a slack muscle and its bulk modulus (K) is calculated using the shear modulus (G) and poisson ratio (v) according to the following equation:

$$K = \frac{2G(1+v)}{3(1-2v)}$$

Table 3.1: Mechanical properties of tissue layers.

Tissue Layer	Young's Modulus, E (MPa)	Poisson's Ratio, v
Soft tissue [34]	2	0.45
Achilles tendon [35]	520.4	0.49
Bone [36]	20,000	0.42
Base (steel) [37]	210,000	0.3
Urethane [33]	0.025	0.0

Table 3.2: Hyperelastic Neo-Hookean properties of tissue layers.

Tissue Layer	Shear Modulus, G (Mpa)	C10: Shear/2 (Mpa)	Bulk Modulus, K (Mpa)	D1: 2/Bulk (1/Mpa)
Soft tissue (fats) [31]	0.000286	0.000143	0.0285	70.175
Panniculus carnosus	0.0071 [38]	0.00355	0.08019 [39]	24.94
Skin [31]	0.039	0.01695	3.179	0.629

After running simulations with a steel base to simulate a sturdy support, we also ran simulations using the properties of urethane foam, a material commonly used in seating and mattresses. When using Neo-Hookean hyperelastic properties, the simulations took 14–24 h to complete on a PC with Intel I7 4 GHz processor and 16 Gb of RAM.

3. Results

After obtaining the final results of the simulation for the model with PC, we then ran the simulation except using the model without PC. Each pair of simulations was done using the same material properties, with the same boundary conditions and interactions in place, and the only difference was the presence or absence of the PC. In each mesh, there were approximately 18,300 elements generated with 47,000 nodes. A comparison of the stress fields from the simulations can be seen in the heat maps of Figs. 3.4 and 3.5. These heat maps use the same ranges of stress, between 0.00,002 Mpa (blue) to 3 Mpa (red). White/gray denotes any stress greater than 3 Mpa.

Figs. 3.4 and 3.5 show a distinctive difference in stress behavior, depending on the presence/absence of the PC. We focus attention only on the regions were PU can occur (not inside the bone and not outside the body). Specifically, we focus on regions with >3 Mpa of stress in soft tissues layers of the fat, skin, and muscle. The model without the PC experiences this extreme level of stress in a broad column between bone and base, passing through all the tissue layers. In contrast, the model with PC has low stress in skin and only a small region of high stress >3 Mpa in the fatty soft tissue.

After obtaining results for a heel resting on a steel base, we next ran simulations using urethane foam as the base. Models with and without PC were simulated using these

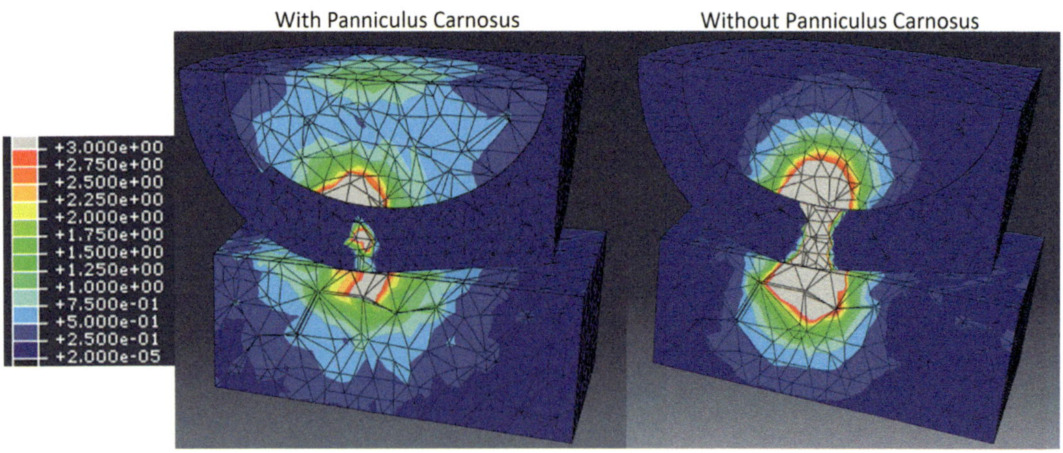

Figure 3.4
Stress heat map of the simulation with steel base. Model is viewed in the X-plane (range is in Mpa).

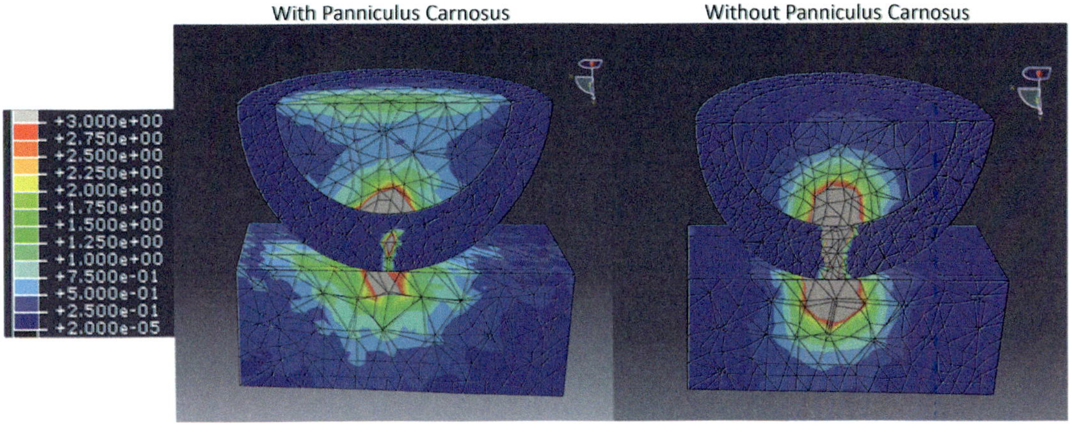

Figure 3.5
Stress heat map of the simulation with steel base. Model is viewed in the Z-plane (range is in Mpa).

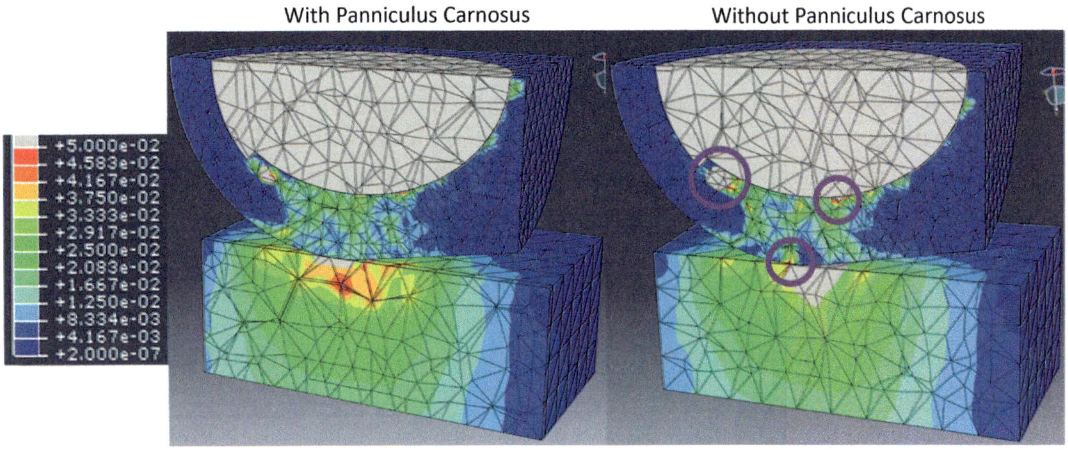

Figure 3.6: Stress heat map of the simulation with urethane foam base. Model is viewed in the X-plane (range is in Mpa).

conditions (i.e., the same boundary conditions and interactions in place), and all material properties unchanged except the soft base. We calculated the effects of applying 20 N force to the heel bone (mimicking the weight of a foot), and the resulting stress fields are shown as heat maps in Figs. 3.6 and 3.7 below. Compared with the results using a steel base, the heat maps for stress on a foam base show a much lower range of stress values, between 0.0000002 Mpa (blue) and 0.05 Mpa (red). White denotes any stress value above 0.05 Mpa. The calcaneus is colored white because it experiences stress above the mapped range, but this heat map is calibrated for a much lower range than Figs. 3.4 and 3.5.

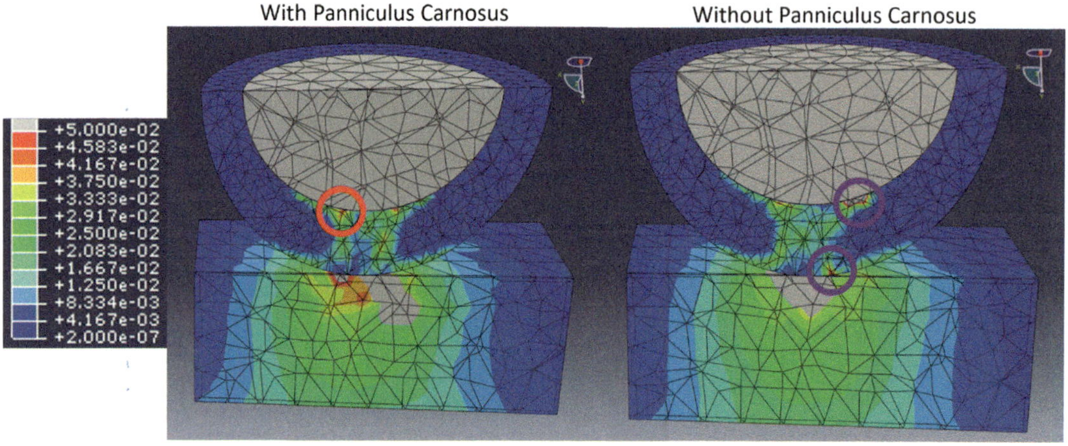

Figure 3.7: Stress heat map of the simulation with urethane foam base.
Model is viewed in the Z-plane (range is in Mpa).

Because the foam surface creates far less stress than steel, the impact of the PC is more subtle. Fig. 3.7 is annotated with circles showing some differences. However, because of meshing and numerical approximation at each node, it is not possible to infer significance from small differences in isolated nodes. Overall, the model without PC still experiences a higher amount of stress in the layers of soft tissue, but this might not have any practical importance when the base support is soft.

4. Discussion

A simplified 3D biomechanical model of heel tissues was developed for determining whether the presence of a firm elastic layer, resembling the PC muscle, would mitigate stresses typical of PU formation. The heel model consists of three large components: a rigid bone, a merged set of nonbone layers (fats, tendon, muscle, and skin; in the model without, the PC will have fats replaced by the muscle layer), and a rigid support surface below. Stress behavior was computed using a biomechanical simulation of the heel on a flat base using the Abaqus software suite.

The results from the biomechanical simulations suggest that there are differences between the two models, when loaded on a hard surface. In the presence of the PC tissue, the heel will experience higher stress in the inner layers of the heel, as seen in the white and red parts in the heat map. In the absence of the PC tissue, we can observe that the heel experiences high stress in a much larger area in surrounding tissues, which spans from the contact at the calcaneus to the contact point at the inferior part of the model. This suggests that the presence of a thin muscle layer can improve the distribution of load during high-stress conditions and motivate us to propose development of PC-like biomimetic dressings for adhesion to the skin surface. When we run the simulation using a soft base of support, the differences become much smaller, suggesting that the PC is less important to the biomechanics in a well-cushioned context. As expected, dressings to prevent pressure injury would be less likely to bring any benefit in low-stress contexts.

There are several limitations to our approach that need to be considered when deciding whether to pursue development of PC-like dressings for PU prevention. The first is that our model is a first-order approximation of the heel geometry and is not faithful to the anatomical details of a real human heel. The approximation was constructed using published dimensions, but without use of clinical imaging. For example, the fat layer has been modeled to entirely fill a volume that should also contain vasculature and nerves. Although these nonfat components make up a very small faction of the soft tissue volume, they have disproportionate importance, which the current model neglects. Secondly, the mechanical properties of the different tissue types were taken from prior publications with macroscopic mechanics represented by bulk parameters. Such simplifications fail to reflect the complexity of microarchitecture and heterogeneous reality. Although we may aspire to comprehensive parameterization of computational models through empirical fitting of extensive measurements (as for biochemical systems [40]), biomechanical models are only starting to approach authenticity to human patients [41]. A third limitation is the set of interactions between the different layers. Our simulations used high friction to create connection between the calcaneus and the surrounding tissue layers, but future work should improve this approximation. Lastly, the simulated location of the PC is just below the skin, due to the PC being a dermal muscle layer. However, the detailed placement of PC in human heels requires further clarification, especially for its continuation from the plantar to the dorsal side of the heel. Simulations can still be done with different positions of the PC with the current conditions and finite element simulation methods, simply by changing the position of the PC in the 3D model (e.g., using Solidworks software). Future simulations, in conjunction with dressing design and development, should demonstrate the effects of PC-like layers on the external surface of the heel.

In conclusion, we used 3D finite element modeling to run simulations of pressure on a heel-like geometry, and we compared the levels of tissue deformation with and without the PC layer. The presence of the PC caused consistent and extensive decrease in stress experienced by soft tissues resting on a firm surface. This could be due to its elasticity and compressibility, allowing redistribution of the applied forces over larger regions. Future work can provide more detailed models with greater anatomical accuracy, based on dissection or clinical imaging of actual heels, and based on engineered PC-like dressings for adhesion to the epidermis, at the dorsal heel and other vulnerable regions. Understanding our native biology for preventing pressure injury will aid in design of safe and effective interventions such as prophylactic dressings.

Acknowledgments

We are grateful to Sandeep Jacob Sebastin, Ang Eng Tat, Julia H. Jenkins, Elson Neo, and Goh Jia Ling for fruitful discussions about the panniculus carnosus.

We gratefully acknowledge support to LTK from the Singapore Ministry of Health's National Medical Research Council (NMRC) under its Open Fund Individual Research Grant scheme (OFIRG15nov062) and also from the National Research Foundation (NRF), Prime Minister's Office, Singapore, under its CREATE program, Singapore-MIT Alliance for Research and Technology (SMART) BioSystems and Micromechanics (BioSyM) IRG.

References

[1] National Pressure Ulcer Advisory Panel. Pressure definition and stages revised by NPUAP. February 2007. Available at: www.npuap.org/documents/PU_Definition_Stages.pdf.

[2] Barton AA. The pathogenesis of skin wounds due to pressure. J Tissue Viability 2006;16(3):12–5.

[3] Cichowitz A, Pan WR, Ashton M. The heel: anatomy, blood supply, and the pathophysiology of PUs. Ann Plast Surg 2009;62(4):423–9. https://doi.org/10.1097/SAP.0b013e3181851b55.

[4] Kalogeris T, Baines CP, Krenz M, Korthuis RJ. Cell biology of ischemia/reperfusion injury. Int Rev Cell Mol Biol 2012;298:229–317. https://doi.org/10.1016/B978-0-12-394309-5.00006-7.

[5] Jagannathan NS, Tucker-Kellogg L. Membrane permeability during pressure ulcer formation: a computational model of dynamic competition between cytoskeletal damage and repair. J Biomech 2016;49(8):1311–20. https://doi.org/10.1016/j.jbiomech.2015.12.022.

[6] Slomka N, Gefen A. Relationship between strain levels and permeability of the plasma membrane in statically stretched myoblasts. Ann Biomed Eng 2012;40(3):606–18. https://doi.org/10.1007/s10439-011-0423-1.

[7] Nguyen BP, Heemskerk H, So PTC, Tucker-Kellogg L. Superpixel-based segmentation of muscle fibers in multi-channel microscopy. BMC Syst Biol 2016;10(Suppl. 5):124. https://doi.org/10.1186/s12918-016-0372-2.

[8] Bahl N, Winarsih I, Tucker-Kellogg L, Ding JL. Extracellular haemoglobin upregulates and binds to tissue factor on macrophages: implications for coagulation and oxidative stress. Thromb Haemostasis 2014;111(1):67–78. https://doi.org/10.1160/TH13-03-0220.

[9] Venkatraman L, Tucker-Kellogg L. The CD47-binding peptide of thrombospondin-1 induces defenestration of liver sinusoidal endothelial cells. Liver Int 2013;33(9):1386–97. https://doi.org/10.1111/liv.12231.

[10] Umoh NA, Walker RK, Millis RM, Al-Rubaiee M, Gangula PR, Haddad GE. Calcitonin gene-related peptide regulates cardiomyocyte survival through regulation of oxidative stress by PI3K/Akt and MAPK signaling pathways. Ann Clin Exp Hypertens 2014;2(1):1007.

[11] Nim TH, Luo L, White JK, Clément M-V, Tucker-Kellogg L. Non-canonical activation of Akt in serum-stimulated fibroblasts, revealed by comparative modeling of pathway dynamics. PLoS Comput Biol 2015;11(11):e1004505. https://doi.org/10.1371/journal.pcbi.1004505.

[12] Shi Y, Mellier G, Huang S, White J, Pervaiz S, Tucker-Kellogg L. Computational modelling of LY303511 and TRAIL-induced apoptosis suggests dynamic regulation of cFLIP. Bioinformatics (Oxford, England) 2013;29(3):347–54. https://doi.org/10.1093/bioinformatics/bts702.

[13] Tucker-Kellogg L, Shi Y, White JK, Pervaiz S. Reactive oxygen species (ROS) and sensitization to TRAIL-induced apoptosis, in Bayesian network modelling of HeLa cell response to LY303511. Biochem Pharmacol 2012;84(10):1307–17. https://doi.org/10.1016/j.bcp.2012.08.028.

[14] Venkatraman L, Chia S-M, Narmada BC, White JK, Bhowmick SS, Forbes Dewey C, So PT, Tucker-Kellogg L, Yu H. Plasmin triggers a switch-like decrease in thrombospondin-dependent activation of TGF-β1. Biophys J 2012;103(5):1060–8. https://doi.org/10.1016/j.bpj.2012.06.050.

[15] Cheah J. Chronic disease management: a Singapore perspective. BMJ 2001;323(7319):990–3.
[16] Fowler E, Scott-Williams S, McGuire JB. Practice recommendations for preventing heel PUs. Ostomy/Wound Manag 2008;54(10):42–57.
[17] Content last reviewed Are we ready for change. Rockville, MD: Agency for Healthcare Research and Quality; October 2014. https://www.ahrq.gov/professionals/systems/hospital/pressureulcertoolkit/putool1.html.
[18] Levy A, Kopplin K, Gefen A. An air-cell-based cushion for pressure ulcer protection remarkably reduces tissue stresses in the seated buttocks with respect to foams: finite element studies. J Tissue Viability 2014;23(1):13–23. https://doi.org/10.1016/j.jtv.2013.12.005.
[19] Byrne J, Nichols P, Sroczynski M, Stelmaski L, Stetzer M, Line C, Carlin K. Prophylactic sacral dressing for pressure ulcer prevention in high-risk patients. Am J Crit Care 2016;25(3):228–34. https://doi.org/10.4037/ajcc2016979.
[20] Walker R, Aitken LM, Huxley L, Juttner M. Prophylactic dressing to minimize sacral pressure injuries in high-risk hospitalized patients: a pilot study. J Adv Nurs 2015;71(3):688–96. https://doi.org/10.1111/jan.12543.
[21] Van Gilder C, Washienko C, Eckstein A, Decker S, MacFarlane G. International pressure ulcer prevalence survey results for long-term acute care facilities in the United States. 2005. Available at: www.sawc.net/ses/sawc/abstracts/06323.
[22] Van Gilder C, Macfarlane GD, Meyer S. Results of nine international pressure ulcer prevalence surveys: 1989 to 2005. Ostomy/Wound Manage 2008;54(2):40–54.
[23] Black J. Preventing PUs occurring on the heel. Wounds Int 2012;3(3):1–4.
[24] Salcido R. Editorial: what is the purple heel? Adv Skin Wound Care 2006;19(1):6.
[25] Gefen A. The biomechanics of heel ulcers. J Tissue Viability 2010;19(4):124–31. https://doi.org/10.1016/j.jtv.2010.06.003.
[26] Greenwood JE. Function of the panniculus carnosus—a hypothesis. Vet Rec 2010;167(19):760. https://doi.org/10.1136/vr.c6210.
[27] Ronald AB, Adel KA, Ryosuke M, Carnosus P. Illustrated encyclopedia of human anatomic variation: opus I: muscular system: alphabetical listing of muscles: P. March 1, 2018. Retrieved From, http://www.anatomyatlases.org/AnatomicVariants/MuscularSystem/Text/P/05Panniculus.shtml.
[28] Spina AA. The plantaris muscle: anatomy, injury, imaging, and treatment. J Can Chiropr Assoc 2007;51(3):158–65.
[29] Moore CW, Rice CL. Structural and functional anatomy of the palmaris brevis: grasping for answers. J Anat 2017;231(6):939–46. https://doi.org/10.1111/joa.12675.
[30] Luboz V, Perrier A, Stavness I, Lloyd JE, Bucki M, Cannard F, et al. Foot ulcer prevention using biomechanical modelling. Comput Methods Biomech Biomed Eng Imag Vis 2014;2(4):189–96. https://doi.org/10.1080/21681163.2013.837410.
[31] Levy A, Frank MB-O, Gefen A. The biomechanical efficacy of dressings in preventing heel ulcers. J Tissue Viability 2015;24(1):1–11. https://doi.org/10.1016/j.jtv.2015.01.001.
[32] Kurihara T, Sasaki R, Isaka T. Mechanical properties of achilles tendon in relation to various sport activities of collegiate athletes. 2012.
[33] Shigekazu I, Keiko I, Mitsuo N. Finite element estimation of pressure distribution inside the trunk on a mattress. 2015. Retrieved April 9, 2018, from https://www.ausmt.org/index.php/AUSMT/article/view/949/462.
[34] Choi PA. Estimation of Young's modulus and Poisson's ratio of soft tissue using indentation. The Hong Kong Polytechnic University; 2009. Retrieved February 20,2018 from http://hdl.handle.net/10397/2777.
[35] Cui L, Maas H, Perreault EJ, Sandercock TG. In situ estimation of tendon material properties: differences between muscles of the feline hindlimb. J Biomech 2009;42(6):679–85. https://doi.org/10.1016/j.jbiomech.2009.01.022.

[36] Rupin F, Saied A, Dalmas D, Peyrin F, Haupert S, Barthel E, Boivin G, Laugier P. Experimental determination of Young modulus and Poisson ratio in cortical bone tissue using high resolution scanning acoustic microscopy and nanoindentation. J Acoust Soc Am 2008;123(No 5) (n.d.). Retrieved February 21, 2018, from https://asa.scitation.org/doi/abs/10.1121/1.2935440.

[37] Beardmore R. Properties of metals. 2010. Retrieved February 21, 2018, from http://www.roymech.co.uk/Useful_Tables/Matter/prop_metals.htm.

[38] Koo TK, Guo J-Y, Cohen JH, Parker KJ. Quantifying the passive stretching response of human tibialis anterior muscle using shear wave elastography. Clin Biomech (Bristol, Avon) 2014;29(1):33−9. https://doi.org/10.1016/j.clinbiomech.2013.11.009.

[39] Hansang K, Lawrence Y, Andrew S, Joseph LD. Determination of poisson ratio of bovine extraocular muscle by computed X-ray tomography. BioMed Res Int 2012;2013. https://doi.org/10.1155/2013/197479. Article ID 197479, 5 pages, 2013.

[40] Nim TH, Luo L, Clément M-V, White JK, Tucker-Kellogg L. Systematic parameter estimation in data-rich environments for cell signalling dynamics. Bioinformatics 2013;29(8):1044−51. https://doi.org/10.1093/bioinformatics/btt083.

[41] Bucki M, Luboz V, Perrier A, Champion E, Diot B, Vuillerme N, Payan Y. Clinical workflow for personalized foot pressure ulcer prevention. Med Eng Phys 2016;38:845−53. https://doi.org/10.1016/j.medengphy.2016.04.017. pdf.

CHAPTER 4

Lower leg compression and its biomechanical effects on the soft tissues of the leg

Fanette Chassagne[1,2], Pierre Badel[1], Jérôme Molimard[1]
[1]Mines Saint-Etienne, Univ Lyon, Univ Jean Monnet, INSERM, U 1059 Sainbiose, Centre CIS, Saint-Etienne, France; [2]Thuasne, Saint-Etienne, France

Chapter Outline

1. Introduction

The role of the venous system is to flow the blood back to the heart. In the lower leg, in standing position, blood flow must counteract gravity. For this, several mechanisms contribute to venous return: the sympathetic nervous activity, the calf pump, and the valves

Innovations and Emerging Technologies in Wound Care. https://doi.org/10.1016/B978-0-12-815028-3.00004-3

in the lower leg veins. If one or several of these mechanisms become deficient, blood stagnation and reflux may occur and lead to venous diseases.

Elastic compression of the lower leg is the traditional preventive and curative treatment of venous insufficiency. This compression, whose clinical efficacy is widely admitted, can either be performed, thanks to stockings or bandages. Compression stockings are made of elastic fabric and are sized to the patient leg. They are prescribed for the treatment of nonacute stages of venous insufficiency and can be put on by the patient or a nontrained person. Bandages consist in a stretched fabric wrapped around the leg and are mainly used to treat the most severe pathologies, such as venous ulcers. As they are not sized to the patient's leg, they are also recommended for the first stages of the treatment, when the leg swelling is reduced rapidly. Moreover, they are prescribed when the patient's pathology prevents the use of stockings (for knee arthroplasty for instance). However, bandages must be applied by experienced care givers, and this treatment modality is very operator dependent.

Both medical devices apply a pressure onto the skin called interface pressure. The basic principle of compression therapy may seem simple: the pressure applied onto the skin is transmitted through soft tissues (muscle and adipose tissues—Fig. 4.1) to the veins. This pressure tends to reduce the diameter of the vein and consequently the blood volume in the leg. However, though the efficacy of this treatment is admitted, the therapeutic objectives are not always reached, underlying the fact that the mechanisms of compression therapy are not yet fully understood.

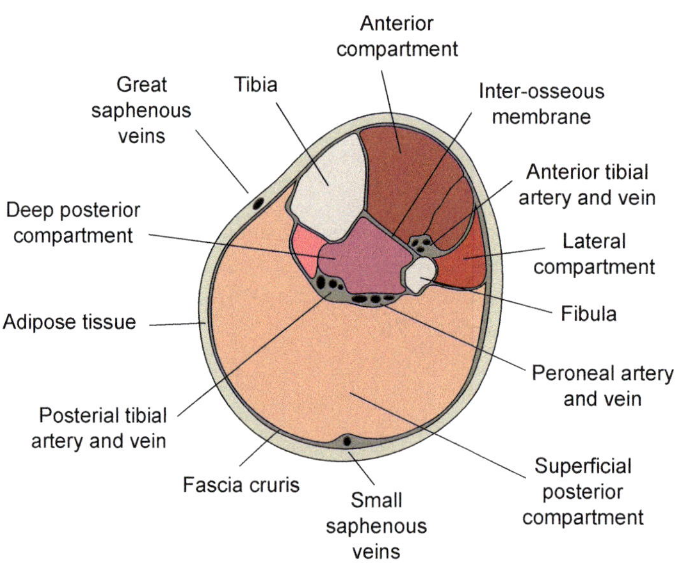

Figure 4.1

Cross section of the lower leg. *Adapted from Braus H, C Elze. Anatomie Des Menschen: Ein Lehrbuch Für Studierende Und Ärzte, vol. 1. J. Springer; 1921.*

A first step toward improved understanding is to investigate how interface pressure is generated, then how it is transmitted to the veins, and how this pressure will enhance venous return. In vivo measurements of interface pressure are a common clinical approach, but investigating pressure transmission, for instance, is more challenging. For this, engineering approaches such as numerical simulation may represent promising tools. Numerical simulation can provide in situ information that is very difficult or impossible to measure and could result in the prediction of patient-specific outcomes of the treatment.

After a description of the medical management of compression treatment, this chapter introduces the current advances in both clinical and engineering approaches designed to study the biomechanical effects of compression therapy. Perspectives and current questions are addressed at the end of the chapter.

2. Medical strategy

2.1 Venous return and chronic venous insufficiency

Although the pathogenesis of venous insufficiency is not fully understood [1], this pathology occurs when venous return becomes less competent. The role of the venous system is to flow deoxygenated blood from the lower extremities to the heart. In standing position, because of gravity, the intravenous pressure increases from the upper leg to the lower leg [2] and can reach 75 mmHg [3]. Thus, venous return must play against pressure gradient and gravity [4].

For this, main mechanisms compose venous return are:

- *The sympathetic activity on venous return*: The venous wall is composed of smooth muscular cells that can contract under the effect of the sympathetic nervous system. When the flow is too low, the cells contract to enhance venous return and counteract gravity [6].
- *The contraction of the calf muscles and the vein valves*: While contracting, the calf muscles compress deep veins. These veins have valves that prevent backflow. The compression of veins by their surrounding muscles results in an increase in intravenous pressure and the opening of the valves. Blood will be propelled to the heart, and valves will close again when the pressure decreases (Fig. 4.2). The calf pump provides about 50% of the energy required for venous return [6].

A lower efficiency of venous return will lead to blood stasis, reflux, and an increased intravenous pressure. This pathology is called chronic venous insufficiency and ranges from "heavy leg" feelings to venous ulcers. The most common conservative and preventive treatment for this pathology is compression therapy, which can be performed, thanks to stockings, bandages, or intermittent pneumatic compression devices.

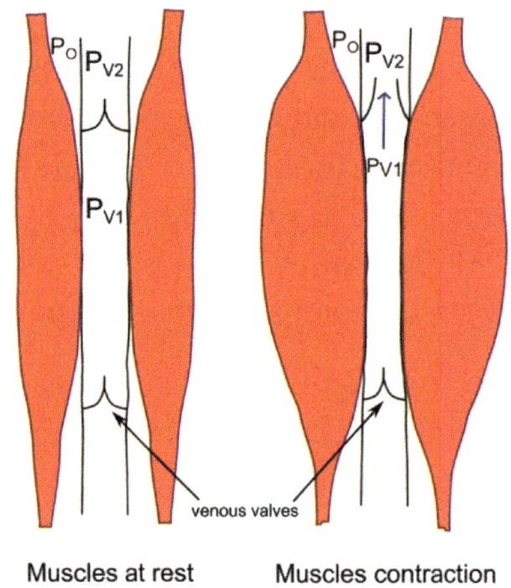

Figure 4.2
Illustration of the calf pump and deep veins valves.

The main objective of compression therapy is to reduce the difference in pressure between the inside and the outside of the venous wall [3], also called transmural pressure, to reduce the caliber of veins and facilitate venous return. The action mechanism of compression consists in applying a pressure onto the leg, which is called interface pressure. This external pressure is then transmitted through soft tissues (adipose and muscle tissues—Fig. 4.1) to the veins [2,7].

Although it is widely admitted that compression enhances venous return, the action mechanisms of compression therapy are not fully understood [8] and neither is their relative influence. Nevertheless, irrespectively of compression modality, interface pressure remains the key aspect of the treatment.

2.2 Interface pressure as a therapeutic dosage

Interface pressure applied by compression devices should be adapted to each patient and his/her pathology [9]. However, the level of pressure recommended for each pathology is not clearly defined. It was observed that even low pressure (about 20 mmHg) can improve venous pumping [10]. Increasing interface pressure (up to 60 mmHg) would result in a larger improvement of venous pumping [10] and was proven to have a better therapeutic effect on ulcers [11]. Higher interface pressure can result in veins narrowing. This would lead to a reduction of venous reflux, [12] venous hypertension [13], and to an increase of calf pump function [14]. However, if the applied pressure is too high, the treatment becomes less bearable for patients and can lead to complications such as pressure ulcers [10]. For the best compromise between treatment observance and efficacy, it is often recommended to apply the highest pressure the patient can tolerate.

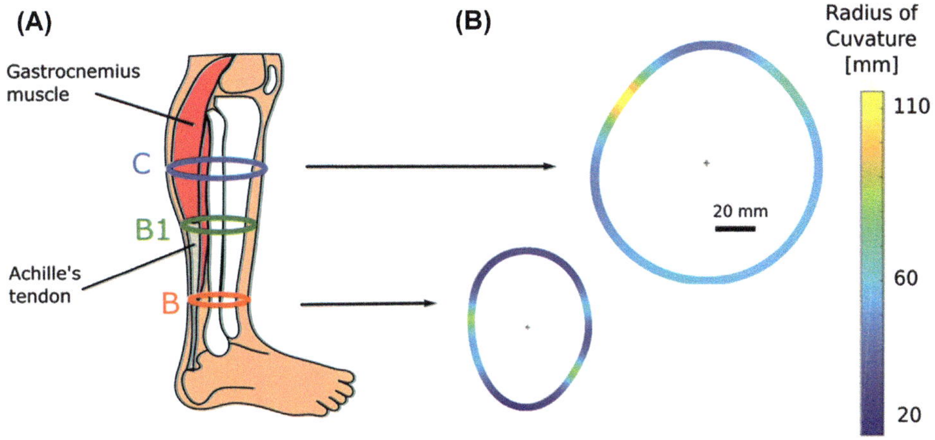

Figure 4.3
(A) Location of standard measurement points; (B) illustration of the variation of local radii along the leg.

To standardize the quantification of compression therapeutic dosage, standard measurement points have been defined for in vivo interface pressure measurement [15] (Fig. 4.3A):

- Measurement point B: right above the ankle where the leg circumference is the smallest
- Measurement point B1: when the Achilles tendon turns into the gastrocnemius muscles
- Measurement point C: at the largest circumference of the calf.

Measurement point B is the reference point for the design of textile compression devices and is also the location where most ulcers occur [16]. Lately, measurement point B1 was considered as the point of interest for in vivo pressure measurements [17]. In this area, the change in leg circumference from supine to standing position (and while walking) is very large. Thus, interface pressure variations in this area are more important [18].

Yet, interface pressure is also influenced by the patient's morphology. Compression stockings are sized to the calf circumference, within the aim of applying similar pressures for all morphologies. To design compression stockings, manufacturers commonly use wooden legs of different sizes (the so called Hohenstein legs) to assess the pressure applied by a stocking at measurement point B [19]. The target pressures are divided into "compression classes" that vary among countries (Table 4.1).

On the other hand, compression bandages are applied with the same stretch for any patients' morphology. The same bandage being applied on different patients' morphology will result in different interface pressures [20].

In addition, pressure may vary between a passive muscular state (relaxed calf muscles) and an active muscular state (active calf muscles, while walking, for instance) [21].

Table 4.1: Pressures (in mmHg) at measurement point B for different compression classes and countries [100].

Compression class	United States	United Kingdom	France	Germany
I	15–20	≤20	10–15	18–21
II	20–30	21–30	15–20	23–32
III	30–40	31–40	20–36	34–46
IV	40	41–60	36	49

These pressures are called resting and working pressures. Thus, interface pressure is also influenced by body position [22].

To reach the appropriate interface pressure, there are several ways of modulating the compression treatment. An additional compression stocking can be worn on top of a stocking to increase interface pressure [23]. For compression bandages, varying the application technique (either the wrapping pattern or the bandage overlapping) would have an impact on the pressure [24]. In addition, it is possible to combine two to four bandages with different mechanical properties to modify the applied interface pressure and its variations with body position [25].

2.3 Progressivity versus degressivity

Most often, compression stockings or bandages apply a degressive profile of pressure: the applied pressure decreases from the ankle to the knee (Fig. 4.4A—left). This profile is mainly the consequence of the conical shape of the leg: the radius of curvature is smaller at the ankle than at the larger circumference of the calf (Fig. 4.3A). In addition, such pressure profile tends to counteract the increase of intravenous pressure from the knee to the ankle because of gravity. The therapeutic efficacy of degressive compression profile is widely admitted [26].

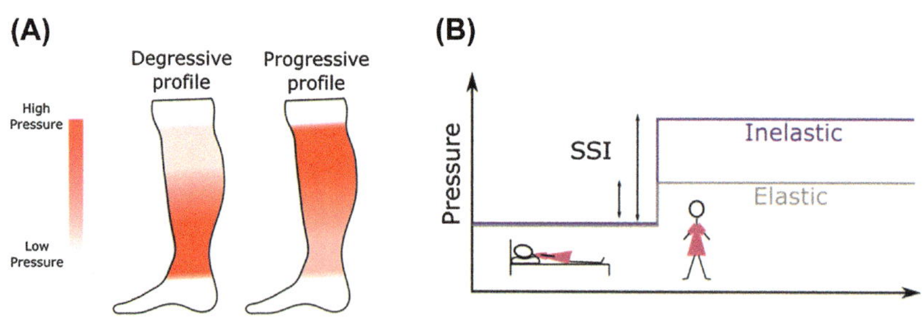

Figure 4.4
(A) Illustration of a degressive and a progressive pressure profile; (B) static stiffness index for an elastic and an inelastic bandage.

Lately, another type of compression profile was tested, called progressive. In the case of a progressive compression device, the pressure applied at the largest circumference of the calf (measurement point C) is higher than that at the ankle (Fig. 4.4A—right). The therapeutic effect of progressive compression was assessed either with stockings [27] or bandages [28]. For the least severe forms of the pathologies, the efficacy of progressive compression was found to be equivalent to degressive compression [29]. Whether progressive compression is performed with stockings [27] or bandages [30], the positive impact on venous return (venous ejection fraction) was found to be higher than with degressive compression. However, these outcomes result from short-term experiments, and the long-term effects of progressive compression still needs to be investigated [31].

2.4 Elasticity and stiffness

An important property of compression therapy with bandages is elasticity. In the case of bandages, elasticity is not defined as the capacity of the material to return to its original shape after being stretched [32]. Instead, bandages are divided in three groups depending of the bandage stretch under a given applied force (10 N/cm, [33]) (Table 4.2).

Usually, elasticity and extensibility are connected [34]: long-stretch bandages are associated with the so-called "elastic" properties and short-stretch bandages with the so-called "inelastic" properties. The main clinical difference between bandages with different mechanical properties is the pressure variation from supine to standing position [35]. Compression bandages are applied on the lower leg while the patient is lying in supine position. From supine to standing position, the leg volume and the calf mechanical properties change. Elastic bandages will tend to adapt to these changes, which will result in a low pressure increase from supine to standing position (Fig. 4.4B). For inelastic bandages, the pressure increase is higher [22,32].

Besides mechanical tensile testing, a clinical method allows assessing the mechanical properties of a bandage: the static stiffness index (SSI). The SSI is the pressure increase from supine to standing position [22]. A bandage is considered as inelastic if the SSI is higher than 10 mmHg [22].

The difference in elasticity is the cornerstone of the difference between compression and contention. *Compression* consists in the effects induced on a leg by an elastic bandage

Table 4.2: Classification of compression bandages.

	Inelastic		Elastic
Maximal stretch at 10 N/cm bandage width	Rigid 0–10	Short-stretch 10–100	Long-stretch >100

(active), whereas *contention* is the passive action of an inelastic fabric on the leg [3]. The inelastic fabric will oppose to a change in leg morphology (from supine to standing position, for instance), whereas elastic fabric will comply with such change and apply quasiconstant pressure on the leg. Different bandage properties will result in different clinical outcomes. It was shown that the elasticity of the bandage may have a positive impact on ulcers healing [36]. Other studies have observed a better improvement of venous return with short-stretch bandages compared with long-stretch bandages [12,14].

2.5 Different schools of compression

The diversity of modalities for compression therapy is reflected by the different schools of compression within countries. For instance, in France, elastic compression is commonly prescribed: mainly elastic stockings and elastic bandages [3]. In Germany and Austria, the standard treatment is also performed with stockings and bandages. However, inelastic compression is usually prescribed. In England, the four-layer bandage is the standard treatment for venous ulcers [37]. This bandage is composed of a wool bandage (padding layer), a creped bandage, an elastic bandage, and a cohesive bandage [38]. In the Netherlands, the treatment of venous ulcers is mainly performed with inelastic compression or intermittent pneumatic compression devices.

Although the kind of compression therapy may differ among countries, it is most often the standard treatment for venous insufficiency. Nevertheless, in the United States, compression stockings are sparingly prescribed for chronic venous insufficiency or after sclerotherapy, and compression bandages are never prescribed.

The prescription of the appropriate compression is a real challenge. It must be adapted to the patient and its pathology while being the best compromise between a high interface pressure and comfort for the patient (and compliance to the treatment).

The large discrepancies among treatment modalities and traditions underline the need of a better understanding of the mechanisms of compression therapy, which was addressed through clinical and engineering studies.

3. Clinical approach

3.1 Pressure measurement

3.1.1 Pressure measurement systems

The main effect of elastic compression is the pressure applied on the skin, which induces a change in depth, at the levels of tissues and veins. Therefore, the pressure applied onto the leg has been measured and/or estimated in many studies as an efficacy indicator.

Pressure sensors are based on four main technologies: pneumatic, electropneumatic, capacitive, or resistive, corresponding to well-known commercial products, for example, PicoPress (Neots Cambridgeshire, United Kingdom), Salzmann (Boston MA, USA), X-Sensor (St. Gallen, Switzerland), and finally Tekscan (Calgary Alberta, Canada) or Novel (Munich, Germany), respectively. Resistive or capacitive sensors may be assembled into a structure that enables the pressure to be measured at several points simultaneously; this structure is often called a pressure mapping system.

In the medical field, many authors have used pressure measurement to evaluate medical devices. They have been employed to prevent bedsores [39–45] and wheelchair ulcers [46], in the study of rigid or soft braces [47–49], or more recently for ulcer prevention in diabetic patients [50,51] or even in the study of bandage on animals [52]. Yet, most work has been developed to measure the interface pressure of compression stockings or compression bandages [17,53–59].

For a given study, the choice of the right pressure sensor has to be made with care. Pneumatic and electropneumatic sensors are easy to use and rather low cost, but the quality of measurements depends on the user. These sensors give a single measurement on a large area compared with the expected pressure spatial variations on the leg. In addition, these sensors have a low (i.e., good) resolution that is well adapted to the pressure range to be measured on a leg. Capacitive sensors usually have a low measurement error but are more expensive, and their drift makes them more relevant for dynamic measurements than static (most usual) measurements. The latest resistive sensors have a good resolution and a reasonable price compared with capacitive ones. However, their sensitivity to surface curvature has been pointed out, even if piezoresistive ones seem to be less affected.

From this brief overview of existing technologies, it is shown that the choice of a pressure measurement device should be driven not only by the expected resolution, measuring range, acceptable error, and cost but also by the size and the curvature of the measured area. Note that sensors with a curvature fitting the curvature of the zone to be measured have even been specifically designed, for example, for rigid bracing [60] or for prosthetic sockets [59]. Last, it should be mentioned that in compression therapies, most of the clinical or experimental work has been done using pointwise pneumatic or electropneumatic sensors [61]. An overview of these studies is presented below.

3.2 Clinical studies

Clinical studies related to elastic compression address different health problems. They have been applied to venous ulcers, varicose veins, deep vein thrombosis, postsurgery, postsclerotherapy, or postphlebitic legs. The use of compression hosiery opened the way to various questions: the compression level, the impact of the hosiery design on the desired level of compression in particular with time, the representativeness of phantom legs, the patient compliance, etc.

Different classifications are used in compression hosiery (Table 4.1). They are based on local practices (Great Britain, France, Germany, USA) and define compression levels and pressure degressivity. As a logical consequence, several studies investigated the pressure effectively applied on the leg. Reich-Schupke [62] showed that 32% were out of the target, and among them, a quarter above could result in clinical complications. Partsch established compression pressure to achieve vein closure on small saphenous vein and concluded that this pressure varies from 15 mmHg in supine position to 70 mmHg in standing position [2].

The quality of the hosiery has been questioned, either at the design level or at the clinical one. For example, Kumar [57] or Troynikov [58] studied the influence of different materials and knitting patterns on the pressure evolution with time; Ma [63] compared the efficacy of different brands, among them the low-cost ones, without significant difference between low-cost and others. It is worth noting anyway that these studies used universal testing machine, and no clinical verification was provided. In a dedicated study, Partsch [17] concluded on the good correlation between in vitro and in vivo measurements.

Variations of pressure were also investigated over time [53,54], or with patient activity [64]. Viscoelasticity effects have been widely demonstrated and can explain part of the pressure loss observed over time. However, Damstra [54] suggested that the pressure loss is mainly due to the leg volume variation coupled with the strong nonlinear elastic behavior of the so-called "inelastic" textiles. In the same way, Partsch [64] established the interaction between the stocking and muscular pump function.

In a review focused on uncomplicated varicose veins, Palfreyman outlined the variety of indicators for the blood return [65]. Authors have measured foot or leg volume variations, venous reflux, skin blood flow, blood volume, or pressure variation. Despite standards already reported, the adequate compression level is still questioning. Even if higher pressures usually lead to better results, the patient compliance is low in this case. Only 37% of patients in initial clinical evaluation maintained their treatment (50% for the one who had history with deep vein thrombosis). As a matter of fact, even if a link between pressure and comfort is intuitive, no correlation was sought.

The case of bandages is specific because these devices fit better to any leg shape and are suitable for severe cases. However, they are more dependent on the practitioner's experience and show a wider variability in the applied pressure. Studies on this type of compression device are less frequent than for stockings or socks. Most of the measurement campaigns that were performed on men and women did not take the sex difference into account [66,67]; however, the leg morphology has an influence on the applied pressure and varies from a subject to another, maybe even more especially from a female subject to a male subject. Other groups investigated the impact of the application technique on the interface pressure [24], and they also measured the stretch of the applied bandage. However, this study was carried out for only one bandage type. Other studies were focused on the influence of bandage mechanical properties and position (supine, standing, sitting) on the interface pressure [68–71]. Of particular interest, Rimaud [71] concluded on the high variability of interface pressure between subjects for some compression bandages (inelastic fabrics, figure-of-eight method). Yet, this variability can be reduced by training nurses, as noted by Hafner who showed that no bandage exceeded the vein collapse pressure after training of the caregiver [56]. Recently, Chassagne added the measurement of bandage stretch in a clinical study [20]. The objective of this study was to perform a complete campaign including bandage stretch and pressure measurements to evaluate the influence of subject morphology, position, bandage mechanical properties, and application technique on interface pressure applied by elastic compression bandage. It revealed a very strong correlation not only between the applied pressure and the bandage mechanical properties but also between the pressure and the application technique, hence the need to control the bandage stretch and the application technique to control interface pressure.

Despite the quantity of literature available on compression hosiery and its capacity in delivering the appropriate therapeutic dosage of pressure, there is still a lack of understanding that has been expressed very recently by Hageman [72]: "In short, a fundamental understanding of the connection between compression therapy, modes of achieving compression, disease onset, and disease progression is needed". Indeed, pressure measurements are usually pointwise measurements on the surface of the leg. Other methodologies measuring blood return are useful global information, but none could describe a mechanism of action from the skin to the veins.

3.3 Modeling pressure using the law of Laplace

For years, using the law of Laplace has been a classical way to estimate the pressure applied by a compression device on the body [73]. It is used in particular to qualify hosiery in the French standard [19] by converting the textile stiffness into pressure (Fig. 4.5).

The law of Laplace was proposed by both Pierre Simon de Laplace and Thomas Young in 1805 and describes the pressure difference between two fluids (for example, droplets in emulsions, capillaries). It relates this pressure difference to the surface tension and the radius of curvature. It has been applied later to fluid/membrane interface and is now well established in medicine: alveoli of the lungs as well as blood vessels can be described by the law of Laplace. It was applied to compression therapies in the 1980s [73].

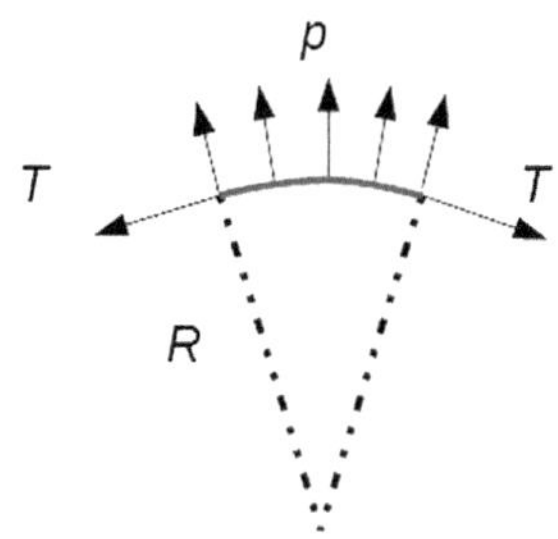

Figure 4.5
Law of Laplace.

According to the law of Laplace, the pressure p applied on the leg by compression hosiery is linked to the radius of curvature R of the leg and the tensile force per unit width T of fabrics as following:

$$p = \frac{T_1}{R_1} + \frac{T_2}{R_2}, T_1, T_2 \geq 0, R_1, R_2 > 0 \tag{4.1}$$

with direction 1 and 2 being the principal radii of curvature.

When getting into details, many limitations affect the law of Laplace. First, the pressure estimated by Eq. (4.1) is valid only for convex surfaces. If the curvature is concave, the pressure is zero as the textile would not be in contact with the body. Second, it is assumed that the two directions of traction correspond to the two main curvature radii; this assumption is reasonable for devices applied on a leg, which is mostly cylindrical, with one prevailing direction. Third, the textile is considered as a membrane, with no compression or bending behavior. Last, the law of Laplace assumes that the textile is sliding on the body without friction. In the specific case of leg compression, this assumption is not fulfilled; otherwise, the garment would fall on the patient's ankle as Chassagne [74] showed.

Because of these limitations, authors tried to modify the law of Laplace to take extra phenomena into account. The most classical presentation includes the number of layers in a bandage [75]. Many authors discussed the law of Laplace's validity by comparing with experiments on phantom legs or on patients with contradictory conclusions [76]. Even if no author directly rejected the law of Laplace, there is still a lack in this analytical description to get predictions with enough precision. According to Chassagne [77], leg deformation—especially during wrapping of bandages—and adhesion are the two main factors for this. A first attempt to overcome this limitation using semianalytical approach has been proposed recently [78], but the only complete way to address the problem is numerical modeling, which is detailed in the following section.

4. Engineering leg compression

As presented in the previous section, a large amount of clinical studies focused on characterizing the interface pressure generated by compression textiles. Still, several questions remain open, which encouraged research work on this topic. An example of such an approach is elastic bandage compression for which pressure generation is much more complex than for socks or stockings and remains to be fully characterized. Engineering approaches were developed in this aim, providing new knowledge and understanding and raising new questions as well, which will be presented in this section.

Deeper in the basic understanding of compression biomechanics, engineering approaches also addressed the more fundamental questions of the mechanisms actually involved in elastic compression and their effects on soft tissues of the leg. This section will present several contributions, which are mostly on the side of modeling and simulation (though directly linked to different sets of measurements) due to the complexity of performing relevant experiments on this topic. These contributions addressed various questions of the problem focusing on aspects such as the direct effect on soft tissues, vein closure, or the development of patient-specific models to study interindividual variability in the observed mechanisms.

4.1 Characterizing and understanding surface pressure generation

The clinical studies presented in Section 3 evidenced the complexity of interface pressure generation, even more markedly for bandages [32]. They revealed that the law of Laplace is not sufficient in spite of being widely used in practice [75,79]. In this regard, several engineering studies were proposed to better understand and characterize how interface pressure is distributed over the leg.

Experimental approaches were developed using other techniques than those presented in Section 3 for clinical use. Using a measurement device previously developed in Ref. [80], the heterogeneity of the interface pressure along the circumference of a leg cross section was quantified showing marked variations in this pressure profile [81] (see Fig. 4.6). Although it was performed on a rigid wooden leg, this type of measurement revealed the influence of the patient's morphology and challenged again the relevance of the law of Laplace from the perspective of individualizing compressive treatments.

Today, an improvement of the experimental characterization of interface pressure faces several issues, which limit novel results:

- Sensors should measure a surface distribution, not just a local or averaged measurement. Current matrix sensors have not demonstrated their reliability on complex and highly curved surfaces such as the leg [82] (Rébecca [48].

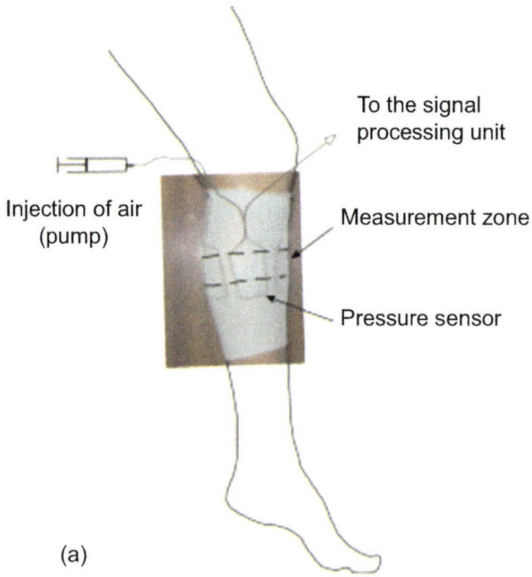

Figure 4.6
Surface pressure measurement device, used on a wooden leg in Ref. [81].

- The influence of the sensor size deserves attention. In addition to possible averaging effects, it was shown that its geometry can alter the true pressure generated by a compressive textile [77].
- The deformation of a leg under the action of a compressive textile is generally not negligible. This highlights the need to perform such studies on soft phantom legs or on real subjects/patients. Furthermore, this phenomenon may induce sensor misplacement, slippage, and/or adhesion, which should not impair the quality of measurement.

In this context, using numerical simulation obviously turns out to be an interesting alternative. As a first step toward improved interface pressure calculation, an interesting numerical approach was presented in Ref. [83], where the law of Laplace was modified to compute interface pressure as a function of the local radius of curvature of the leg obtained from magnetic resonance imaging (MRI) acquisition of the whole leg. This method—though not validated in terms of pressure values, but rather in terms of leg deformation under elastic compression—was able to assess pressure heterogeneities over the leg, hence under- or overpressure areas that could be important from the perspectives of treatment efficacy and observance (possible link with comfort). Yet, this approach requires an MRI measurement of the leg shape that is still not easily applicable in practice. Using similar inputs, the method proposed by Ref. [84] successfully determined interface pressure by directly modeling the motion of sock application onto the limb (see Fig. 4.7).

The above studies were performed with compression socks, which consist in a "simple" device as it is made of a single textile layer whose stretch can be properly assessed.

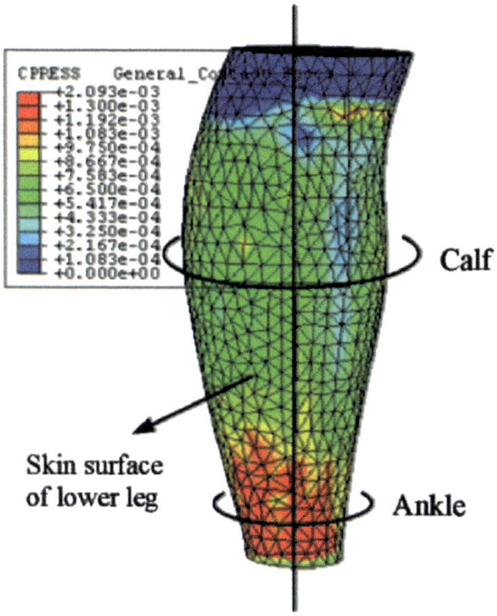

Figure 4.7
Skin pressure distribution obtained from a finite element simulation of a stocking being put on Ref. [84].

When studying compression bandages, complexity significantly increases as the textile layer undergoes superimposition with itself or other bandages. At least, additional frictional effects between the layers or with skin are involved in pressure generation with such bandages.

Clinical studies were numerous (see Section 3) on studying this complex configuration where each bandage can also have its own properties. Similarly, part of the research effort was dedicated to addressing this more complex case. Among the most recent works, numerical models derived from the modeling techniques used for socks/stockings first showed the need for refined approaches in which friction and application process (i.e., bandage wrapping) must be considered [77]. Therefore, they led to the development of a specific modeling and simulation approach in which the actual wrapping movement was simulated to properly integrate the effects of friction on the skin and between layers [74] (see Fig. 4.8). It is interesting to note that the current technological limitations (simulation time) led the authors to develop a model reduction approach, enabling a fast assessment of pressure on any three-dimensional (3D) leg geometry.

The main finding regarding interface pressure generation was the respective influence of parameters possibly involved in the mechanism of pressure generation. First, although their deformability must be taken into account, the elastic properties of soft tissues have little influence when their values are in the range of those measured in vivo. It is important

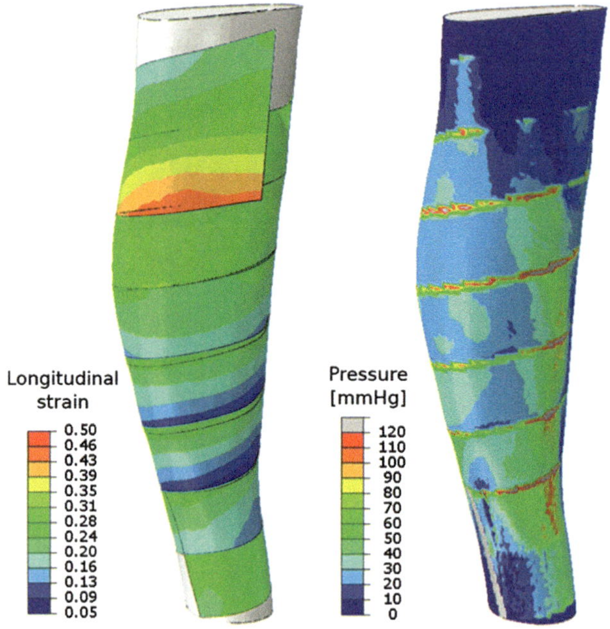

Figure 4.8
Finite element simulation of bandage wrapping (left: bandage longitudinal strain, right: interface pressure).

to notice that this claim is only valid for interface pressure. Indeed, these properties influence the transmission of mechanical efforts through soft tissues [7]. Second, bandage-to-bandage friction was also a low-influence parameter because the frictional properties (obtained from experimental measurements) were in a range that prevented any sliding between layers in this purely static case (Coulomb fiction coefficient from 0.5 to 0.7). This statement may be different when considering dynamic conditions, but this remains to be explored. Finally, skin-to-bandage friction was found to be very sensitive because bandages would tend to slide down and then to unstretch if these frictional effects are too low. Obviously, bandage tension was the most influencing parameter overall.

In spite of these advances, several aspects that may influence interface pressure have not been fully characterized yet. Among them, short-term (period of a walking cycle) and long-term (period of a day) dynamic effects were not detailed. It is likely that slippage effects occur over time, but this has not been characterized yet; it is worth reminding that the objective is a decrease of the limb volume, which should have an incidence on the medical device tension and pressure [54]. In addition to frictional effects and possible stabilization over time, it is sensible to expect effects of hygrometry/hydration variations that could influence the mechanics of the compressive textile as well as its mechanical/biological interactions with skin [85]. The latter direction of research would indirectly address both the mechanical efficacy of the treatment and comfort that is supposed to be a key to improve observance.

4.2 Effects on soft tissues of the leg

Because one of the main supposed effects of compression is a mechanical effect on the underlying soft tissues, an important effort of research was focused on this aspect, addressing the effects on soft tissues (adipose and muscular tissues) and the effect on veins that are in fine directly related to venous return intensity, the main objective of compression treatments.

Although venous hemodynamics improvement is the target, it is almost surprising to notice that literature is very scarce on this topic. Based on acquisitions of MR imaging and pulsed echo-Doppler data with and without elastic compression stockings, the hemodynamics effects of elastic compression were assessed via the computation of wall shear stresses in one superficial (great saphenous) vein and one deep (peroneal or posterior tibial) vein [86]. This study showed that stockings were effective in raising venous wall shear stresses in prone subjects, with a much more pronounced effect in deep vessels. The latter claim may be explained by the fact that deep veins collect the resulting effects from all superficial/perforating veins and microvessels. Similar data were used by the same group to compute the reduction of vein caliber using a finite element model fed from MRI data [87]. The authors also mentioned that tissue compressibility was a key parameter in this phenomenon of vein area reduction, suggesting that the effect of compression on different subjects could be very different if mechanical properties of tissues are different.

Implementing relevant experimental approaches for this question of the mechanical effects of compression on internal soft tissues is a challenge. Apart from the previously mentioned studies that mainly address veins and not other soft tissues, other works are very scarce in the literature, to our knowledge. This statement probably arises from the complexity to setup experimental studies that would require measuring internal mechanical quantities in opaque, highly deformable materials. If in vivo measurements could be implemented with MR elastography, manufacturing phantoms is another alternative that faces the issue of being representative enough of the real problem.

As a consequence, the biomechanical effects of elastic compression on underlying soft tissues were addressed mainly with numerical approaches, most of which are based on finite element simulations.

Targeting this objective, two-dimensional (2D) (cross-sectional) models were developed as a first attempt to study the effects in superficial adipose tissue and deep muscular tissues. Among them, Avril et al. [88] developed a model using MRI data from one patient for the geometry, in which each type of tissue (adipose and muscular) was considered as homogeneous. The application of elastic compression was modeled using the law of Laplace taking the local curvature into account and enabled computing the whole pressure field in the soft tissues. Discrepancies of more than 35% could be observed from one location to another, showing that the same compression garment may not be suitable for treating deficiencies of both the deep venous system and the large superficial veins (see Fig. 4.9). Moreover, it was shown that the internal morphology of the human leg can play an important role, in particular, when considering the ratio of adipose to muscular tissue.

Toward more complexity, Refs. [87,89,90] introduced veins in their 2D models to directly evaluate the effects on the venous network, target of the treatment. The caliber reduction of one or several veins could be quantified. Although useful from a didactic point of view, several assumptions were made such as a uniform interface pressure and homogeneous mechanical properties of tissues, which were found to be a source of major sensitivity in the abovementioned papers. One more recent work addressed these aspects and proposed two patient-specific models obtained from MR and ultrasound (US) images to study the effect of medical compression stockings on vein closure, both for superficial and deep veins [7,91]. The effect on superficial varicose veins was quantified using the venous transmural pressure variation induced by the compression garment, which was found to decrease as a result of the increase of adipose tissue pressure. The transmural pressure is known to be related to the tension of the vein wall according to the law of Laplace and probably to alterations in the vein wall. Another consequence is a greater stability with respect to axial buckling and tortuosity development [92,93]. In the second model where muscle compartments, the crural fascia, and deep veins were included, the combined effects of the elastic compression device and muscular contraction were investigated (see Fig. 4.9). The contribution of the compression stocking to the deep vein diameter reduction was rather small and in fact negligible, compared with that of the contracting muscle: 3% and 9% decrease in the vein cross-sectional area were obtained with a grade II compression stocking in the supine and standing positions, respectively, while complete collapse was obtained at the end of muscle activation.

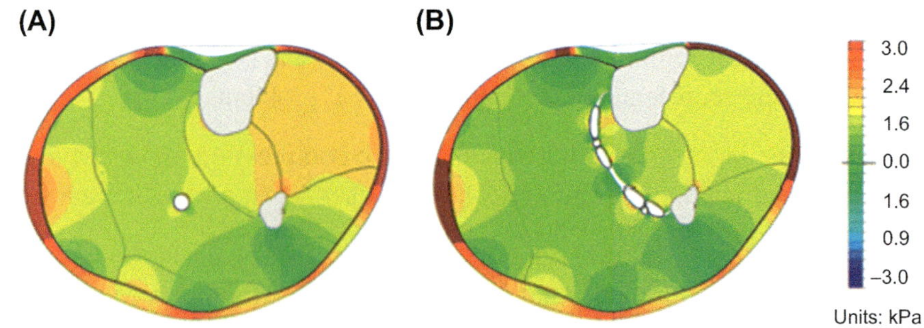

Figure 4.9
Hydrostatic pressure distribution in two leg cross sections.
(A) Intramuscular model, (B) models including compartments, fasciae, and some veins [7].

An approach that aimed at developing a 3D patient-specific model of a complete leg under elastic compression should be mentioned [83]. This model was (i) patient-specific in geometry (MRI acquisition), (ii) contained two types of soft tissues, subcutaneous adipose tissue and muscular tissue, whose properties were also patient-specific (the approach is detailed in the following section), and (iii) the applied pressure was computed using a local law of Laplace. The results of this model confirmed the important heterogeneity of the pressure field in the internal soft tissues, hence different veins being subjected to different pressure levels. They also showed, via a study on six subjects, a very large intersubject variability with, for example, variation about 50% of the pressure computed in the area of deep veins in spite of an identical grade compressive sock.

The 3D analysis of the internal pressure field is crucial to detail the actual pressure gradients induced by the elastic compression, hence the possible fluid flows, their direction, and their intensity. The development of such models represents an interesting direction, in particular if the main (superficial, perforating, deep) veins and the main different components (muscle, fasciae) could be included. For now, this perspective is a challenge from a technical point of view, as a finite element mesh of a structure like this would contain a huge number of elements to properly discretize the geometry of the venous network. The question of the accuracy of imaging data required for that purpose, and reasonable acquisition conditions of these, may also be raised. Poroelastic modeling is another alternative that has to be mentioned here.

All works above were performed in the frame of studies on various external compression devices, the question of interface pressure generation being not the main problem (as shown in Section 3, this question is a topic of research in itself). Yet, the effects of morphology and mechanical properties of internal tissues were highlighted, and quantitative studies on vein closure were proposed. These works brought significant advances in the current knowledge of the mechanisms involved in compression therapy and more specifically how the interface pressure applied by the compressive treatment impacts the mechanical state of internal tissues and structure.

4.3 Patient-specific approaches

In previous paragraphs, several approaches were developed on a patient-specific basis. This aspect is crucial in understanding the variability that can be observed in terms of targeted effects. The present section aims at presenting how patient specificity was addressed in this field.

The first, and probably easiest, element of a patient-specific model that can be addressed is geometry. Indeed, current techniques that can be used for that purpose are numerous and various. Furthermore, many of them are already widely used in the medical sector.

They allow a wide range of scales, accuracy of details, ease of use, and ease of access. Eventually, the choice of a geometry acquisition technique often consists of a trade-off between ease of implementation and anatomical accuracy/detail requirements.

The richest imaging modalities are probably MRI, which provides a detailed description of all tissues, and in a lesser extent, computed tomography (CT) scanner. Access to these modalities may be difficult as it is limited by other clinical demands, the irradiating nature of CT scanner, and the need to obtain authorization from competent institutions. In addition, it is important to notice that such equipment only provide supine position geometries (standing position equipment being very rare). Much simpler to use and with easier access, US echography enables to image local details but not a complete geometry.

The principle to create a geometric model from the obtained data is classical. It consists in segmenting the images and meshing the obtained result. In practice, image quality and resolution are a challenge for automatic methods, which most often led to developing semiautomatic methods [87,90].

Multimodality methods are an interesting alternative, which was used in Ref. [91] to merge the advantage of MRI and US echography. The former was suitable to segment the exterior surface and internal tissues (bones, adipose tissue, muscular compartments, vessel locations) and the latter was suitable to obtain details such as vein caliber and their precise geometry. Such approach requires image registration methods that are typically based on landmark registration.

Another approach presented in Ref. [94] was proposed to obtain a complete 2D cross section of a leg taking advantage of the ease of use/access of US echography. This approach is based on a sequential acquisition of several US-echo images along the circumference of the leg using a specific mounting device (see Fig. 4.10) and a registration method including the minimization of landmark-based criteria. Nevertheless, this approach is not able to provide deep geometric details.

It must be acknowledged that all previous methods are not suitable for use in a daily clinical routine. Accordingly, when only interface pressure is of interest, the external surface may be sufficient and the whole leg surface should be acquired. In this context, the use of a 3D optical scanner turns out to be a relevant alternative: it is nonirradiating, commercially available, and very fast to obtain the surface of a leg in any position. Such a device was used in a previous (clinical) study, thanks to its versatility [95]. It should be noted that assumptions on the internal structures/morphology must be made, though their sensitivity is low with respect to interface pressure.

The methods to ensure geometrically patient-specific models are numerous, and their implementation relies on the compromise between available equipment and targeted objectives. With regard to patient-specific material properties for the internal soft tissues,

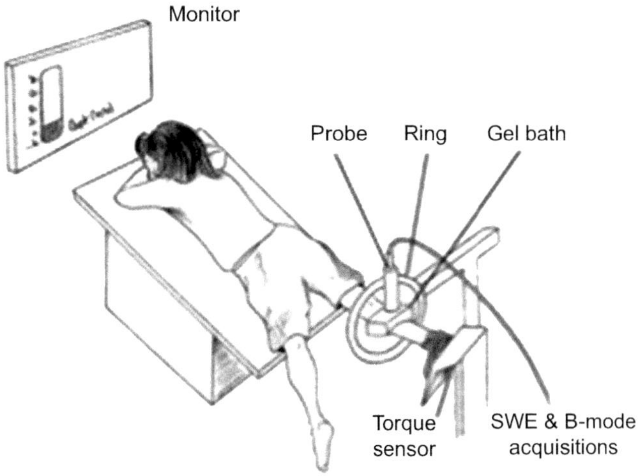

Figure 4.10
Experimental setup developed in Ref. [94]: the probe is mounted on a ring, which allows collecting images around the leg. The foot is fixed through a shoe to a torque sensor. A feedback of the torque is given to the subject with the monitor. Data, including B-mode scans and SWE maps, are acquired with Aixplorer (SuperSonic Imagine, Aix-en-Provence, France).

approaches and methods are not as straightforward. Again, the complexity of the considered method should be adjusted to the targeted objective. Most of the models presented above used linear or quasilinear (hyper)elastic constitutive equations. The problematics then turned into (i) defining the number of different constituents and (ii) identifying the parameters of the material models for each of them.

For instance, in studies focusing on interface pressure characterization, all soft tissues were considered as a single homogeneous material, whereas complexity could be increased by distinguishing adipose and muscular tissue when focusing on internal mechanical fields or even including other heterogeneities such as fasciae, distinct muscle compartments, veins, and arteries when focusing on the specific response of constituent such as veins.

Various approaches were developed to identify the constitutive parameters of the constituents involved in those models. Identifying a single homogeneous material for the leg soft tissues is possible, thanks to external testing devices. Indentation, or localized compression, of the calf was used in several studies to this [96-98]. In these studies, the parameters were identified using an inverse method based on a finite element model of the test on the patient-specific leg. The relative simplicity of this approach makes it a good candidate for clinical applications [95]. Other mechanical tests were used to identify these mechanical properties. In particular, imaging the deformation induced by the compressive stocking or cuff is a way to obtain these data. In this manner, MR images with/without compression were used as input for inverse (or similar) approaches in Refs. [88,90,99]. A similar approach (with/without compression) was implemented using US-echo imaging of superficial varicose veins to identify their tangent elastic moduli in standing and supine positions [91]. Because venous tissue behaves very nonlinearly, the latter distinction was considered essential in this work.

Recently, shear wave elastography has become more and more used to assess soft tissue elastic moduli. Although implementing this technique to obtain a complete map of stiffness in the leg is complex, a method was proposed in this aim in Ref. [94]. In addition to the underlying assumptions (for moduli reconstruction in anisotropic muscular tissue, with several heterogeneities), the applicability of this technique in a clinical context remains an issue.

Implementing patient-specific approaches mainly relies on gathering geometric and mechanical properties. Simple as well as complex methods were proposed in the existing literature. Yet, not all of them are applicable in clinical routine, and important data, such as fasciae mechanical properties and initial tension, which may play a significant role, are still not properly explored. Other characteristics, such as arterial pressure (easily measured and extrapolated at the location of interest, if the patient is at rest), muscle activity, skin friction properties, and reflex contractions, which could be influenced via proprioceptive effects, for instance, were not addressed in this section. These would require other equipment that were not used, to our knowledge, for consideration in biomechanical studies of elastic compression.

All presented studies and approaches were useful in understanding trends in the effects of elastic compression that are directly related to the interindividual variability measured in these studies. Among interesting conclusions that could be drawn, it is worth noting two examples of them. First, the mechanical properties (stiffness) of calf soft tissue have negligible influence on the interface pressure value, provided they are in a reasonable range, i.e., deformable. Second, geometry and tissue mechanical properties have strong (spatial) impact on the internal distribution of pressure, and in particular on the effect on veins.

Such conclusions are interesting for clinical practice and will potentially have implications in designing specific care strategies. The objective of the following section is to specifically address and detail new knowledge and results that were obtained from clinical and research studies, along with their possible implications for the clinical practice.

5. Contribution to the clinical practice

The present section sums up the essential findings and new relevant knowledge arising from the previously presented engineering and clinical research works.

Regarding internal mechanisms of elastic compression, which can be related to treatment efficacy, the following can be retained:

- Although they have little effect on interface pressure, the mechanical properties of soft tissues (e.g., elasticity) influence pressure transmission to veins and internal soft tissues.
- The internal distribution of pressure in soft tissues is also influenced by the external and internal geometry of the leg. In particular, the pressure observed at the location of deep veins may be very different from one patient to another.
- When focusing on the effects of compression on veins, it can be summarized that the response of superficial veins to compression is subject to three main factors: the vein size, the local radius of curvature of the leg, and the fat stiffness. On the contrary, although the effect on deep veins may be noticeable in supine position, previous studies showed that, in standing position, the effect of compression is marginal compared with that of muscular contraction.

Interface pressure is on the one hand the condition to obtain the expected internal effects, but on the other hand it may compromise the compliance to the treatment. Although the latter is not addressed, the following should be mentioned about interface pressure:

- Bandage application is a sensitive process that highly depends on the operator and the application technique. As a consequence, a marked variability may be observed in the interface pressure applied onto skin, and this should be carefully considered. Improving the control of bandage stretch and application technique are probably the keys to a better control of interface pressure.
- Bandage to skin friction plays a key role in compression by bandages. Indeed, it is essential in maintaining the wrapped bandage and preventing it from falling down. Its effect on pressure generation remains to be clarified, as the law of Laplace is questioned in this context.
- For compression stockings or socks, the stiffness of the sock, the size of the sock, and the size of the leg are the main parameters influencing interface pressure. They are also the parameters used for choosing the appropriate treatment. However, studies presented in Section 4 highlighted different pressure distribution patterns for different morphologies. In particular, very homogeneous pressure distributions are observed in cases of round leg cross sections, while high pressure spots may be found in other cases. This additional aspect is usually not considered though it may have important impact on comfort, hence observance of the treatment. It is important to note that these results were obtained for purely static simulation and should be confirmed in dynamic conditions.

Beyond the understanding of elastic compression mechanisms, it is believed that these findings can help enhance care management.

An important impact of these research-level results would be to head toward an improved personalization of compression treatment. The complete external geometry of the leg could be used to estimate personalized interface pressure distributions and improve the choice of the right size of a device. In the long term, it is believed that it will be possible to assess the impact of a treatment on the venous return, which would help in targeting the right treatment. For instance, the influence of bandage application order—in the case of multilayer bandages—was highlighted, and this specificity would deserve to be considered.

The presented results also underline the needs to standardize practices among health-care professional. For example, specific or improved training of practitioners willing to apply compression bandages could be encouraged. On the same aspect, bandage design could be addressed for a better control of application.

6. Prospects

As already mentioned, clinical studies gave limited results but could at least conclude that the maximum applied pressure should not exceed 70 mmHg in standing position and that training medical staff is necessary to reduce complications related to overpressure; the correlation between phantom and real legs is established in the case of static pressure measurement.

Clinical studies opened the way to more questions that are important perspectives in this research area.

Firstly, the applied pressure varies with time. Several authors studied the time dependence of the applied pressure to the knitting process, the material viscoelasticity, the leg volume variation, or the patient activity, but these studies considered only one parameter at a time, and it is still impossible to rank these phenomena or even consider the effects of interactions among them. No numerical approach could describe time-dependent phenomena yet. Of course, compression level is crucial for the efficacy of the treatment; therefore, the evolution of the tension or the leg volume with time and/or patient activity is a major prospect.

Indeed, from a clinical point of view, such prospect cannot be achieved in a short term: a multiparametric study should lead to a too large cohort, and unwanted variabilities related both to patients and operators should make variations statistically nonsignificant. A lot of work is still to be done to extract the relevant parameters from the mechanical behavior of the textiles and from the patient daily life activity and also to reduce the interoperator variability.

From a numerical modeling point of view also, studying the time dependence of hosiery compression is very ambitious; it should include new developments on the textile behavior modeling and on the fluid–structure interactions in the leg, may be using poroelastic modeling. Blood return is also dependent on active processes. The muscular contraction in the calf and in the foot should be put in a dynamic biomechanical model of the limb, including the foot.

For now, computer time has been a limitation, and innovative numerical methods should be found; because modeling the hosiery to leg contact is computationally expensive, one way to reduce the global computing time should be the development of an analytical contact model based on the law of Laplace and including the tangential effects (friction or adhesion).

Second, as it has already been mentioned, the numerical models could describe the mechanisms of action of the hosiery onto the veins. Even if the applied pressure poorly depends on the soft tissue properties, these properties have a great influence on the load transfer from the skin to the veins, so another limitation lies in the knowledge of the patient-specific soft tissue properties (skin, fat, muscles, aponeurosis) and the mechanical interaction between them. Therefore, the second prospect proposed here should be the development of experimental methods to quantify relevant material properties of the leg. These methods should ideally be available in the clinical context to evaluate the efficiency of a treatment for a specific patient.

Last, the skin–textile interface has been poorly studied in the context of compression therapies. Skin, which is the natural mechanical, chemical, and thermal protection of the body, is highly stressed by compression or contention devices, and it might be the cause of the low level of compliance to the compression-based treatments.

Two main risks are related to the skin: a high compression, even during short times, can lead to ulcers, and irritation should be related to skin shear. The latter has been partially addressed by medical doctors, but studies from biophysicists all focused on ulcers or prosthesis complications. The study of the skin–textile interface in the context of compression therapies should be developed both from medical and engineering angles of view. Clinical studies should be designed to establish the conditions leading to irritation. On the other hand, experimental studies should better describe the skin behavior under pressure and shear according to various parameters of the skin (moisture, hydration, etc.), the hosiery (textile microstructure, roughness, etc.), or the interaction of both (load, shear distance, etc.) to understand basic mechanisms leading to skin diseases and to prevent them.

References

[1] Santler B, Goerge T. Chronic venous insufficiency – a review of pathophysiology, diagnosis, and treatment. J Dtsch Dermatol Ges 2017;15(5):538–56. https://doi.org/10.1111/ddg.13242.

[2] Partsch B, Partsch H. Calf compression pressure required to achieve venous closure from supine to standing positions. J Vasc Surg 2005;42(4):734–8. https://doi.org/10.1016/j.jvs.2005.06.030.

[3] Gardon-Mollard C, Ramelet A-A. La compression médicale. Elsevier Masson; 2008.

[4] Meissner MH, Moneta G, Burnand K, Gloviczki P, Lohr JM, Lurie F, Mattos MA, et al. The hemodynamics and diagnosis of venous disease. J Vasc Surg 2007;46(6):S4–24. https://doi.org/10.1016/j.jvs.2007.09.043.
[5] Braus H, Elze C. Anatomie Des Menschen: Ein Lehrbuch Für Studierende Und Ärzte, Vol. 1. J. Springer; 1921.
[6] Marieb EN, Hoehn K. Anatomie et Physiologie Humaines. ERPI; 2010.
[7] Rohan P-Y, Badel P, Lun B, Rastel D, Avril S. Prediction of the biomechanical effects of compression therapy on deep veins using finite element modelling. Ann Biomed Eng 2015;43(2):314–24. https://doi.org/10.1007/s10439-014-1121-6.
[8] Lattimer CR, Kalodiki E, Kafeza M, Azzam M, Geroulakos G. Quantifying the degree graduated elastic compression stockings enhance venous emptying. Eur J Vasc Endovasc Surg 2014;47(1):75–80. https://doi.org/10.1016/j.ejvs.2013.10.020.
[9] Vicaretti M. Compression therapy for venous disease. Aust Prescr 2010;33:186–90.
[10] Mosti G, Partsch H. Is low compression pressure able to improve venous pumping function in patients with venous insufficiency? Phlebology 2010;25(3):145–50. https://doi.org/10.1258/phleb.2009.009023.
[11] Milic DJ, Zivic SS, Bogdanovic DC, Jovanovic MM, Jankovic RJ, Milosevic ZD, Stamenkovic DM, Trenkic MS. The influence of different sub-bandage pressure values on venous leg ulcers healing when treated with compression therapy. J Vasc Surg 2010;51(3):655–61. https://doi.org/10.1016/j.jvs.2009.10.042.
[12] Partsch H, Menzinger G, Mostbeck A. Inelastic leg compression is more effective to reduce deep venous refluxes than elastic bandages. Dermatol Surg 1999;25(9):695–700.
[13] Partsch H. Improving the venous pumping function in chronic venous insufficiency by compression as dependent on pressure and material. VASA. Zeitschrift für Gefässkrankheiten 1984;13(1):58–64.
[14] Mosti G, Mattaliano V, Partsch H. Inelastic compression increases venous ejection fraction more than elastic bandages in patients with superficial venous reflux. Phlebology 2008;23(6):287–94. https://doi.org/10.1258/phleb.2008.008009.
[15] Partsch H, Partsch B, Braun W. Interface pressure and stiffness of ready made compression stockings: comparison of in vivo and in vitro measurements. J Vasc Surg 2006;44(4):809–14. https://doi.org/10.1016/j.jvs.2006.06.024.
[16] Callam MJ, Harper DR, Dale JJ, Ruckley CV. Chronic ulcer of the leg: clinical history. Br Med J 1987;294(6584):1389–91.
[17] Partsch H, Clark M, Bassez S, Benigni J-P, Becker F, Blazek V, Caprini J, et al. Measurement of lower leg compression in vivo: recommendations for the performance of measurements of interface pressure and stiffness: consensus statement. Dermatol Surg 2006;32(2):224–32. Discussion 233.
[18] Chassagne F, Helouin-Desenne C, Molimard J, Convert R, Badel P, Giraux P. Superimposition of elastic and nonelastic compression bandages. J Vasc Surg Venous Lymphat Disord 2017;5(6):851–8. https://doi.org/10.1016/j.jvsv.2017.07.006.
[19] Afnor. NF G30-102 – articles de Bonneterie – détermination de La pression de Contention. 1986.
[20] Chassagne F, Martin F, Badel P, Convert R, Giraux P, Molimard J. Experimental investigation of pressure applied on the lower leg by elastic compression bandage. Ann Biomed Eng 2015;43(12):2967–77. https://doi.org/10.1007/s10439-015-1352-1.
[21] Dissemond J, Assenheimer B, Bültemann A, Gerber V, Gretener S, Kohler-von Siebenthal E, Koller S, et al. Compression therapy in patients with venous leg ulcers: compression in leg ulcers. J Dtsch Dermatol Ges 2016;14(11):1072–87. https://doi.org/10.1111/ddg.13091.
[22] Partsc H. The use of pressure change on standing as a surrogate measure of the stiffness of a compression bandage. Eur J Vasc Endovasc Surg 2005;30(4):415–21. https://doi.org/10.1016/j.ejvs.2005.06.002.
[23] Cornu-Thenard A, Boivin P, Carpentier PH, Courtet F, Ngo P. Superimposed elastic stockings: pressure measurements. Dermatol Surg 2007;33(3):269–75. Discussion 275, https://doi.org/10.1111/j.1524-4725.2007.33061.x.

[24] Coull A, Tolson D, McIntosh J. Class-3c compression bandaging for venous ulcers: comparison of spiral and figure-of-eight techniques. J Adv Nurs 2006;54(3):274–83. https://doi.org/10.1111/j.1365-2648.2006.03816.x.

[25] Hanna R, Bohbot S, Connolly N. A comparison of interface pressures of three compression bandage systems. Br J Nurs 2008;17(20):S16–24.

[26] Amsler F, Willenberg T, Blättler W. In search of optimal compression therapy for venous leg ulcers: a meta-analysis of studies comparing divers bandages with specifically designed stockings. J Vasc Surg 2009;50(3):668–74. https://doi.org/10.1016/j.jvs.2009.05.018.

[27] Mosti G, Partsch H. Compression stockings with a negative pressure gradient have a more pronounced effect on venous pumping function than graduated elastic compression stockings. Eur J Vasc Endovasc Surg 2011;42(2):261–6. https://doi.org/10.1016/j.ejvs.2011.04.023.

[28] Mosti G, Partsch H. High compression pressure over the calf is more effective than graduated compression in enhancing venous pump function. Eur J Vasc Endovasc Surg 2012;44(3):332–6. https://doi.org/10.1016/j.ejvs.2012.06.001.

[29] Couzan S, Assante C, Laporte S, Mismetti P, Pouget J-F. Étude booster : évaluation comparative d'un nouveau concept de compression élastique dans l'insuffisance veineuse chronique légère et modérée. La Presse Médicale 2009;38(3):355–61. https://doi.org/10.1016/j.lpm.2008.09.019.

[30] Mosti G, Iabichella ML, Partsch H. Compression therapy in mixed ulcers increases venous output and arterial perfusion. J Vasc Surg 2012;55(1):122–8. https://doi.org/10.1016/j.jvs.2011.07.071.

[31] Shepherd J. Progressive compression versus graduated compression for the management of venous insufficiency. Br J Community Nurs September 2016;21(Suppl. 9). S13–18, https://doi.org/10.12968/bjcn.2016.21.Sup9.S13.

[32] Clark M. Compression Bandages : principles and definition. Eur Wound Manag Assoc 2003:5–7.

[33] DIN 61632. Verbandmittel – idealbinden. 2009.

[34] Partsch H, Flour M, Smith PC, International Compression Club. Indications for compression therapy in venous and lymphatic disease consensus based on experimental data and scientific evidence. Under the auspices of the IUP. Int Angiol 2008;27(3):193–219.

[35] Partsch H. The static stiffness Index: a simple method to assess the elastic property of compression material in vivo. Dermatol Surg 2005;31(6):625–30.

[36] Danielsen L, Munk Madsen S, Henriksen L, Sindrup J, Petersen LJ. Subbandage pressure measurements comparing a long-stretch with a short-stretch compression bandage. Acta Derm Venerol. 1998;78:201–4.

[37] Blair SD, Wright DD, Backhouse CM, Riddle E, Mc Collum CN. Sustained compression and healing of chronic venous ulcers. BMJ November 1988;297:1159–61.

[38] Moffatt C. Four-layer bandaging: from concept to practice. Int J Low Extrem Wounds 2002;1(1):13–26.

[39] Fenety PA, Putnam C, Walker JM. In-chair movement: validity, reliability and implications for measuring sitting discomfort. Appl Ergon 2000;31(4):383–93.

[40] Hamanami K, Tokuhiro A, Inoue H. Finding the optimal setting of inflated air pressure for a multi-cell air cushion for wheelchair patients with spinal cord injury. Acta Med Okayama 2004;58(1):37–44. https://doi.org/10.18926/AMO/32114.

[41] Reenalda J, Van Geffen P, Nederhand M, Jannink M, IJzerman M, Rietman H. Analysis of healthy sitting behavior: interface pressure distribution and subcutaneous tissue oxygenation. J Rehabil Res Dev 2009;46(5):577–86.

[42] Rithalia SV, Heath GH, Gonsalkorale M. Assessment of alternating-pressure air mattresses using a time-based pressure threshold technique and continuous measurements of transcutaneous gases. J Tissue Viability 2000;10(1):13–20.

[43] Shelton F, Wright Lott J. Conducting and interpreting interface pressure evaluations of clinical support surfaces. Geriatr Nurs 2003;24(4):222–7. https://doi.org/10.1016/S0197-4572(03)00212-X.

[44] Tam EW, Arthur FM, Wai Nga L, Evans JH, Chow YY. Pelvic movement and interface pressure distribution during manual wheelchair Propulsion. Arch Phys Med Rehabil 2003;84(10):1466–72. https://doi.org/10.1016/S0003-9993(03)00269-7.

[45] Vaisbuch N, Meyer S, Weiss PL. Effect of seated posture on interface pressure in children who are able-bodied and who have myelomeningocele. Disabil Rehabil 2000;22(17):749–55.

[46] Meffre R, Gehin C, Dittmar A. MAPI: active interface pressure sensor integrated into a seat. IEEE; 2007. p. 1358–61. https://doi.org/10.1109/IEMBS.2007.4352550.

[47] Aubin C-E, Labelle H, Cheriet F, Villemure I, Mathieu PA, Jean D. Tridimensional evaluation and optimization of the orthotic treatment of adolescent idiopathic scoliosis. Med Sci 2007;23(11):904–9. https://doi.org/10.1051/medsci/20072311904.

[48] Bonnaire R, Verhaeghe M, Molimard J, Paul C, Convert R. Characterization of a pressure measuring system for the evaluation of medical devices. Proc Inst Mech Eng H J Eng Med 2014;228(12):1264–74. https://doi.org/10.1177/0954411914562871.

[49] van den Hout Ja aM, van Rhijn LW, van den Munckhof RJH, van Ooy A. Interface corrective force measurements in boston brace treatment. Eur Spine J 2002;11(4):332–5. https://doi.org/10.1007/s00586-001-0379-1.

[50] Bus SA, van Deursen RW, Armstrong DG, Lewis JEA, Caravaggi CF, Cavanagh PR, International Working Group on the Diabetic Foot. Footwear and offloading interventions to prevent and heal foot ulcers and reduce plantar pressure in patients with diabetes: a systematic review. Diabetes Metabol Res Rev January 2016;32(Suppl. 1):99–118. https://doi.org/10.1002/dmrr.2702.

[51] Perrier A, Vuillerme N, Luboz V, Bucki M, Cannard F, Diot B, Colin D, Rin D, Bourg J-P, Payan Y. Smart diabetic socks: embedded device for diabetic foot prevention. IRBM 2014;35(2):72–6. https://doi.org/10.1016/j.irbm.2014.02.004.

[52] Taroni M, Molimard J, Cachon T, Viguier E, Carozzo C. Robert Jones bandage pressure range assessment using a pressure mapping system and application to band calibration. Comput Methods Biomech Biomed Eng 2015;18(Suppl. 1):2068–9. https://doi.org/10.1080/10255842.2015.1069560.

[53] Candy LHY, Li-Tsang Wai Ping C, Zheng Yong P. Effect of different pressure magnitudes on hypertrophic scar in a Chinese population. Burns 2010;36(8):1234–41. https://doi.org/10.1016/j.burns.2010.05.008.

[54] Damstra RJ, Els R B, Hugo P. Controlled, comparative study of relation between volume changes and interface pressure under short-stretch bandages in leg lymphedema patients. Dermatol Surg 2008;34(6):773–8. Discussion 778–779, https://doi.org/10.1111/j.1524-4725.2008.34145.x.

[55] Dumbleton T, Arjan WP, Buis AMF, McHugh BF, McKay G, Murray KD, Sexton S. Dynamic interface pressure distributions of two transtibial prosthetic socket concepts. J Rehabil Res Dev 2009;46(3):405–15.

[56] Hafner J, Lüthi W, Hänssle H, Kammerlander G, Burg G. Instruction of compression therapy by means of interface pressure measurement. Dermatol Surg 2000;26(5):481–8. https://doi.org/10.1046/j.1524-4725.2000.99257.x.

[57] Kumar B, Das A, Alagirusamy R. Effect of material and structure of compression bandage on interface pressure variation over time. Phlebology 2013;29(6):376–85. https://doi.org/10.1177/0268355513481772.

[58] Troynikov O, Wardiningsih W, Koptug A, Watson C, Oggiano L. Influence of material properties and garment composition on pressure generated by sport compression garments. Procedia Eng 2013;60:157–62. https://doi.org/10.1016/j.proeng.2013.07.054.

[59] Wheeler JW, Dabling JG, Chinn D, Turner T, Filatov A, Anderson L, Rohrer B. MEMS-based bubble pressure sensor for prosthetic socket interface pressure measurement. IEEE; 2011. p. 2925–8. https://doi.org/10.1109/IEMBS.2011.6090805.

[60] Lou E, Zbinden D, Mosberger P, Hill DL, Raso VJ. A wireless personal wearable network system to understand the biomechanics of orthotic for the treatment of scoliosis. Conf Proc IEEE Eng Med Biol Soc 2008:3426–9. https://doi.org/10.1109/IEMBS.2008.4649942.

[61] Partsch H, Mosti G. Comparison of three portable instruments to measure compression pressure. Int Angiol 2010;29(5):426–30.
[62] Reich-Schupke S, Gahr M, Altmeyer P, Stücker M. Resting pressure exerted by round knitted moderate-compression stockings on the lower leg in clinical practice–results of an experimental study. Dermatol Surg 2009;35(12):1989–97. https://doi.org/10.1111/j.1524-4725.2009.01318.x.
[63] Ma H, Blebea J, Malgor RD, Taubman KE. Variability in leg compression provided by gradient commercial stockings. J Vasc Surg Venous Lymphat Disord 2015;3(4):431–7. https://doi.org/10.1016/j.jvsv.2015.07.001.
[64] Partsch H. Efficacy of sport stockings can Be improved. Ann Sports Med Res 2016;3(8).
[65] Palfreyman SJ, Michaels JA. A systematic review of compression hosiery for uncomplicated varicose veins. Phlebology 2009;24(Suppl. 1):13–33. https://doi.org/10.1258/phleb.2009.09s003.
[66] Mosti G, Partsch H. Inelastic bandages maintain their hemodynamic effectiveness over time despite significant pressure loss. J Vasc Surg 2010;52(4):925–31. https://doi.org/10.1016/j.jvs.2010.04.081.
[67] Mosti G, Partsch H. Bandages or double stockings for the initial therapy of venous oedema? A randomized, controlled pilot study. Eur J Vasc Endovasc Surg 2013;46(1):142–8. https://doi.org/10.1016/j.ejvs.2013.04.015.
[68] Benigni JP, Uhl JF, Cornu-Thénard A, Blin E. Compression bandages: influence of techniques of use on their clinical efficiency and tolerance. Int Angiol 2008;27(1):68–73.
[69] Danielsen L, Madsen SM, Henriksen L. Healing of venous leg ulcers. A randomized prospective study of a long-stretch versus short-stretch compression bandage. Phlebology 1998;13(2):59–63.
[70] Hirai M, Niimi K, Iwata H, Sugimoto I, Ishibashi H, Ota T, Nakamura H. A comparison of interface pressure and stiffness between elastic stockings and bandages. Phlebology 2009;24(3):120–4. https://doi.org/10.1258/phleb.2008.008057.
[71] Rimaud D, Convert R, Calmels P. In vivo measurement of compression bandage interface pressures: the first study. Ann Phy Rehabil Med 2014;57(6–7):394–408. https://doi.org/10.1016/j.rehab.2014.06.005.
[72] Hageman DJ, Wu S, Kilbreath S, Rockson SG, Wang C, Melissa L, Tate K. Biotechnologies toward mitigating, curing, and ultimately preventing edema through compression therapy. Trends Biotechnol 2018;36(5):537–48. https://doi.org/10.1016/j.tibtech.2018.02.013.
[73] Thomas S. The use of the Laplace equation in the calculation of sub-bandage pressure. Eur Wound Manag Assoc 2003;3:21–3.
[74] Chassagne F, Molimard J, Convert R, Pascal G, Badel P. Numerical model reduction for the prediction of interface pressure applied by compression bandages on the lower leg. IEEE Trans Biomed Eng 2018;65(2):449–57. https://doi.org/10.1109/TBME.2017.2774598.
[75] Schuren J, Mohr K. The efficacy of Laplace's equation in calculating bandage pressure in venous leg ulcers. Wounds U K 2008;4(2):38–47.
[76] Aghajani M, Jeddi AAA, Amani Tehran M. Investigating the accuracy of prediction pressure by Laplace law in pressure-garment applications. J Appl Polym Sci 2011;121(5):2699–704. https://doi.org/10.1002/app.33640.
[77] Chassagne F, Molimard J, Convert R, Giraux P, Badel P. Numerical approach for the assessment of pressure generated by elastic compression bandage. Ann Biomed Eng 2016;44(10):3096–108. https://doi.org/10.1007/s10439-016-1597-3.
[78] Bonnaire R, Molimard J, Han WS, Convert R, Calmels P. Predicting strategy of lumbar belts mechanical performance at design stage. In: Lyon, France. ⟨hal-01406590⟩; 2016.
[79] Basford JR. The law of Laplace and its relevance to contemporary medicine and rehabilitation. Arch Phys Med Rehabil 2002;83(8):1165–70.
[80] Blazek V, Schultz-Ehrenburg U. A new pneumatic-electronic procedure for the determination of pressure of medical compression stockings. Phlebologie 1997;26:143–9.

[81] Gaied I, Drapier S, Lun B. Experimental assessment and analytical 2D predictions of the stocking pressures induced on a model leg by medical compressive stockings. J Biomech 2006;39(16):3017–25. https://doi.org/10.1016/j.jbiomech.2005.10.022.
[82] Baldoli I, Mazzocchi T, Paoletti C, Ricotti L, Salvo P, Dini V, Laschi C, Di Francesco F, Menciassi A. Pressure mapping with textile sensors for compression therapy monitoring. Proc Inst Mech Eng H J Eng Med 2016;230(8):795–808. https://doi.org/10.1177/0954411916655184.
[83] Dubuis L, Avril S, Debayle J, Badel P. Identification of the material parameters of soft tissues in the compressed leg. Comput Methods Biomech Biomed Eng 2002;15(1):3–11. https://doi.org/10.1080/10255842.2011.560666.
[84] Dai XQ, Lui R, Abel M, Kwok YL. Numerical simulation and prediction of skin pressure distribution applied by graduated compression stockings. Comput Textile 2007;55:301–9.
[85] Gerhardt L-C, Strassle V, Lenz A, D Spencer N, Derler S. Influence of epidermal hydration on the friction of human skin against textiles. J R Soc Interface 2008;5(28):1317–28. https://doi.org/10.1098/rsif.2008.0034.
[86] Downie SP, Raynor SM, Firmin DN, Wood NB, Thom SA, Hughes AD, Parker KH, Wolfe JHN, Xu XY. Effects of elastic compression stockings on wall shear stress in deep and superficial veins of the calf. Am J Physiol Heart Circ Physiol 2008;294(5):H2112–2120. https://doi.org/10.1152/ajpheart.01302.2007.
[87] Wang Y, Downie S, Wood N, Firmin D, Xu XY. Finite element analysis of the deformation of deep veins in the lower limb under external compression. Med Eng Phys 2013;35(4):515–23. https://doi.org/10.1016/j.medengphy.2012.06.019.
[88] Avril S, Bouten L, Dubuis L, Drapier S, Pouget J-F. Mixed experimental and numerical approach for characterizing the biomechanical response of the human leg under elastic compression. J Biomech Eng 2010;132(3):031006. https://doi.org/10.1115/1.4000967.
[89] Dai G, Gertler JP, Kamm RD. The effects of external compression on venous blood flow and tissue deformation in the lower leg. J Biomech Eng 1999;121(6):557.
[90] Narracott AJ, John GW, Morris RJ, Woodcock JP, Hose DR, Lawford PV. A validated model of calf compression and deep vessel collapse during external cuff inflation. IEEE Trans Biomed Eng 2009;56(2):273–80. https://doi.org/10.1109/TBME.2008.2005964.
[91] Rohan CP-Y, Badel P, Lun B, Rastel D, Avril S. Biomechanical response of varicose veins to elastic compression: a numerical study. J Biomech 2013;46(3):599–603. https://doi.org/10.1016/j.jbiomech.2012.10.043.
[92] Badel P, Christian P-YR, Avril S. Finite element simulation of buckling-induced vein tortuosity and influence of the wall constitutive properties. J Mech Behav Biomed Mater October 2013;26:119–26. https://doi.org/10.1016/j.jmbbm.2013.05.006.
[93] Martinez R, Fierro C, Shireman P, Han H-C. Mechanical buckling of veins under internal pressure. Ann Biomed Eng 2010;38(4):1345–53. https://doi.org/10.1007/s10439-010-9929-1.
[94] Frauziols F, Molimard J, Navarro L, Badel P, Viallon M, Testa R, Avril S. Prediction of the biomechanical effects of compression therapy by finite element modeling and ultrasound elastography. IEEE Trans Biomed Eng 2015;62(4):1011–9. https://doi.org/10.1109/TBME.2014.2378553.
[95] Chassagne F. Biomechanical study of the action of compression bandages on the lower leg. Université de Lyon – Mines Saint-Etienne; 2017.
[96] Frauziols F, Chassagne F, Badel P, Navarro L, Molimard J, Curt N, Avril S. In vivo identification of the passive mechanical properties of deep soft tissues in the human leg: in vivo identification of passive mechanical properties of leg soft tissues. Strain 2016;52(5):400–11. https://doi.org/10.1111/str.12204.
[97] Sengeh DM, Moerman KM, Petron A, Herr H. Multi-material 3-D viscoelastic model of a transtibial residuum from in-vivo indentation and MRI data. J Mech Behav Biomed Mater June 2016;59:379–92. https://doi.org/10.1016/j.jmbbm.2016.02.020.

[98] Vannah WM, Childress DS. Indentor tests and finite element modelling of bulk muscular tissue in vivo. J Rehabil Res Dev 1996;33(3):239–52.

[99] Dubuis L, Rohan CP-Y, Avril S, Badel P, Debayle J. Patient-specific computational models: tools for improving the efficiency of medical compression stockings. In: Adam W, Miller K, Nielsen PMF, editors. Computational biomechanics for medicine, vols. 25–37. New York, NY: Springer New York; 2013. https://doi.org/10.1007/978-1-4614-6351-1_4.

[100] Rabe E, Partsch H, Jünger M, Abel M, Achhammer I, Becker F, Cornu-Thenard A, Flour M, Hutchinson J, Ißberner K, Pannier C Moffatt &F. Guidelines for clinical studies with compression devices in patients with venous disorders of the lower limb. European Journal of Vascular and Endovascular Surgery 2008;35(4):494–500. https://doi.org/10.1016/j.ejvs.2007.08.006.

CHAPTER 5

Real-time computer modeling in prevention of foot pressure ulcer using patient-specific finite element model and model order reduction techniques

S. Niroomandi[1], A. Perrier[1,2], M. Bucki[2], Yohan Payan[1]
[1]Laboratoire TIMC-IMAG, CNRS & University Grenoble Alpes, France;
[2]TexiSense, Montceau-les-Mines, France

Chapter Outline

1. Introduction

This chapter summarizes some technologies that could be used to help with prevention of diabetic foot pressure ulcer. The latest work from the authors is also presented. The first component is a fully wireless, customizable, and washable smart sock to monitor the pressure applied to the foot. The second component is a patient-specific finite element (FE) model of the foot including soft tissues and bony prominences, which is specifically built and used to compute the internal strains and stresses from the pressure data provided by the smart sock sensors. To simulate such complex model, the ArtiSynth platform (artisynth.org) was used. Although ArtiSynth is powerful in combining rigid bodies and soft tissues, simulation time is in the order of hours, which is not suitable for a monitoring device. To overcome this issue, we could use a reduced order model (ROM) to decrease

Innovations and Emerging Technologies in Wound Care. https://doi.org/10.1016/B978-0-12-815028-3.00005-5

computation time to less than a second and thus make the simulations run in real time. In the following sections, foot pressure ulcer and associated complications are briefly reviewed while the smart diabetic socks are described. Then a patient-specific foot model from the authors is presented. Afterward, main ideas of model order reduction techniques that are used, namely proper orthogonal decomposition (POD) and POD with interpolation (PODI), are reviewed. After the ROM is constructed, a displacement field for evaluation points was simulated in a time frame of about 0.1s. Despite such short computation time, the results of the ROM show good accuracy in comparison to the full ArtiSynth solution and the maximum relative error norm was 4.6%, which seems acceptable from a clinical point of view. This technique could equally be extended to use an online version of the foot model by applying pressure values measured by the smart sock.

2. Foot ulcer and smart diabetic socks

Foot ulcers are mostly the consequence of a trauma or excessive pressures associated to diabetes and resulting from repetitive high stresses and strains, ill-fitting footwear, or sometimes an external object inside the shoe. Diabetic peripheral neuropathy and angiopathy are chronic complications of diabetes affecting the lower limbs. When both complications are associated to an external trivial trauma, they determine the progression of the pathology from a nonulcerated condition of a foot at risk to an acute syndrome characterized by the classic diabetic foot ulcer, followed by a chronic condition in the postulcerative phase, which can lead to amputation [1]. Healthy individuals with undamaged somatosensory function would experience pain and avoid the offending pressures if such external trauma occurred. On the contrary, in a person who suffers from diabetic neuropathy there is no warning of excessive pressure or tissue damage, which could lead to skin breakdown or ulceration. Previous studies suggested that an efficient care could reduce foot ulcer [2]. This monitoring strategy is largely based on renewed vigilance of the diabetic patient, which mainly consists in making a daily inspection of his/her feet. Unfortunately, studies show that patient vigilance decreases over time and complications which appear after some time are directly correlated with the increasing lack of vigilance. In the case of diabetic patients, it is precisely when the first foot ulcers appear that serious complications begin. This is mainly due to recurrent disease in the diabetic angiopathy, which severely limits the healing process. It becomes therefore essential to propose new technologies and services dedicated to the monitoring and assistance of the diabetic patient to help maintain his/her alertness and by extension his/her autonomy in society. The external trauma that causes tissue breakdown could be intrinsic, such as repetitive stresses from high pressure and/or callus, or extrinsic, such as from ill-fitting footwear rubbing on the skin or an object inside the shoe. It hence appears very suitable to be able (1) to monitor the external pressure applied all around the foot and (2) if possible, to estimate the corresponding internal stresses and strains.

To monitor the external pressures, a fully wireless customizable and washable smart sock was designed in a collaboration between two research laboratories (TIMC-IMAG and AGIM), three companies (TexiSense, IDS, and IFTH), and a clinical center (Center de l'Arche) in France [3].This smart sock is made of three main components: a 100% textile sock that collects the foot external pressures, a central unit, and an external device that receives the wireless information sent by the central unit and estimates the risks for pressure ulcer. A schematic of this system and details of the sock and its embedded sensors are shown in Figs. 5.1 and 5.2, respectively.

It is now well established that measuring pressures at the skin surface is not sufficient to prevent the most dangerous foot ulcers that start in deep tissues and progress outward rapidly, causing substantial subcutaneous damage underneath intact skin [22,4]. Indeed, such surface measurements are not sufficient to predict ulcer formation caused by internal tissue loading [5,6]. For example, a very similar pressure distribution could be observed under the heel of a thin person with blunt calcaneus bone and of a heavy diabetic person with sharp calcaneus bone, whereas in that case, the diabetic person has obviously much higher risks for foot pressure ulcers. The likelihood of a pressure ulcer forming is

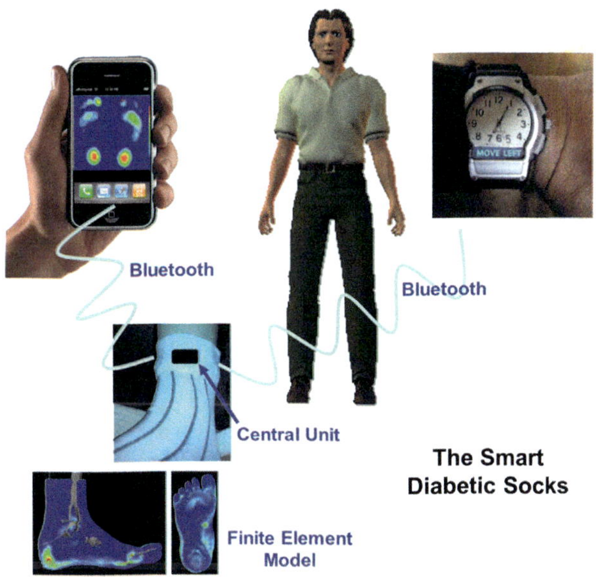

Figure 5.1

This smart sock is made of three main components: a 100% textile sock that collects the foot external pressures, a central unit, and an external device that receives the wireless information sent by the central unit and estimates the risks for pressure ulcer. *From Perrier A, Vuillerme N, Luboz V, Bucki M, Cannard F, Diot B, Colin D, Rin D, Bourg J-P, Payan Y. Smart Diabetic Socks: embedded device for diabetic foot prevention. IRBM April 2014; 35:72–6.*

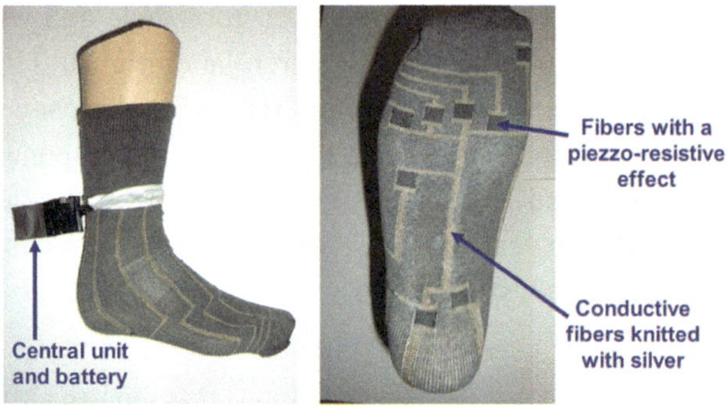

Figure 5.2

The sock knitted with fibers coated with silver (current conduction) and with fibers having a piezoresistive effect. The central unit (left) is connected to the silver-coated fibers and to a soft battery. *From Perrier A, Vuillerme N, Luboz V, Bucki M, Cannard F, Diot B, Colin D, Rin D, Bourg J-P, Payan Y. Smart Diabetic Socks: embedded device for diabetic foot prevention. IRBM April 2014; 35:72–6.*

therefore highly patient-specific because it depends on the anatomical properties of the patient foot such as the bones curvatures (calcaneus, metatarsal heads, toes) as well as the thicknesses of the soft tissues layers (skin, fat, and muscles). As measuring in vivo and in an embedded way the internal strains of the foot tissues is not possible, we have proposed to quantitatively estimate the internal stresses and strains from the measured external pressures. To achieve this goal, a patient-specific FE model of the foot including soft tissues and bony prominences is specifically built and used to compute the internal strains and stresses (Fig. 5.3).

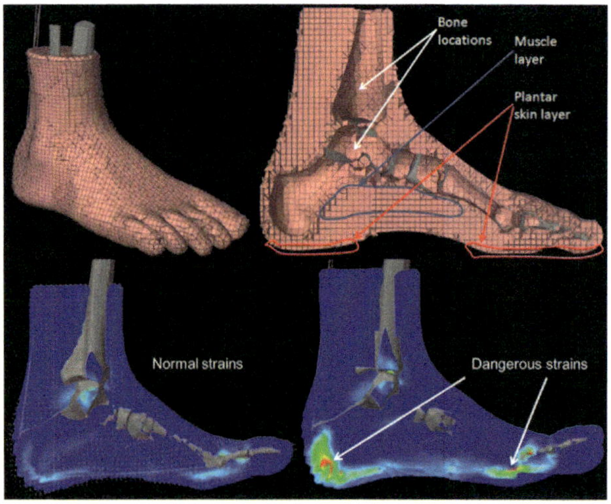

Figure 5.3

Finite element model of the foot: a 3D mesh (top) is used to compute internal strains (bottom). If strain values become too high (as it is illustrated in the lower right panel), an alert is launched. *From Perrier A, Vuillerme N, Luboz V, Bucki M, Cannard F, Diot B, Colin D, Rin D, Bourg J-P, Payan Y. Smart Diabetic Socks: embedded device for diabetic foot prevention. IRBM April 2014; 35:72–6.*

3. A patient-specific foot model

A detailed biomechanical model of the foot has been developed by Perrier et al., 2015. For this, a high-resolution computed tomography (CT) volume acquired on the unloaded right foot of a volunteer was used to segment the 28 bones and reconstruct their 3D shapes. Rigid body contact constraints were implemented in the ArtiSynth framework to model joint interactions. All 33 joint motions were further constrained by 210 ligaments modeled by cables and inserted on the bones using CT images of the subject. Such contacts between bones and ligaments attached to the bones were thus used to guide the foot kinematics. The aponeurosis was modeled using five ligaments linked by transverse structures. Finally, 15 Hills model muscles were positioned according to their anatomical course and can be independently activated. Soft tissues were modeled by an FE mesh comprising three subdomains representing skin, fat, and muscle tissues. The bones were modeled as rigid bodies to accelerate the simulation. The behavior of the bony rigid bodies coupled with FE models of the soft tissue domains was computed using a semiimplicit integration scheme in the ArtiSynth open source framework [7]. To increase the realism from a physiological and a dynamic point of view, the masses of the different structures were estimated using the tables from Ref. [8], which yielded a foot mass of 1.37% of the subject mass. The density and mass of the bones and soft tissues are shown in Table 5.1.

The muscles and part of the ligament insertions were localized within the medical images and the other part of the ligament insertions by resorting to the literature (Fig. 5.4).

The soft tissue FE mesh was generated using a meshing tool provided by the Texisense Company. In this application, modeling domains are defined by the 3D surfaces recovered from medical image segmentation. The meshing algorithm tends to maximize the number of hexahedrons in the generated mesh to avoid the locking effect inherent to linear tetrahedrons in nearly incompressible materials. The final mesh is composed of 23,298 elements and 13,087 nodes. It is composed of four domains: skin, heel fat pad, other fatty tissues, and muscles (see Fig. 5.5). As bones are modeled as rigid bodies, they do not need to be discretized.

To replicate a nonlinear hyperelastic behavior, an isotropic neo-Hookean material is assumed to describe the behavior of the soft tissues. This material behavior is characterized by the strain density energy function given by

$$\Psi = C_{10}(I_1 - 3) + (J - 1)^2 / D \tag{5.1}$$

Table 5.1: The density and mass of the bones and soft tissues.

	Density (kg/m^3)	Mass (kg)
Bones	1850	0.433
Soft tissues	1200	1.13

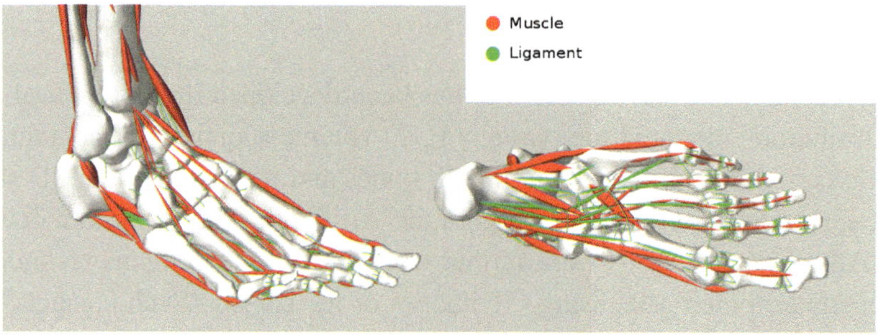

Figure 5.4
The musculoskeletal model with 28 bones, 15 activated muscles (in red), and 210 ligaments (in green). The insertions of those structures have been determined from the MRI scan and from literature.

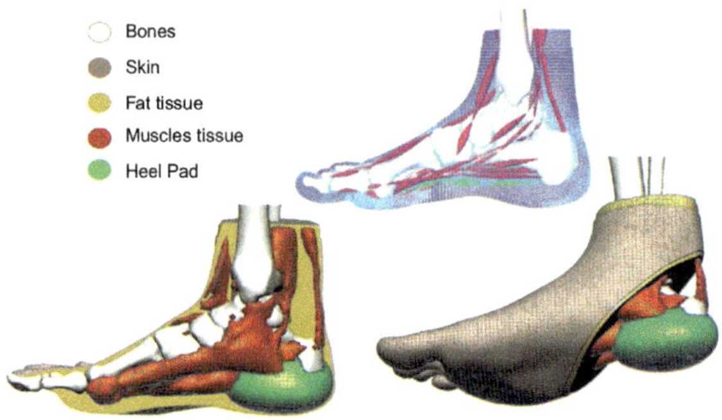

Figure 5.5
The complete biomechanical foot model with bony structures represented as rigid bodies, and soft tissues modeled as a finite element mesh. Soft tissues are divided in four layers: skin (in beige color), heel fat pad (in green), other fatty tissue (in yellow), and muscles (in red). *From Perrier et al., 2017 [3].*

where I_1 is the first invariant of the right Cauchy–Green deformation tensor, C_{10} is a material parameter, J is the determinant of the deformation gradient, and D is a material incompressibility parameter. The material properties of each soft tissue layer (or domain) are shown in Table 5.2 [9]. More details about this model can be found in Perrier et al., 2015, 2017 [20,21].

4. Model order reduction

The technique we used in this work goes by many names as it was employed in different branches of science and engineering among which Karhunen–Loeve (KL) decomposition, POD, or singular value decomposition (SVD) can be mentioned [10–12]., [13,14]. POD was successfully applied to real-time simulation of soft tissue [15–17]. In this work a variation of that method called PODI was used to construct a ROM with a parameter, which is the muscle activation. Such a model makes it possible to have a fast computation of the field of interest for any value of the parameter [18,19].

4.1 Proper orthogonal decomposition

In this technique we assume that the evolution of a certain field $u(x,t)$is known for example displacements. In practical applications (assuming that some numerical simulations were performed offline, for instance), this field is expressed in a discrete form which is known at the nodes $i$of a spatial mesh at time instancet^m. Thus, we consider that $u(\boldsymbol{x}_i, t^m) \equiv u_i^m$is known. We can also write $\boldsymbol{u}^m$for the vector containing the nodal displacements at time instancet^m. The main idea of KL decomposition is to obtain the most typical or characteristic structure $\phi(\boldsymbol{x})$among these $u^m(\boldsymbol{x})$, $\forall m$. Here ϕ is a scalar field defined at all the nodes of the FE mesh. The characteristic structure is defined in the KL approach as the function that maximizes.

$$\frac{\sum_{m=1}^{M}\left[\sum_{i=1}^{N}\phi(\boldsymbol{x}_i)u^m(\boldsymbol{x}_i)\right]^2}{\sum_{i=1}^{N}(\phi(\boldsymbol{x}_i))^2} \tag{5.2}$$

where N represents the number of nodes of the complete model and M the number of computed time steps. The maximization is equivalent to finding for any direction $\tilde{\phi}$ the zero of the Gâteaux derivative of the right-hand side in the above equation, which leads to

$$\sum_{m=1}^{M}\left[\left(\sum_{i=1}^{N}\tilde{\phi}(\boldsymbol{x}_i)u^m(\boldsymbol{x}_i)\right)\left(\sum_{j=1}^{N}\phi(\boldsymbol{x}_j)u^m(\boldsymbol{x}_j)\right)\right] = \alpha\sum_{i=1}^{N}\tilde{\phi}(\boldsymbol{x}_i)\phi(\boldsymbol{x}_i); \quad \forall\tilde{\phi} \tag{5.3}$$

which can be rewritten in the form

$$\sum_{i=1}^{N}\left\{\sum_{j=1}^{N}\left[\sum_{m=1}^{M}u^m(\boldsymbol{x}_i)u^m(\boldsymbol{x}_j)\phi(\boldsymbol{x}_j)\right]\tilde{\phi}(\boldsymbol{x}_i)\right\} = \alpha\sum_{i=1}^{N}\tilde{\phi}(\boldsymbol{x}_i)\phi(\boldsymbol{x}_i); \ \forall\tilde{\phi} \tag{5.4}$$

Defining the vector ϕ such that its i-th component is $\phi(\boldsymbol{x}_i)$, Eq. (5.4) takes the following matrix form

$$\tilde{\boldsymbol{\phi}}^T\boldsymbol{c}\boldsymbol{\phi} = \alpha\boldsymbol{\phi}^T\boldsymbol{\phi}; \ \forall\tilde{\boldsymbol{\phi}} \Rightarrow \boldsymbol{c}\boldsymbol{\phi} = \alpha\boldsymbol{\phi} \tag{5.5}$$

Table 5.2: Material properties of the neo-Hookean model for each soft tissue.

	C_{10}	Poisson's ratio
Skin	33.3	0.485
Heel fat pad	16.6	0.499
Other fat tissues	5.0	0.490
Muscles	10.0	0.495

where the two-point correlation matrix is given by

$$c_{ij} = \sum_{m=1}^{M} u^m(\boldsymbol{x}_i) u^m(\boldsymbol{x}_j) \Leftrightarrow \boldsymbol{c} = \sum_{m=1}^{M} \boldsymbol{u}^m (\boldsymbol{u}^m)^T \tag{5.6}$$

which is symmetric and positive definite. If we define the matrix $\boldsymbol{Q}$ containing the discrete field history:

$$\boldsymbol{Q} = \begin{pmatrix} u_1^1 & u_1^2 & \cdots & u_1^M \\ u_2^1 & u_2^2 & \cdots & u_2^M \\ \vdots & \vdots & \ddots & \vdots \\ u_N^1 & u_N^2 & \cdots & u_N^M \end{pmatrix} \tag{5.7}$$

then it is easy to verify that the matrix c in Eq. (5.5) results

$$c = \boldsymbol{Q}\boldsymbol{Q}^T \tag{5.8}$$

If some simulations were carried out, we can determine $\boldsymbol{u}_i^m$, $\forall i \in [1, \cdots, N]$and $\forall m \in [1, \cdots, M]$, and from these solutions the r modes related to the r-highest eigenvalues that are expected to contain the most important information about the problem solution. For this purpose we solve the eigenvalue problem defined by Eq. (5.5) retaining the modes corresponding to the eigenvalues that sum up to 99% of the variance. In practice, r is much lower than N, and this constitutes the main advantage of the technique. Thus, we can try to use these r eigenmodes $\phi_k$for approximating the solution of a problem *slightly* different to the one that has served to define u_j^m. For this purpose we define the matrix $\boldsymbol{A} = [\phi_1 \cdots \phi_r]$, which is an orthonormal basis

$$\boldsymbol{A} = \begin{pmatrix} \phi_1(\boldsymbol{x}_1) & \phi_2(\boldsymbol{x}_1) & \cdots & \phi_r(\boldsymbol{x}_1) \\ \phi_1(\boldsymbol{x}_2) & \phi_2(\boldsymbol{x}_2) & \cdots & \phi_r(\boldsymbol{x}_2) \\ \vdots & \vdots & \ddots & \vdots \\ \phi_1(\boldsymbol{x}_N) & \phi_2(\boldsymbol{x}_N) & \cdots & \phi_r(\boldsymbol{x}_N) \end{pmatrix} \tag{5.9}$$

Now, assuming that the displacement vector contains the nodal degrees of freedom, it can be expressed as

$$\boldsymbol{u}^m = \sum_{i=1}^{r} \zeta_i^m \phi_i = \boldsymbol{A}\,\zeta^m \tag{5.10}$$

in which ζ^m are the coordinates with respect to the POD basis. They are sometimes referred to as the reduced coordinates. It is worth reiterating that as r is much smaller than N, the displacement vector could be expressed with a few reduced coordinates.

4.2 Proper orthogonal decomposition with interpolation

In the previous section, POD procedure was explained considering a time-varying field, although that procedure can be applied in parameter space for example computing displacements by varying a parameter [18,19]. In this work, muscle activations were used as parameters and the displacement field was obtained for a range of physiological values. The procedure for fast prediction of the displacement field u for any value of the parameter γ can be summarized as

1. Perform simulations for each parameter value $\gamma_i$to obtain snapshots $\boldsymbol{u}^{\gamma i}$.
2. Apply POD technique to obtain the matrix $\boldsymbol{A}$ in Eq. (5.9), which is the orthonormal basis.
3. Compute the reduced coordinates for each snapshot using the inner product as

$$\zeta_j^{\gamma_i} = \left(\boldsymbol{u}^{\gamma_i}, \phi_j\right) \tag{5.11}$$

4. The reduced coordinates for intermediate values of γ that were not included in constructing POD basis are calculated by cubic spline interpolation of $\boldsymbol{\zeta}^{\gamma i}$, which is the vector containing the reduced coordinates corresponding to the parameter value γ_i. Now, the displacement field for this parameter value can be calculated using Eq. (5.4) as

$$\boldsymbol{u}^{\gamma} = \sum_{i=1}^{r} \zeta_i^r \phi_i = \boldsymbol{A}\boldsymbol{\zeta}^{\gamma}$$

Remark 1: Often, one is interested in finding rather an approximating affine subspace than a linear subspace. Hence the snapshots are centered by subtracting its average $\overline{\boldsymbol{u}}$ and collect the centered vectors as columns of the matrix $\boldsymbol{Q}$.

Remark 2: POD modes can also be computed by SVD of $\boldsymbol{Q}$ $\boldsymbol{Q}(\boldsymbol{U}\sum \boldsymbol{V}^T)$. In that case the POD modes are the columns of $\boldsymbol{U}$.

Remark 3: In many applications N is very large and calculation of the POD modes would require solving a large eigenvalue problem, which is not feasible. In such cases the modes can be computed by the method of snapshots proposed by Ref. [11].

5. Results and discussion

As a preliminary result, to investigate the ability of the model order reduction technique to run our foot model in an interactive way, one example is considered. In this example, the tibialis anterior muscle was activated and the displacement field was achieved using the ArtiSynth program. Afterward, the POD was applied to these solutions to construct a ROM with the muscle activation as a parameter. The force exerted by the tibialis anterior

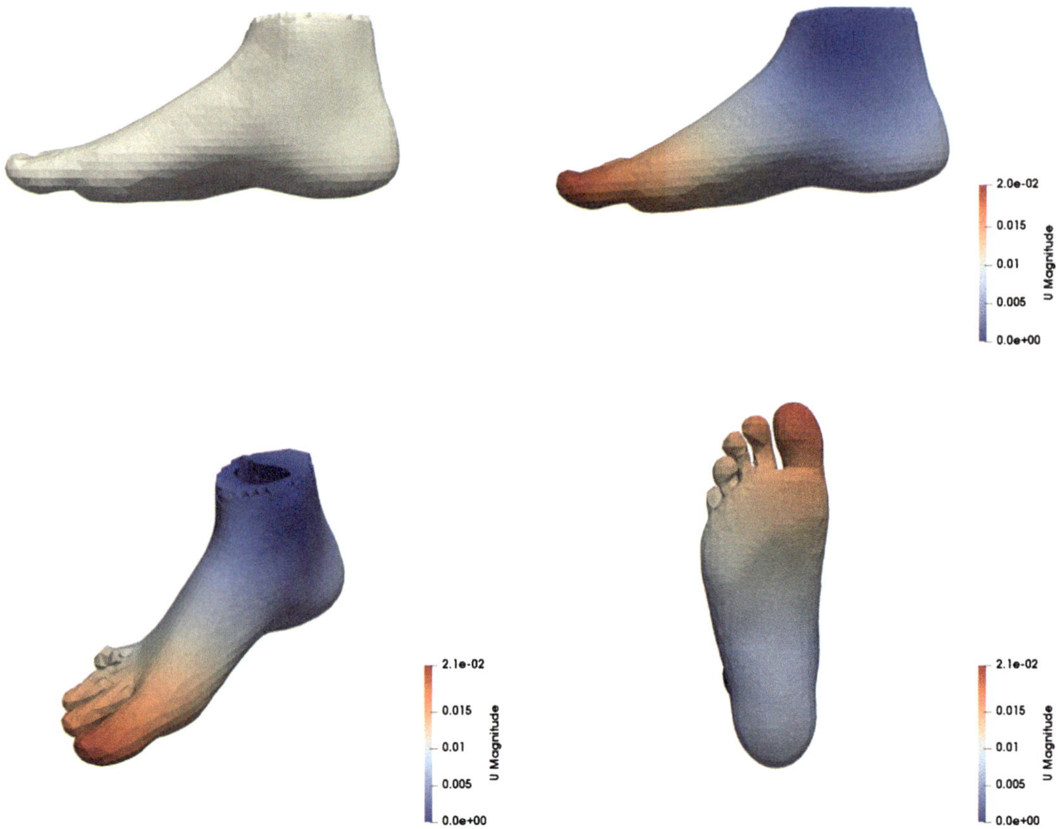

Figure 5.6
The nodal displacement magnitude for the first eigenmode.

muscle is increased from 0% to 70% of its maximum value and the displacement field of the FE model is computed by ArtiSynth and saved at each interval along with the muscle activation value (with a total of 200 simulated results). If we consider the muscle activation as a parameter, we can use PODI technique to have a fast prediction of the displacement field for any parameter value between 0% and 70%. We used 101 snapshots to construct the ROM and 99 snapshots to evaluate the ROM for intermediate values of muscle activation. The nodal displacement magnitudes for the first three eigenmodes are shown in Figs. 5.6–5.8.

The displacement field due to the tibialis anterior muscle 70% activation and computed by ArtiSynth from one side and by the ROM from the other side are depicted in Figs. 5.9 and 5.10, respectively.

As it can be seen, the ROM solution has a good agreement with the ArtiSynth solution, which is considered as the reference solution. The evolution of the relative error norm with muscle activation is plotted in Fig. 5.11 and the error in displacement

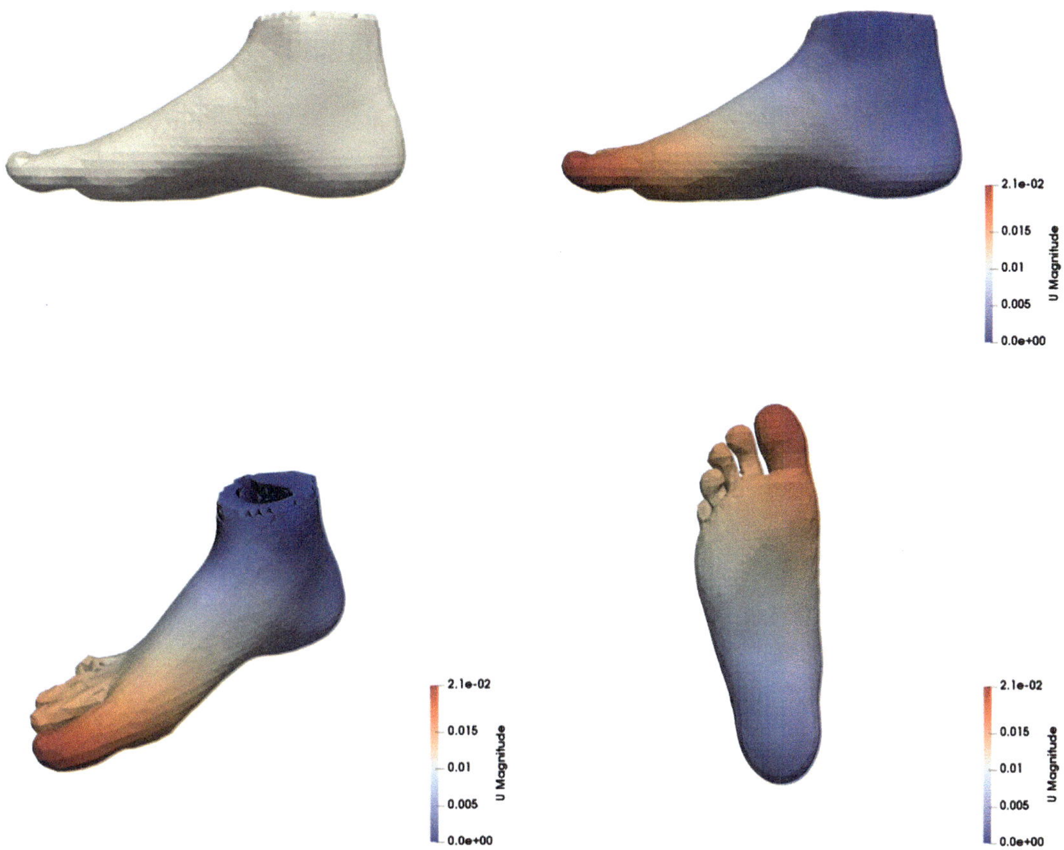

Figure 5.7
The nodal displacement magnitude for the second eigenmode.

$\|\boldsymbol{U}_{PODI} - \boldsymbol{U}_{Art}\|$was computed in the whole foot volume to determine the zones with less accurate results, which are around the ball of the foot and the tip of the big toe (Fig. 5.12).

The nodes around the big toe have large displacements and the node with maximum displacement was picked. The evolution of this value with the muscle activation calculated using the ArtiSynth program and the ROM is plotted in Fig. 5.13.

The computation time using ArtiSynth was about 12 h on a PC running Linux Mint with Intel Xeon CPU E5-1607 at 3 GHz; only four cores were used. Once the ROM is constructed, the displacement field generated by any activation of the tibialis anterior muscle requires 0.1 s to be computed. Despite such short computation time, the results of the ROM show good accuracy in comparison to the ArtiSynth solution because the maximum relative error norm was 3.8%, which seems acceptable in clinical routine. This technique could equally be extended to use an online version of the foot model by applying pressure values measured by the smart sock.

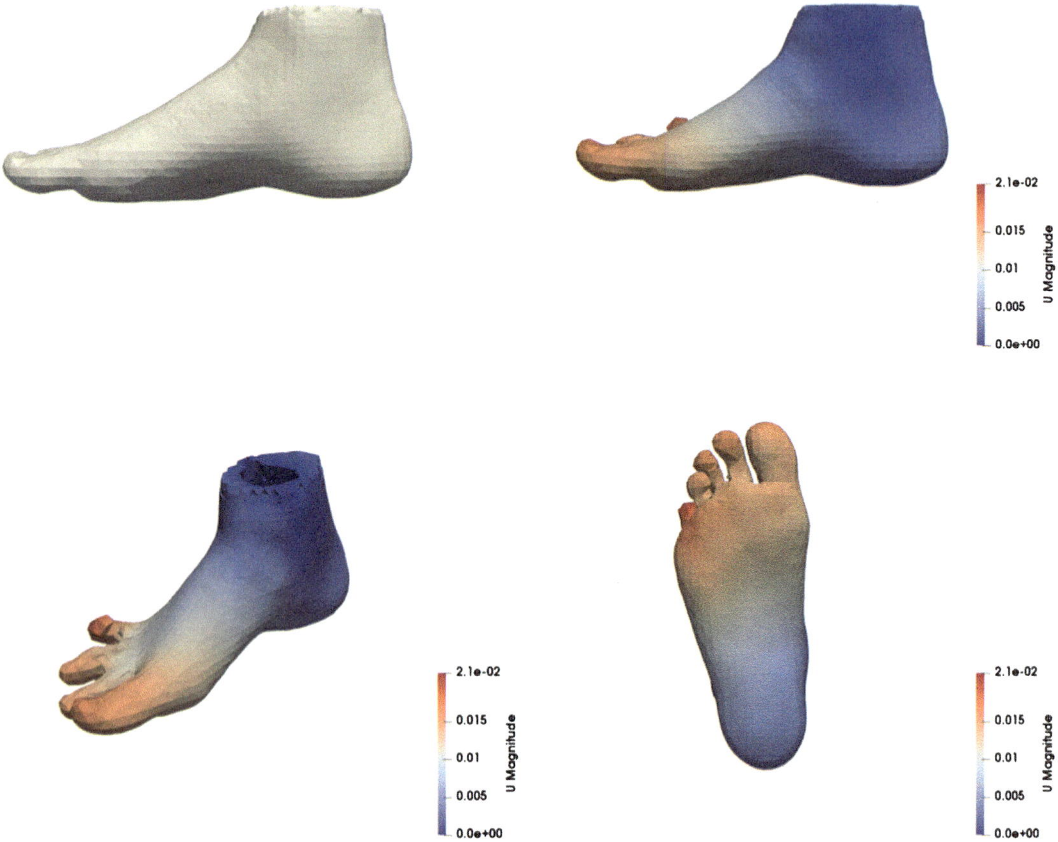

Figure 5.8
The nodal displacement magnitude for the third eigenmode.

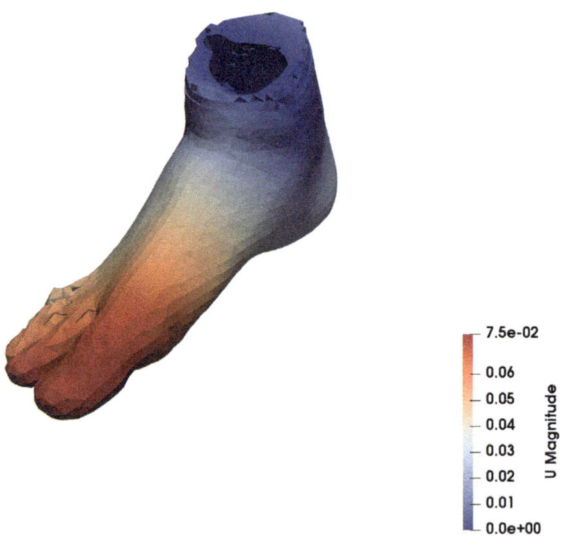

Figure 5.9
The displacement field computed by the ArtiSynth program.

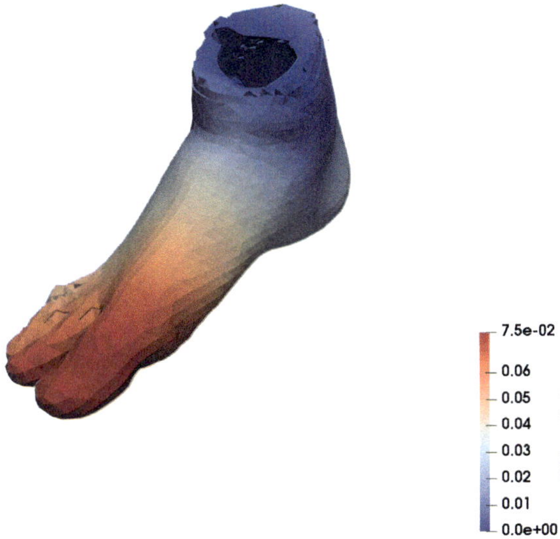

Figure 5.10
The displacement field computed by the reduced order model.

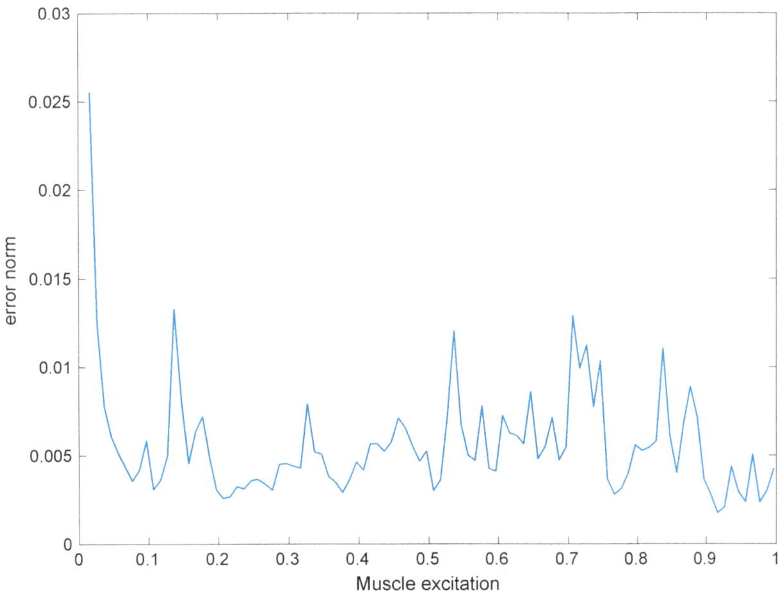

Figure 5.11
The relative error norm evolution with muscle activation.

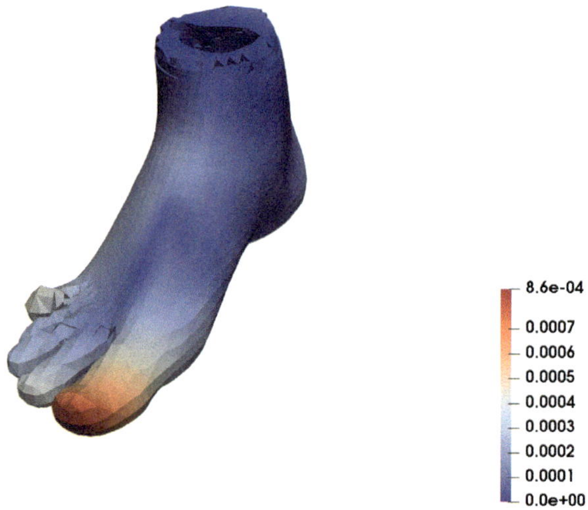

Figure 5.12
The error in displacement for the whole foot.

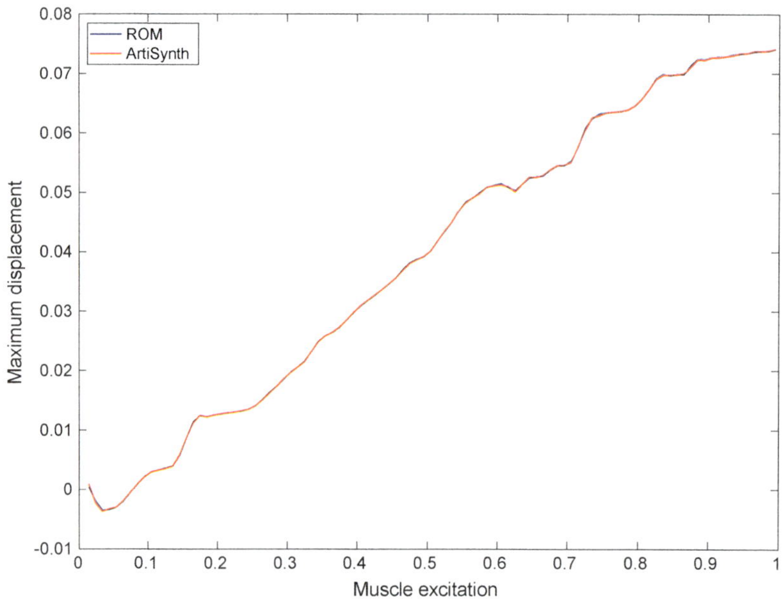

Figure 5.13
The maximum displacement evolution with muscle activation. *ROM*, reduced order model.

Acknowledgments

This work was funded by the *Fondation pour la Recherche Médicale* under the project FRM DIC20161236448.

References

[1] Shaw JE, Boulton AJM. The pathogenesis of diabetic foot problems: an overview. Diabetes 1997;46(Suppl. 2):S58–61. https://doi.org/10.2337/diab.46.2.S58. ISSN 0012-1797.
[2] Boulton AJM, Vileikyte L, Ragnarson-Tennvall G, Jan A. The global burden of diabetic foot disease. Lancet 2005;366:1719–24.
[3] Perrier A, Vuillerme N, Luboz V, Bucki M, Cannard F, Diot B, Colin D, Rin D, Bourg J-P, Payan Y. Smart Diabetic Socks: embedded device for diabetic foot prevention. IRBM April 2014;35:72–6.
[4] Atlas E, Yizhar Z, Khamis S, Slomka N, Shlomo H, Amit G. Utilization of the foot load monitor for evaluating deep plantar tissue stresses in patients with diabetes: proof-of-concept studies. Gait Posture 2009;29(3):377–82. ISSN 0966-6362.
[5] Linder-Ganz E, Shabshin N, Itzchak Y, Yizhar Z, Siev-Ner I, Gefen A. Strains and stresses in sub-dermal tissues of the buttocks are greater in paraplegics than in healthy during sitting. J Biomech 2008;41(3):567–80.
[6] Loerakker S, Manders E, Strijkers GJ, Nicolay K, Baaijens FPT, Bader DL, Oomens CWJ. The effects of deformation, ischemia, and reperfusion on the development of muscle damage during prolonged loading. J Appl Physiol 2011;111(4):1168–77. https://doi.org/10.1152/japplphysiol.00389.2011. PMID: 21757578.
[7] Lloyd JE. Ian Stavness, and Sidney Fels. Artisynth: a fast interactive biomechanical modeling toolkit combining multibody and finite element simulation. In: Payan Y, editor. Soft tissue biomechanical modeling for computer assisted surgery. Springer Berlin Heidelberg; 2012. p. 355–94.
[8] de Leva P. Adjustments to zatsiorsky-seluyanovs segment inertia parameters. J Biomech 1996;29(9):1223–30.
[9] Luboz V, Perrier A, Bucki M, Diot B, Cannard F, Vuillerme N, Payan Y. Influence of the calcaneus shape on the risk of posterior heel ulcer using 3d patient-specific biomechanical modeling. Ann Biomed Eng February 2015;43(2):325–35. https://doi.org/10.1007/s10439-014-1182-6. ISSN 1573-9686.
[10] Karhunen K. Uber lineare methoden in der wahrscheinlichkeitsrechnung. Ann Acad Sci Fennicae, Ser Al Math Phys 1946;37.
[11] Sirovich L. Turbulence and the dynamics of coherent structures. i - iii. Q Appl Math 1987;45(3):561–1590.
[12] P. Krysl, S. Lall, and J. E. Marsden. Dimensional model reduction in non-linear finite element dynamics of solids and structures. Int J Numer Methods Eng 51(4):479–504. https://doi.org/10.1002/nme.167.
[13] Liang YC, Lee HP, Lim SP, Lin WZ, Lee KH, Wu CG. Proper orthogonal decomposition and its applications-part I: Theory. J Sound Vib 2002;252(3):527–44. ISSN 0022-460X, https://doi.org/10.1006/jsvi.2001.4041.
[14] Pinnau R. Model reduction via proper orthogonal decomposition. In: Model order reduction: theory, research aspects and applications, vol. 13. Springer-Verlag Berlin Heidelberg; 2008. p. 95–109.
[15] Niroomandi S, Alfaro I, Cueto E, Chinesta F. Real-time deformable models of non-linear tissues by model reduction techniques. Comput Methods Progr Biomed 2008;91(3):223–31. https://doi.org/10.1016/j.cmpb.2008.04.008. ISSN 01692607.
[16] Niroomandi S, Alfaro I, Cueto E, Chinesta F. Accounting for large deformations in real-time simulations of soft tissues based on reduced-order models. Comput Meth Prog Bio 2010. https://doi.org/10.1016/j.cmpb.2010.06.012. ISSN 0169-2607.
[17] Cueto E, Chinesta F. Real time simulation for computational surgery: a review. Adv Model Simul Eng Sci April 2014;1(1):11. https://doi.org/10.1186/2213-7467-1-11. ISSN 2213-7467.

[18] Ly HV, Tran HT. Modeling and control of physical processes using proper orthogonal decomposition. Math Comput Model 2001;33(1):223–36. ISSN 0895-7177, https://doi.org/10.1016/S0895-7177(00)00240-5. Computation and control VI proceedings of the sixth Bozeman conference.

[19] Epureanu BI, Dowell EH, Hall KC. A parametric analysis of reduced order models of potential flows in turbomachinery using proper orthogonal decomposition. In: ASME TURBO EXPO; 2001. 2001.

[20] Perrier A, Bucki M, Vincent L, Vuillerme N, Payan Y. 3D musculoskeletal finite element analysis of the foot kinematics under muscle activation with and without ankle arthrodesis. Comput Methods Biomech Biomed Eng 2015;0(0):1–2. https://doi.org/10.1080/10255842.2015.1069605. PMID: 26273957.

[21] Antoine P, Vincent L, Bucki M, Cannard F, Vuillerme N, Payan Y. Biomechanical modeling of the foot. In: Biomechanics of living organs: hyperelastic constitutive laws for finite element modeling. Academic Press; 2017. p. 545–63.

[22] Gefen A. Plantar soft tissue loading under the medial metatarsals in the standing diabetic foot. Med Eng Phys 2003;25(6):491–9. ISSN 1350-4533.

CHAPTER 6

Bioengineering studies of cell migration in wound research

Daphne Weihs

Faculty of Biomedical Engineering, Technion-Israel Institute of Technology, Haifa, Israel

Chapter Outline

1. *Introduction*

Wounds, and especially chronic wounds, remain one of the most challenging medical problems and are responsible for a substantial fraction of medical system costs and constitute a large burden [1], while also causing significant patient discomfort. The two general types of wounds are traumatic, e.g., acquired at surgery or acute injury, and chronic wounds, such as diabetic or pressure ulcers (PUs). PU prevention is the only strategy to reduce resulting morbidity and eventual mortality while avoiding the substantial health-care costs of 10s of thousands USD for each serious case; the problem exists at all types of health-care facilities, e.g., hospitals, clinics, long-term care facilities, and homes. Thus, PUs have a prevalence of 5.7%−12.8% in hospital patients and 6.2%−12.8% in surgical intensive care units, and their incidence is as high as 25%−33% in community settings [2,3]. Beyond the substantial costs and burden associated with wound treatment, such wounds reduce the patients' quality of life and can often be life

Innovations and Emerging Technologies in Wound Care. https://doi.org/10.1016/B978-0-12-815028-3.00006-7

threatening. Many serious PUs (category 3 and 4) become chronic and can eventually lead to mortality from complications, e.g., sepsis, osteomyelitis, renal, or organ system failure.

Prevention is the only effective approach to reduce the formation, development, and serious implications of PUs. For clinical applications, cost-effective and efficient prevention and treatment approaches are critically required. Current prevention is focused on nutrition and support surfaces to redistribute tissue loads; however, PU prevalence remains high. Many approaches exist for treatment of PUs, yet those are not immediately applicable for prevention requiring novel preventative approaches.

The accepted knowledge in the international scientific and clinical PU communities [4], based on rigorous scientific and clinical evidence [5–8], has shown that PUs initiate at sites of elevated tissue deformations in deep tissues (e.g., muscle and fat), near bony prominences because of bodyweight. At those sites, the bodyweight of bedridden or chair-bound patients causes cell-level sustained deformations, which can lead to plasma membrane (PM) permeabilization, loss of cell homeostasis, necrosis, and apoptosis. Increasing the adaptability and resistance of cells to tissue deformations caused by the bodyweight at susceptible body sites may reduce PU formation and progression and extend the safe time for a patient at a certain posture (e.g., lying supine or sitting motionless).

1.1 Cell death due to mechanical loading

The dynamic cytoskeleton of a cell typically adapts to withstand changing environmental conditions. Under external forces and sustained deformations, changes in cytoskeleton structure and dynamics likely prevent it from maintaining the cell shape and the overall cell integrity and function.

Local loads, such as those at sites of boney prominences, are applied to the surrounding tissues. Tissues are composed of an extracellular matrix (ECM) in which the cells of that tissue layer adhere, grow and proliferate, or migrate; migration is typical for skin cells (keratinocytes or fibroblasts) or undifferentiated stem cells. The mechanical loads are distributed in the tissue, potentially damaging the matrix and the cells [9]. The mechanical loads can induce strains (in [%], i.e., an induced size change relative to the original size) and mechanical stresses (forces per area) that can damage tissue. In the ECM, the strains can induce anisotropy or breakages, which can affect the mechanical stability as well as the function of the cells within; damage to the ECM may be repaired or remodeled locally by cells (e.g., fibroblasts) on local cues [10]. In contrast, damage incurred by the cells may result in cell death or loss of function; dead cells affect neighboring cells and must also be cleared to allow repopulation.

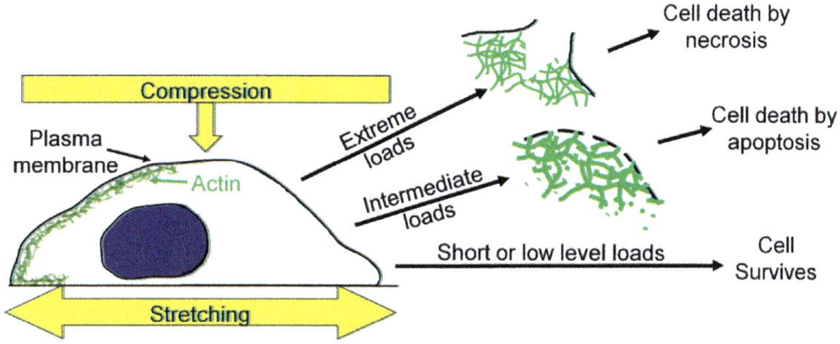

Figure 6.1

The mechanisms of possible cell responses under mechanical loads (e.g., external compression or stretching) are cell survival, apoptosis, or necrosis for, respectively, short- or low-level, intermediate, or extreme extents of loads. Qualitatively, the extreme loads will likely lead to immediate rupture and breakdown of the plasma membrane and cytoskeleton (typically the actin), respectively. The intermediate loads may lead to local failure of the cytoskeleton, consequently causing poration of the plasma membrane which then becomes leaky; homeostasis is gradually lost and the cell eventually dies by apoptosis. Cells can withstand short- or low-level loads, e.g., by self-repair. *Reprinted from Gefen A, Weihs D. Cytoskeleton and plasma-membrane damage resulting from exposure to sustained deformations: a review of the mechanobiology of chronic wounds. Med Eng Phys 2016;38:828–33. https://doi.org/10.1016/j.medengphy.2016.05.014 with permission.*

The damage that mechanical loads can induce in living cells may be classified by level of effect on cell viability and function (Fig. 6.1):

(a) *Low mechanical loads*: The loads are physiological scale and are tolerable by (initially undamaged) cells for prolonged times, relative to the life span of the cell. The tolerable load magnitude will depend on the loading mode (e.g., sustained/dynamic) and the cell type (e.g., skin, fat, or muscle), and cells will function normally throughout the loading period. Such loads occur in daily, real-life scenarios, for example, as skin and subcutaneous tissue cells are able to survive under the weight of a wrist watch or a bracelet.

(b) *Intermediate mechanical loads:* The loads are either larger magnitude or at physiological magnitudes but delivered continuously and without relief for a prolonged period. The viability of the cells is affected in that they will not survive a normal cell life span, yet cells will not fail immediately. The loading will affect the mechanostructure and function of cells, particularly on the dynamics, and the integrity of the cytoskeleton and PM [5] will be such that cells will gradually and slowly die, i.e., by apoptotic pathways.

(c) *High mechanical loads:* The magnitude of the applied load is large enough that cells will immediately and catastrophically fail. The loading will irreversibly disrupt the cell structure and integrity, causing its death. This is typically caused by a traumatic injury, e.g., a knife cut killing cells at the blade location.

2. Cell migration during gap closure in the context of wound healing

Wound healing depends strongly on the ability of cells from sites surrounding the damage to migrate into the gap. The migratory cells involved in gap closure are typically fibroblasts, keratinocytes (in the skin), or stem cells; undifferentiated stem cells are highly migratory even after commitment to a specific lineage. Once in the gap, the cells can proliferate to reconstitute the tissue or can differentiate to restore specific functions or structures. For example, adipose-derived stem cells migrate and also affect the migration of neighboring fibroblasts [10–12]. Cell migration is typically triggered through mechanical or biochemical cues following gap (wound) formation and may also occur naturally in highly migratory cells, e.g., fibroblasts, as part of physiological processes [13]. Thus, migration of highly motile keratinocytes, fibroblasts, and various types of stem cells has been studied extensively.

Cell migration occurs as a natural, physiological process [13] and also in tumorigenesis. Cells migrate either collectively, in groups, as in embryonic development or in cancer metastasis, or as single cells, for example, immune cells or invading cancer cells [14,15]. The substrate stiffness can affect migration of cells and their applied forces [16,17] and can also affect cell differentiation and fate [18].

2.1 Single cell migration: morphology and forces

The highly migratory fraction of cells includes many types of stem cells, which are also able to differentiate to fill specific requirements at the wound site. Those stem cells can maintain their migratory capacity even after committing to a specific lineage, e.g., mesenchymal stem cells and committed preadipocytes are migratory, but adipocytes are not. The fibroblast-like preadipocytes are a good example of highly migratory cells. This characteristic allows them to be called to damage sites, where they can increase the available energy by increasing the fat storage capacity. To migrate, proliferate, and eventually differentiate, the cells require adhesion to an ECM with which they may dynamically interact. We have recently studied the dynamic mechanobiological interactions of migrating preadipocytes with their environment [19]. We have evaluated the forces applied by single, migrating preadipocytes on a two-dimensional (2D) substrate. We demonstrate below the typical approach to determine the in vitro time-dependent forces applied by migrating preadipocytes.

Cell migration includes several steps requiring morphological and mechanostructural changes in the cells, such as cell polarization, formation of a protrusion at the direction of motion, i.e., at the cells' leading edge, combined with contraction, and retraction of the cell through adhesive forces applied to the surrounding matrix [20,21]. As cells interact with their microenvironment, they apply forces and induce deformations, e.g., to the ECM.

Traction force microscopy (TFM) has widely been used to evaluate the forces applied by single cells [17,19,22–24] or groups [25–28] of cells that are adhered to a substrate. The substrate is a synthetic or biological (typically 2D) gel with fiducial markers embedded into it, typically being fluorescent beads; tractions can also be evaluated in three dimensions using different approaches [29,30]. The cell-applied forces induce deformation in the gels, resulting in displacement of the markers; biopolymer fibers may also be fluorescently labeled [31] or imaged with reflectance microscopy [32,33] to reveal local fiber reorientation.

The marker displacements are determined by image correlation, e.g., between time points or before/after treatment, which provides the mechanical strains. The strains can then be used to estimate the stresses or forces applied on the gel, as the gels typically used for TFM are (linear) elastic. The stresses are determined through the linear elastostatics, assuming zero body forces and an infinite half-space elastic substrate. Under those conditions, the Boussinesq solution states that the displacement (u) at every position (x) is related to the traction stress (T) through the Green's functions (G) as a convolution integral:

$$u(x) = \int G(x, x')T(x')dx' \tag{6.1}$$

Butler et al. [34] have solved Eq. (6.1) in the Fourier field, where convolutions reduce to a simple multiplication:

$$\widetilde{u}(k) = \widetilde{G}(k)\widetilde{T}(k) \tag{6.2}$$

with the Green's function in the Fourier space being

$$\widetilde{G}(k) = \frac{2(1+v)}{Ek^3}\begin{pmatrix} (1-v)k^2 + vk_y^2 & -vk_xk_y \\ -vk_xk_y & (1-v)k^2 + vk_x^2 \end{pmatrix} \tag{6.3}$$

Here the gels' Young's Modulus, E, and Poisson's ratio, v, are used over radial wave vector, k, to calculate the displacement from an applied traction force. However, in TFM, the measurable quantity is the displacement field and the strains at each location, and the calculated value is the traction stress; this requires solving the inverse problem. The 2D inverse Fourier transform is applied on the product of the displacements and the inverse Green's function:

$$\widetilde{T}(k) = FT_2^{-1}\left(\widetilde{G}(k)^{-1}\widetilde{u}(k)\right) \tag{6.4}$$

The traction stresses and forces are calculated within the constrained boundaries of the cell location, recognizing that while deformations may occur at various locations in the continuous gel, stresses are only applied where a cell is present; the boundaries of the cells need to be accurately identified. Thus, the tractions are calculated by iterating to convergence the traction forces determined directly from the displacements within the cell boundaries, while concurrently setting the tractions outside the boundaries to be zero [34].

We demonstrate TFM evaluation of the forces applied by migrating preadipocytes on a 2D, synthetic polyacrylamide gel; the gel includes fluorescent beads as markers and has a monolayer of collagen to facilitate biochemical attachment of the cells. Using the TFM approach delineated above, we have determined the forces applied by migrating, single preadipocytes on a physiological stiffness 2.44 kPa collagen-coated polyacrylamide gel. We have imaged cells on the gels every 20 min, evaluating changes in forces and directionality [19]. Cells exhibited fibroblast-like, elongated morphologies with mean migration speeds of 0.27 μm/min; 3T3 fibroblasts on a 2.8 kPa polyacrylamide gel have previously demonstrated speeds ranging 0.19–0.31 μm/min [35].

The migrating preadipocytes exhibited an elongated morphology, with local widening at the leading edge before retraction of the rear end (Fig. 6.2A,B). The speed of the migrating cells was calculated between successive 20 min frames, using the locations of the geometrical center of mass. The distribution of the speeds (magnitude without direction) demonstrated a log-normal distribution (Fig. 6.2C), i.e.,

$$f(x) = \frac{1}{x\sigma\sqrt{2\pi}}\exp\left[(\ln x - \mu)^2 / 2\sigma^2\right] \tag{6.5}$$

Such distributions have previously been observed in biological context, e.g., in intracellular particle dynamics [36,37]. The average migration speed of the preadipocytes on the 2.44 kPa polyacrylamide was 0.27 ± 0.09 μm/min. Previous studies with 3T3 fibroblasts migrating on a 2.8 kPa polyacrylamide gel demonstrated a similar range of speeds, 0.19–0.31 μm/min [35].

We have also evaluated the time-dependent traction forces that the cells apply during migration (Fig. 6.2B), typically including breaking of cell symmetry in terms of force application at the front (leading edge) and the rear [20,21,38]. The sum of the force vectors (magnitude and direction) are expected to be zero, due to force conservation of the forces applied by the cell against the counteracting gel according to Newton's third law. Thus, we focus on the magnitudes of the cell-applied forces, i.e., the sum of the magnitudes of the traction stress vectors (nondirectional) over a specified cells area. For example, the total traction force applied by a cell is in fact its mechanical energy output [22] and is likely the sum of forces applied at all cell–substrate connection points, e.g., focal adhesions.

As each cell is migrating in a random direction (no attractants present), we separate the two sides of the cell, its poles which are expected to be the migratory front, and rear ends. We have used the cell orientation and morphology to generate a local cell-specific axis,

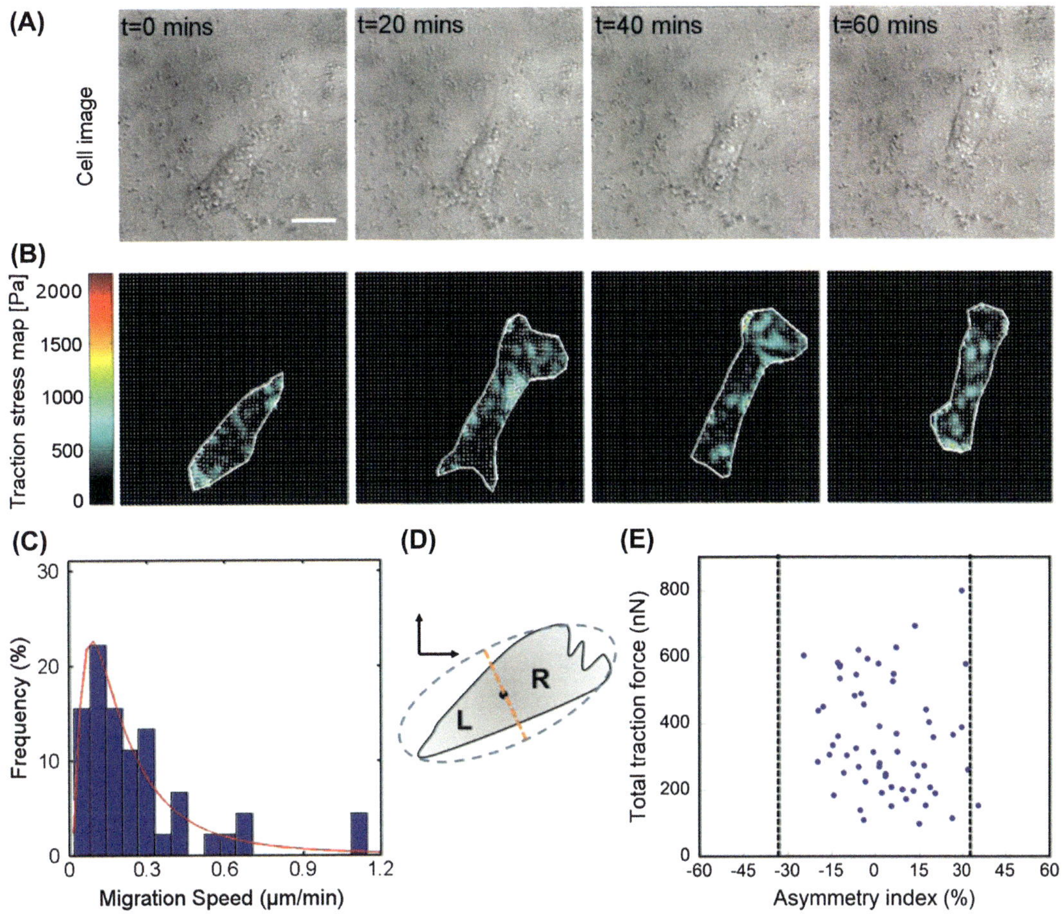

Figure 6.2

(A–B) Single preadipocytes migrating on 2.44 kPa stiffness polyacrylamide gel; the cell is migrating from bottom left to top right corners. Top row shows the DIC images of the cells and bottom row provides the maps of the traction stresses applied by the cell at time point with 20 min intervals. Stresses are confined within the manually marked cell boundaries (*white line*). Scale bar is 20 μm and color bar indicates traction stresses. (C) Distribution of preadipocyte migration speeds at 20 min intervals, a total of 45 images of single cells were included in the distribution. The distribution was fit to a log-normal (*solid red line*) with a mean speed of 0.27 ± 0.09 μm/min. (D) The cell morphologies are estimated with a bounding ellipse, where the minor axis defines left/right (L/R) sides of the cell on its main axis; the total applied traction forces (nondirectional magnitudes) in each side of the cell or within the entire bounded area may be calculated. (E) The total traction force versus the asymmetry index, calculated by the percent difference between the left sum and the right sum of the traction force magnitudes divided by the total traction force applied by the whole cell. The preadipocytes applied a wide range of total traction forces; however, the asymmetry index was bounded to ±33%. *Reprinted from Abuhattum S, Weihs D. Asymmetry in traction forces produced by migrating preadipocytes is bounded to 33%. Med Eng Phys 2016;38:834–8. https://doi.org/10.1016/j.medengphy.2016.05.013 with permission.*

wherein at each time point we can calculate the forces applied at each side of each cell and identify the currently active site. A similar approach has previously been used to define a polar coordinate system using the typically oval-shaped morphology of the nucleus of cells adhered on a 2D substrate [39].

To identify directional preferences in the total forces, we have defined an asymmetry index for the cells. The asymmetry of the traction forces, F, is calculated by

$$\text{Assymetry}\,[\%] = \frac{F_{totL} - F_{totR}}{F_{totL} + F_{totR}} \tag{6.6}$$

being the difference of the total forces applied at the left relative to the right sides of the cell divided by the total traction forces applied by the entire cell. An asymmetry index of zero indicates balanced magnitudes of traction forces in the left and right sides of the cell. Similarly, a positive or negative asymmetry index indicates, respectively, stronger forces on the left or right sides of the cell. Although the total traction force [22] applied by each cell varies widely (100–800 nN), likely as cells are captured at different stages of migration, interestingly the asymmetry between the sides is bounded to 33% for all forces (Fig. 6.2E), that is, at all captured stages of the migration, the force applied on one side of the cell (front/rear) was only up to 33% more than the other side.

2.2 En masse cell migration kinematics determination

Efficient migration is typically achieved when cells migrate as a collective or en masse. Collectively, migrating cells moving in groups or monolayers [40] have been observed in the context of embryonic development [41,42] morphogenesis and regeneration [42] and of cancer metastasis [43] and is also required for wound healing [44–46]. In wound healing, collective cell migration is a major mechanism driving closure of formed gaps.

The kinematics of en masse migration, e.g., during closure of a damage-induced gap, can be quantitatively determined [44,47]. Using a custom image-processing algorithm, we obtain the denuded area [48], devoid of cells at each imaged time point (Fig. 6.3A). The algorithm segments the denuded and cell-populated areas through spatial and temporal local texture homogeneity measures [44]. The onset of en masse cell migration is typically observed when 10% of the gap area has been covered and ends at 90% of coverage (Fig. 6.3B). The linear slope between the start and end of the en masse migration is the average migration rate (AMR). We note that the time of start–end of en masse migration may vary, which could lead to inaccuracies in the AMR calculations (as demonstrated in Fig. 6.3B); hence, a fit is used to more accurately quantify the kinematics and provide a time-independent parameter.

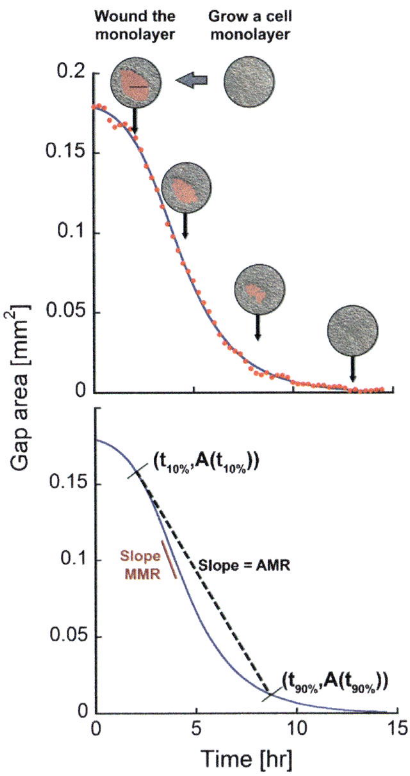

Figure 6.3

Gap formation and analysis of closure kinematics. (A) Cells in a confluent monolayer (here murine 3T3 fibroblasts) are damaged (i.e., by crush or scratch) to form a gap. The gap area is automatically imaged and analyzed using custom MATLAB modules to detect the denuded areas during the gap closure process [47,48]. The experimental results (*red dots*) are fit with the Richard's function (*blue line*). Scale bar is 200 μm. (B) Using the Richard's function fit, the average migration rate (AMR) is calculated as slope of the dashed line connecting the onset ($t_{10\%}$) and end ($t_{90\%}$) times of en masse cell migration. The maximum migration rate (MMR) is calculated as the maximum slope (*in red*) of the Richard's function that is fitted to the experimental data; the MMR is typically normalized to the initial gap size. *Reproduced from Toume S, Gefen A, Weihs D. Low-level stretching accelerates cell migration into a gap. Int Wound J 2017;14:698–703. https://doi.org/10.1111/iwj.12679 with permission.*

To analyze the kinematics of the cell migration during gap closure, we use the Richards function (asymmetric sigmoid). This function was originally developed from the theory of restricted growth [49] and was adapted in an inverse (reducing) representation to describe the reduction of gap area during cell migration [47]. Specifically, we fit the time-dependent gap area plots with the following function:

$$\mathrm{Area}(t) = \frac{\alpha}{\left(1 + ve^{\frac{t-t_0}{\Delta T}}\right)^{\frac{1}{v}}} \tag{6.7}$$

The fit parameters, obtained by minimizing the mean squared error objective function [44], are the upper plateau, α; the "starting time," t_0; the measure of the growth/reduction rate, $1/\Delta T$; and the υ that is correlated to the maximum migration rate. The normalized maximum migration rate (NMMR, in %/h) is the maximum of the derivative (dA/dt), within the linear regime, of the Richards fit (Eq. 6.7), which is normalized to the initial wound area a_0:

$$\text{NMMR} = \frac{1}{a_0} \cdot \frac{\alpha}{\Delta T (1 + \upsilon)^{\frac{1+\upsilon}{\upsilon}}} \tag{6.8}$$

2.3 Effects of honey wound care additive on gap closure kinematics

Honey has long been known to have antiinflammatory and antimicrobial properties, leading to its wide use in skin wound care. It has also been proposed to have cell-stimulating properties, and thus we have evaluated the direct effect of honey application on cell migration. We have evaluated the kinematics of cell migration toward gap closure, under varying concentrations of medical-grade honey [50].

Mouse fibroblasts (NIH3T3) were grown to confluency, forming a cell monolayer, and then a gap was induced by crushing injury using a truncated, 300 μm optic fiber. Following induction of damage, the cell growth medium was replaced with medium containing varying concentrations of medical honey. The kinematics or time dependence of the gap closure was monitored by time lapse microscopy and analyzed using custom image-processing modules [47,48] in MATLAB 2012b. We have generated a wide range of small-scale gaps to identify the effect of initial gap area and observe the typical sigmoidal reduction in gap area (Fig. 6.4A–C). The kinematics of gap closure were calculated by fitting a Richards' function and determining the maximal migration rate. The maximal migration rate scales linearly with the initial gap area (Fig. 6.4D) as observed in other similar experiments [51,52]. The maximal migration rate was normalized by the initial gap area for comparative analysis. We show that the migration rate under 0.02% (w/v) honey supplement was indistinguishable from control, yet significantly reduced ($P < 0.02$) under 0.2% (w/v) honey (Fig. 6.4E).

The effects of other natural medications, such as curcumin, aloe vera, and ginger, on gap closure kinematics have not demonstrated any change in the migration rate [53]. Thus, these natural medications do not directly affect cell migration, and their beneficial effects in wound healing stem from other processes, e.g., antiinflammatory and antimicrobial capacities of honey. We note that while medical honey is widely used to treat various types of wounds (including PUs), it is cannot be used for prevention at the early stages of wound formation, i.e. microscopic gap closure. Moreover, using laboratory *in vitro* models

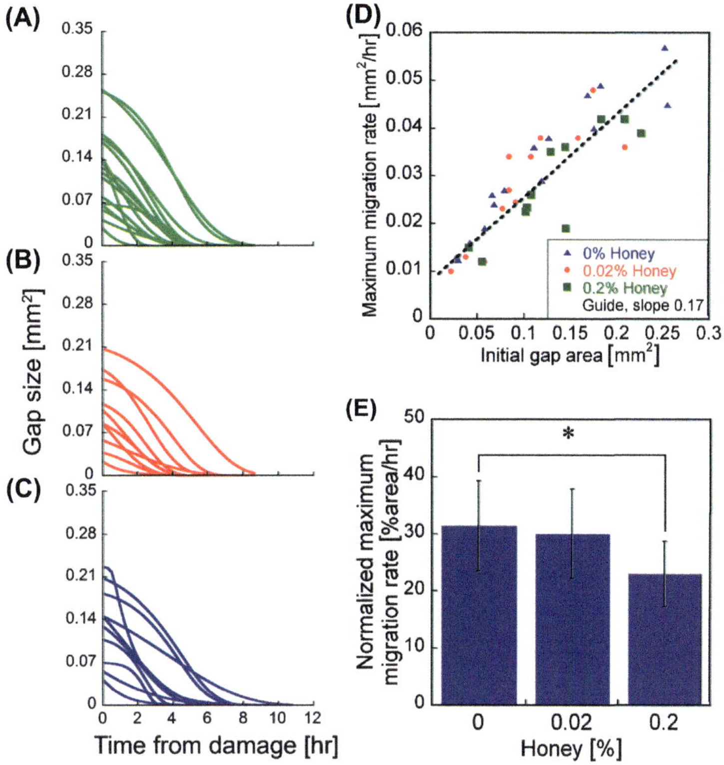

Figure 6.4

Effects of medical honey on the migration kinematics of murine NIH3T3 fibroblasts; gap size was monitored over time for cells with (A) no added honey control; (B) 0.02%w/v honey; and (C) 0.2% w/v honey. (D) The maximal migration rate increases linearly with the initial gap area for all evaluated honey concentrations; *dashed line* is a guide to the eye with slope of 0.17. (E) The normalized migration rate reduces with honey concentration ($^*P < 0.02$).

we show that addition of medical honey, in fact, reduces prevention-associated closure of small gaps that form during initial damage.

3. Accelerating gap closure by combining low-level nondamaging strains and exogenous sodium pyruvate supplement

Cell function and migration is affected by the stiffness of their microenvironment and external forces transmitted through it. Mechanical stresses may be transmitted to the wound site and its cells naturally by daily routine of the patient or care givers, e.g., by application of dressings, or may be applied by contemporary therapies such as negative pressure wound therapy; in the latter, the vacuum applied to the wound site induces macro- and microscale deformations to the tissues and cells [54]. The direct effects of locally applied strains and induced deformations are known to damage cells [5] and may cause PM poration [55–57]; loss of membrane integrity can lead to loss of cell homeostasis and eventual death.

Cells are, however, designed to withstand a certain level of strain, as explained above, and may even benefit from it. Macro- and microscale deformations have been shown to constructively influence, in various types of cells, cell morphology, proliferation, apoptosis, differentiation, and development, as well as adhesive interactions, cell orientation, and direction of migration [54,58–63]; stretching changed expression of proteins and genes controlling these processes [54,59]. As mechanical strains have such far-reaching effects on intracellular composition and structure, cell shape, and fate, we expected a significant effect also on the kinematics of cell migration during gap closure in the context of wound healing.

We have therefore evaluated potentially beneficial effects of low-level, nondamaging strains that can be controllably applied to cells at sites of initial cell damage to halt wound development; initial damage including death of several cells may spiral into large-scale cell death, inflammation, immune response, and become a vicious, self-intensifying cycle that progresses to a PU [64]. Previous works have shown a large increase in poration of the PMs of individual cells, under strain-induced damage (as in Fig. 6.1) above 10% strain [56,57]. In contrast, under 6% strain and even 3% strain, PM poration was, respectively, smaller or insignificant. We have also shown that fibroblast morphology, adherence, and viability are unchanged following application of 6% radial stretching applied either statically or dynamically [65]. Consequently, we have constrained the applied stretching to 6%. As a result, we have shown that low-level, nondamaging strains of 3%–6% applied directly to cells may accelerate cell migration during gap closure, in the context of wound healing [51].

Combined treatments have great potential for synergistic, beneficial effects, and thus we have opted to affect the biochemical composition of the wound site environment, concurrently with external mechanical stretching. Our chosen supplement is sodium pyruvate (NaPy), a natural metabolic intermediate in the Krebs' cycle (the glycolysis pathway) of the cell. NaPy (typically 1 mM) is added in standard cell culture procedures as a supplement to support rapidly proliferating cells [66], such as skin and connective tissue fibroblasts as well as cancer cells. Pyruvates have a double role in cells, as endogenous energy provider and as a powerful antioxidants [67]; NaPy has been shown to have antioxidant and antiinflammatory capacities [68–73]. Being native to the cells in the living body, pyruvates are expected to be a safe supplement; safety is an important feature for any potential in-human application. In this context, pyruvates have been used commercially in skin ointments and treatments [74] as well as orally consumed food supplements and weight loss aids [75].

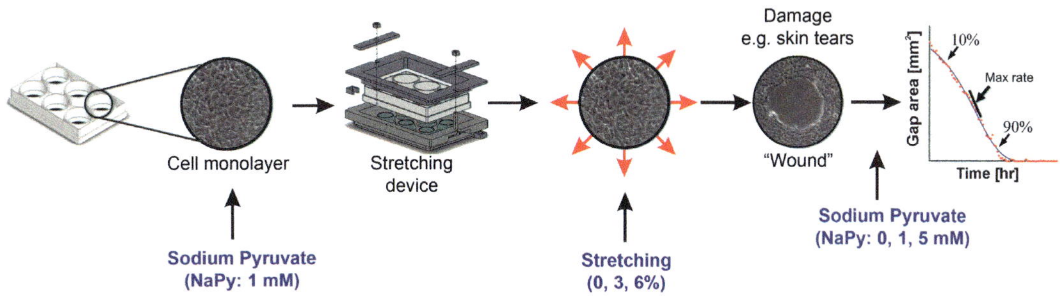

Figure 6.5

Experimental procedure of stretching combined with sodium pyruvate supplement. From left, cells were seeded on stretchable-substrate 6-well plates and grown in 1 mM sodium pyruvate (NaPy) until attaining a confluent monolayer. Well plates were then placed in a custom three-dimensional printed stretching device [65], based on a previously machined design [56,86]. Immediately before damage induction, low- or medium-level, nondamaging radial stretching (0%, 3%, or 6%) was applied and then a small crushing wound was induced in the monolayer. Following local damage, the cell growth media was replaced with medium containing 0, 1, or 5 mM NaPy. The time-dependent gap closure was monitored by determining the denuded area in images collected every 15 min until full gap closure. The results are fit to a Richard's function, which is used to obtain the time for 90% of gap closure (end of collective migration) and the maximal migration rate. *Adapted from Marom A, Berkovitch Y, Toume S, Alvarez-Elizondo MB, Weihs D. Non-damaging stretching combined with sodium pyruvate supplement accelerate migration of fibroblasts and myoblasts during gap closure. Clin Biomech 2019;62:96–103. https://doi.org/10.1016/J.CLINBIOMECH.2019.01.009 with permission.*

We have tested the combined treatment on murine fibroblasts or myoblasts, respectively, connective tissue, and precursor muscle cells, which are typically involved in tissue repair. Each type of cell was (separately) cultured in media containing 1 mM NaPy to a monolayer, on the stretchable-substrate plate (Fig. 6.5). We induce radial, controlled-level stretching using our custom 3D-printed cell stretcher, by lowering the stretchable-substrate plate onto open, cylindrical fixtures to varying depths [52,65]. The depth that the plate is lowered is controlled by a screw that brings the top and bottom plates of the stretcher closer, wherein the distance between the plates defines the stretching. Immediately following stretching, a small gap (0.05–0.5 mm^2) is induced and the time-dependent gap closure kinematics are evaluated.

We observe statistically significant acceleration of gap closure under specific combinations of nondamaging stretching levels and exogenous NaPy supplement concentrations (Fig. 6.6) [52]. In both myoblasts and fibroblasts, gap closure acceleration by low- and mid-level stretching depends strongly on the postinjury NaPy concentration, given preexposure to the NaPy; all the cells were cultured in 1 mM NaPy up to the time of induced injury. Specifically, in damaged myoblast monolayers, when NaPy supplement is added postinjury (at 1 or 5 mM), gap closure is significantly accelerated specifically when combined with low-level stretching (3%). In parallel, we demonstrate in both fibroblasts

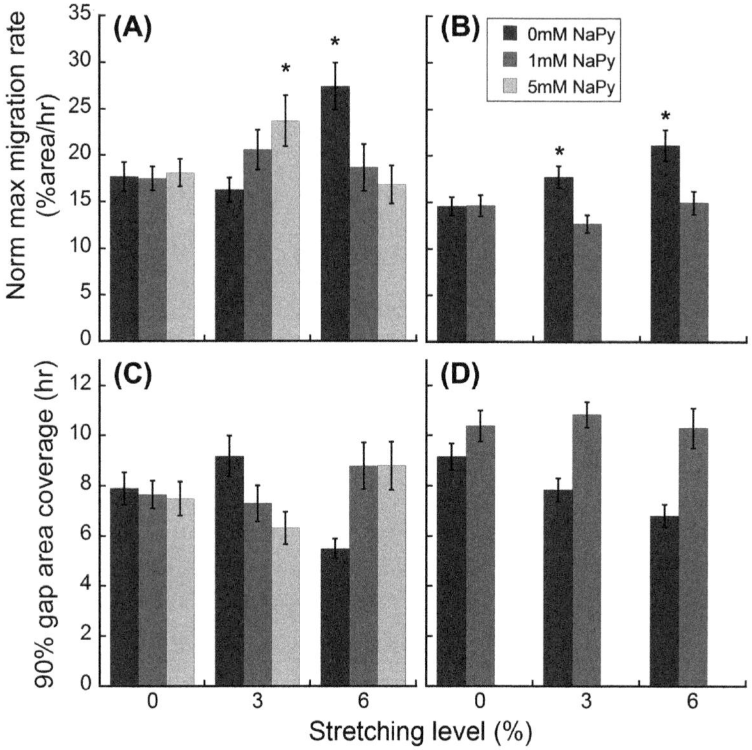

Figure 6.6

Kinematics of gap closure of C2C12 myoblasts (left) and NIH3T3 fibroblasts (right) for varying stretching levels and sodium pyruvate (NaPy) supplement. (A, B) Normalized maximum migration rate; (C, D) time for closure of 90% of initial gap area. Error bars are standard errors. Asterisks mark statistically significant results ($P < 0.05$) as compared with values with the same stretching level or with the same NaPy concentration, in the same cell type; the C2C12 case of 0 mM NaPy + 6% stretch is not statistically different from the 5 mM NaPy + 3% stretch. *Reproduced from Marom A, Berkovitch Y, Toume S, Alvarez-Elizondo MB, Weihs D. Non-damaging stretching combined with sodium pyruvate supplement accelerate migration of fibroblasts and myoblasts during gap closure. Clin Biomech 2019;62:96–103. https://doi.org/10.1016/J.CLINBIOMECH.2019.01.009 with permission.*

and myoblasts that exposure to NaPy before damage induction allows mid-level stretches (here 6% strain) to compensate for the postinjury deficiency in exogenous NaPy, and gap closure is still accelerated in a statistically significant manner. Thus, a potential preventative treatment against PUs could include preexposure to NaPy of cells in various tissues at the site of potential damage, before expected mechanical loading (i.e., before extended periods of immobility, e.g., in an operation), followed with low-level nondamaging stretching applied together with NaPy to accelerate closure of any formed yet undetectable gaps; this would prevent the PU cascade of progression [64].

4. Perspectives on prevention and accelerated healing

PUs and wounds in general remain a worldwide problem. Innovative approaches are required to prevent PU formation as well as to accelerate wound healing, e.g., by accelerating closure of gaps in tissues. Prevention strategies, i.e., in the context of PUs, are all based on macroscale interactions of the patient body with support surfaces. Traditionally, the focus was on mattresses and cushions, which in recent years have been technologically augmented through collaborative bioengineering research in academia and industry [76–80]. Recently, prophylactic (prevention) dressings were introduced in the market, aimed at cushioning vulnerable body sites and removing load from tissues, with high biomechanical efficacy [81]. Both support surfaces and prophylactic dressings are intended to minimize PUs by reducing tissue deformations. In contrast, using the research methodologies highlighted in the current chapter, we demonstrate development of innovative mechanobiology-based approaches to directly influence the cells within loaded or damaged tissues. These innovative mechanobiology-based approaches can lead to development of technologies to prevent PU formation and accelerate healing [82], which should, of course, be applied concurrently with the existing approaches to minimize tissue deformations.

Tissues are a complex environment including a flexible and modifiable ECM and various types of cells, e.g., muscle, fat, and connective tissue cells alongside undifferentiated or committed stem cells, endothelial cells in blood vessels, as well as visiting immune cells. While in vitro monolayer systems do no account for the entire complexity of an in vivo wound, the migration aspects do correlate with in vivo migration [83,84]. As we have demonstrated, nondamaging mechanical interventions (e.g., sustained strains) combined with native cell supplements may hold the key to this complex issue. Unique options to interact with and affect this complex system on the cell level are provided by the growing body of knowledge on mechanobiology and the interplay between cell, tissue, and microenvironment mechanics with the cell biology and function (Fig. 6.7).

Specifically, as we have demonstrated that combining controlled application of small (nondamaging) strains with exogenous NaPy supplement can accelerate closure of

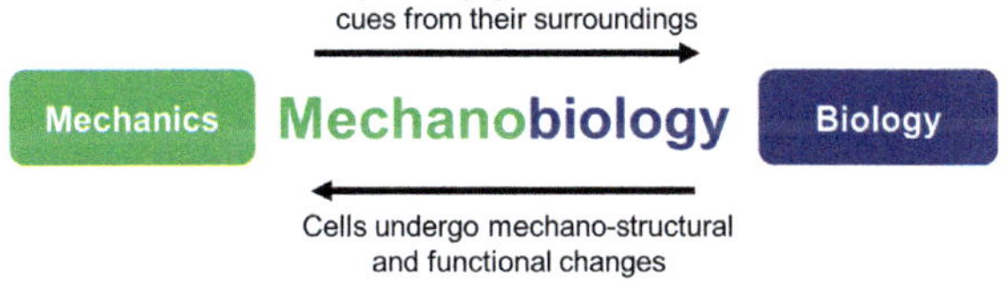

Figure 6.7
Mechanobiology is the interplay between the biological function and capacity of the cells and the mechanics of the cells and tissues and of their environment in micro- and macroscales. For examples, applying mechanical forces to the cells may change their mechanics and function, leading to cell damage/death or conversely to improved performance.

microscale gaps (the initial stages of wound formation), being especially efficacious in deep tissues, e.g., muscle. This combined treatment supplements the cytoskeleton dynamics, concurrently enhancing cell adaptability and stability under applied deformations and accelerating cell migration associated with small gap closure; the gap closure acceleration is relevant to both prevention of wound development and treatment of already existing wounds. We further note that honey, a medical substance widely used for treatment of various types of wounds including PUs, is not applicable for prevention of those wounds, as it is, in fact, slowing gap closure; this may even promote wound progression. It is unsurprising therefore that not all compounds or approaches to treat wounds, or PUs in particular, are applicable also for their prevention. Thus, our approach of pyruvate pre-treatment combined or independently of application of low-level, non-damaging strains as a preventative approach is innovative and unique. Modeling has recently shown that advanced dressings may be designed to release pyruvates to potential damage sites [85]. If such treatments were to be applied both preinjury and postinjury, they may prevent formation or development of wound as well as accelerate closure of existing gaps. These approaches can be developed into clinical technologies [82], where one may envision a method to impart nondamaging strains while releasing required supplements, e.g., through a dressing.

Acknowledgments

The work was partially supported by the Israeli Ministry of Science, Technology and Space in the Program for Science, Technology and Innovation for Third Age Populations (grant no. 3−12804, awarded jointly to Drs. Daphne Weihs and Amit Gefen in 2016).

References

[1] Padula WV, Mishra MK, Makic MBF, Sullivan PW. Improving the quality of pressure ulcer care with prevention. Med Care 2011;49:385−92. https://doi.org/10.1097/MLR.0b013e31820292b3.

[2] NPUAP. Pressure ulcers in America: prevalence, incidence, and implications for the future. An executive summary of the national pressure ulcer advisory panel monograph. Adv Skin Wound Care 2001;14:208−15.

[3] Raghavan P, Raza WA, Ahmed YS, Chamberlain MA. Prevalence of pressure sores in a community sample of spinal injury patients. Clin Rehabil 2003;17:879−84.

[4] EPUAP NPUAP. International pressure ulcer clinical practice guideline. 2nd ed. 2014. http://internationalguideline.com/.

[5] Gefen A, Weihs D. Cytoskeleton and plasma-membrane damage resulting from exposure to sustained deformations: a review of the mechanobiology of chronic wounds. Med Eng Phys 2016;38:828−33. https://doi.org/10.1016/j.medengphy.2016.05.014.

[6] Linder-Ganz E, Shabshin N, Itzchak Y, Yizhar Z, Siev-Ner I, Gefen A. Strains and stresses in sub-dermal tissues of the buttocks are greater in paraplegics than in healthy during sitting. J Biomech 2008;41:567−80. https://doi.org/10.1016/j.jbiomech.2007.10.011.

[7] Gefen A. How much time does it take to get a pressure ulcer? Integrated evidence from human, animal, and in vitro studies. Ostomy Wound Manage 2008;54(26−28):30−5.

[8] Peko Cohen L, Gefen A. Deep tissue loads in the seated buttocks on an off-loading wheelchair cushion versus air-cell-based and foam cushions: finite element studies. Int Wound J 2017;14:1327–34. https://doi.org/10.1111/iwj.12807.
[9] Shoham N, Gefen A. Deformations, mechanical strains and stresses across the different hierarchical scales in weight-bearing soft tissues. J Tissue Viability 2012;21:39–46. https://doi.org/10.1016/j.jtv.2012.03.001.
[10] Kim WS, Park BS, Sung JH, Yang JM, Park SB, Kwak SJ, et al. Wound healing effect of adipose-derived stem cells: a critical role of secretory factors on human dermal fibroblasts. J Dermatol Sci 2007;48:15–24. https://doi.org/10.1016/j.jdermsci.2007.05.018.
[11] Hassan WU, Greiser U, Wang W. Role of adipose-derived stem cells in wound healing. Wound Repair Regen 2014;22:313–25. https://doi.org/10.1111/wrr.12173.
[12] Cherubino M, Rubin JP, Miljkovic N, Kelmendi-Doko A, Marra KG. Adipose-derived stem cells for wound healing applications. Ann Plast Surg 2011;66:210–5. https://doi.org/10.1097/SAP.0b013e3181e6d06c.
[13] Ridley AJ, Schwartz MA, Burridge K, Firtel RA, Ginsberg MH, Borisy G, et al. Cell migration: integrating signals from front to back. Science 2003;302:1704–9. https://doi.org/10.1126/science.1092053.
[14] Friedl P, Wolf K. Plasticity of cell migration: a multiscale tuning model. J Cell Biol 2010;188:11–9. https://doi.org/10.1083/jcb.200909003.
[15] Chen J, Weihs D, Van Dijk M, Vermolen FJ. A phenomenological model for cell and nucleus deformation during cancer metastasis. Biomech Model Mechanobiol 2018. https://doi.org/10.1007/s10237-018-1036-5.
[16] Vermolen FJ, Meijden RP, Es MV, Gefen A, Weihs D. Towards a mathematical formalism for semi-stochastic cell-level computational modeling of tumor initiation. Ann Biomed Eng 2015;43:1680–94. https://doi.org/10.1007/s10439-015-1271-1.
[17] Massalha S, Weihs D. Metastatic breast cancer cells adhere strongly on varying stiffness substrates, initially without adjusting their morphology. Biomech Model Mechanobiol 2017;16:961–70. https://doi.org/10.1007/s10237-016-0864-4.
[18] Engler AJ, Sen S, Sweeney HL, Discher DE. Matrix elasticity directs stem cell lineage specification. Cell 2006;126:677–89. https://doi.org/10.1016/j.cell.2006.06.044.
[19] Abuhattum S, Weihs D. Asymmetry in traction forces produced by migrating preadipocytes is bounded to 33%. Med Eng Phys 2016;38:834–8. https://doi.org/10.1016/j.medengphy.2016.05.013.
[20] Nabi IR. The polarization of the motile cell. J Cell Sci 1999;112:1803. LP – 1811.
[21] Cramer LP. Forming the cell rear first: breaking cell symmetry to trigger directed cell migration. Nat Cell Biol 2010;12:628–32. https://doi.org/10.1038/ncb0710-628.
[22] Abuhattum S, Gefen A, Weihs D. Ratio of total traction force to projected cell area is preserved in differentiating adipocytes. Integr Biol 2015;7:1212–7. https://doi.org/10.1039/c5ib00056d.
[23] Kraning-Rush CM, Califano JP, Reinhart-King CA. Cellular traction stresses increase with increasing metastatic potential. PLoS One 2012;7:e32572. https://doi.org/10.1371/journal.pone.0032572.
[24] Mierke CT, Rosel D, Fabry B, Brabek J. Contractile forces in tumor cell migration. Eur J Cell Biol 2008;87:669–76. https://doi.org/10.1016/j.ejcb.2008.01.002.
[25] Teo A, Lim M, Weihs D. Embryonic stem cells growing in 3-dimensions shift from reliance on the substrate to each other for mechanical support. J Biomech 2015;48:1777–81. https://doi.org/10.1016/j.jbiomech.2015.05.009.
[26] Trepat X, Wasserman MR, Angelini TE, Millet E, Weitz DA, Butler JP, et al. Physical forces during collective cell migration. Nat Phys 2009;5:426–30. https://doi.org/10.1038/Nphys1269.
[27] Krishnan R, Klumpers DD, Park CY, Rajendran K, Trepat X, van Bezu J, et al. Substrate stiffening promotes endothelial monolayer disruption through enhanced physical forces. Am J Physiol Cell Physiol 2011;300:C146–54. https://doi.org/10.1152/ajpcell.00195.2010.
[28] Mierke CT, Zitterbart DP, Kollmannsberger P, Raupach C, Schlotzer-Schrehardt U, Goecke TW, et al. Breakdown of the endothelial barrier function in tumor cell transmigration. Biophys J 2008;94:2832–46. https://doi.org/10.1529/biophysj.107.113613.

[29] Legant WR, Miller JS, Blakely BL, Cohen DM, Genin GM, Chen CS. Measurement of mechanical tractions exerted by cells in three-dimensional matrices. Nat Methods 2010;7:969–71. https://doi.org/10.1038/nmeth.1531.

[30] Koch TM, Munster S, Bonakdar N, Butler JP, Fabry B. 3D Traction forces in cancer cell invasion. PLoS One 2012;7:e33476. https://doi.org/10.1371/journal.pone.0033476.

[31] Lesman A, Notbohm J, Tirrell DA, Ravichandran G. Contractile forces regulate cell division in three-dimensional environments. J Cell Biol 2014;205:155. https://doi.org/10.1083/JCB.201309029.

[32] Weiner GI. Cellular mechanics in the initial stages of 3D cancer cell invasion. Technion-Israel Institute of Technology; 2015.

[33] Kraning-Rush CM, Carey SP, Califano JP, Reinhart-King CA. Quantifying traction stresses in adherent cells. Comput Methods Cell Biol 2012;110:139–78. https://doi.org/10.1016/B978-0-12-388403-9.00006-0.

[34] Butler JP, Tolic-Norrelykke IM, Fabry B, Fredberg JJ. Traction fields, moments, and strain energy that cells exert on their surroundings. Am J Physiol Cell Physiol 2002;282:C595–605.

[35] Munevar S, Wang YL, Dembo M. Traction force microscopy of migrating normal and H-ras transformed 3T3 fibroblasts. Biophys J 2001;80:1744–57.

[36] Gal N, Weihs D. Intracellular mechanics and activity of breast cancer cells correlate with metastatic potential. Cell Biochem Biophys 2012;63:199–209.

[37] Gal N, Lechtman-Goldstein D, Weihs D. Particle tracking in living cells: a review of the mean square displacement method and beyond. Rheol Acta 2013;52:425–43. https://doi.org/10.1007/s00397-013-0694-6.

[38] Sheetz MP, Felsenfeld D, Galbraith CG, Choquet D. Cell migration as a five-step cycle. Cell Behav Control Mech Motil 1999:233–43.

[39] Abuhattoum S, Weihs D. Location-dependent intracellular particle tracking using a cell-based coordinate system. Comput Methods Biomech Biomed Eng 2013;16:1042–9. https://doi.org/10.1080/10255842.2012.761694.

[40] Kim JH, Serra-Picamal X, Tambe DT, Zhou EHH, Park CY, Sadati M, et al. Propulsion and navigation within the advancing monolayer sheet. Nat Mater 2013;12:856–63. https://doi.org/10.1038/Nmat3689.

[41] Weijer CJ. Collective cell migration in development. J Cell Sci 2009;122:3215–23. https://doi.org/10.1242/jcs.036517.

[42] Friedl P, Gilmour D. Collective cell migration in morphogenesis, regeneration and cancer. Nat Rev Mol Cell Biol 2009;10:445–57. https://doi.org/10.1038/Nrm2720.

[43] Friedl P, Wolf K. Tube travel: the role of proteases in individual and collective cancer cell invasion. Cancer Res 2008;68:7247–9. https://doi.org/10.1158/0008-5472.CAN-08-0784.

[44] Topman G, Lin FH, Gefen A. The influence of ischemic factors on the migration rates of cell types involved in cutaneous and subcutaneous pressure ulcers. Ann Biomed Eng 2012;40:1929–39. https://doi.org/10.1007/s10439-012-0545-0.

[45] Monsuur HN, Boink MA, Weijers EM, Roffel S, Breetveld M, Gefen A, et al. Methods to study differences in cell mobility during skin wound healing in vitro. J Biomech 2016;49:1381–7. https://doi.org/10.1016/j.jbiomech.2016.01.040.

[46] Vermolen FJ, Arkesteijn EC, Gefen A. Modelling the immune system response to epithelial wound infections. J Theor Biol 2016;393:158–69. https://doi.org/10.1016/j.jtbi.2015.12.030.

[47] Topman G, Sharabani-Yosef O, Gefen A. A standardized objective method for continuously measuring the kinematics of cultures covering a mechanically damaged site. Med Eng Phys 2012;34:225–32. https://doi.org/10.1016/j.medengphy.2011.07.014.

[48] Topman G, Sharabani-Yosef O, Gefen A. A method for quick, low-cost automated confluency measurements. Microsc Microanal 2011;17:915–22. https://doi.org/10.1017/S1431927611012153.

[49] Richards FJ. A flexible growth function for empirical use. J Exp Bot 1959;10:290–301. https://doi.org/10.1093/jxb/10.2.290.

[50] Berkovitch Y, Toume S, Gefen A, Weihs D. Effect of natural honey treatment and external stretching on kinematics of cell migration during gap closure. In: Gefen A, Weihs D, editors. Comput. Methods Biomech. Biomed. Eng. Proc. 14th Int. Symp. C. Tel Aviv, Isr. 2016. Cham: Springer International Publishing; 2018. p. 75–80. https://doi.org/10.1007/978-3-319-59764-5_9.

[51] Toume S, Gefen A, Weihs D. Low-level stretching accelerates cell migration into a gap. Int Wound J 2017;14:698–703. https://doi.org/10.1111/iwj.12679.
[52] Marom A, Berkovitch Y, Toume S, Alvarez-Elizondo MB, Weihs D. Non-damaging stretching combined with sodium pyruvate supplement accelerate migration of fibroblasts and myoblasts during gap closure. Clin Biomech 2019;62:96–103. https://doi.org/10.1016/J.CLINBIOMECH.2019.01.009.
[53] Topman G, Lin F-H, Gefen A. The natural medications for wound healing – Curcumin, Aloe-Vera and Ginger – do not induce a significant effect on the migration kinematics of cultured fibroblasts. J Biomech 2013;46:170–4. https://doi.org/10.1016/j.jbiomech.2012.09.015.
[54] Huang C, Leavitt T, Bayer LR, Orgill DP. Effect of negative pressure wound therapy on wound healing. Curr Probl Surg 2014;51:301–31. https://doi.org/10.1067/j.cpsurg.2014.04.001.
[55] Slomka N, Gefen A. Relationship between strain levels and permeability of the plasma membrane in statically stretched myoblasts. Ann Biomed Eng 2012;40:606–18. https://doi.org/10.1007/s10439-011-0423-1.
[56] Leopold E, Gefen A. Changes in permeability of the plasma membrane of myoblasts to fluorescent dyes with different molecular masses under sustained uniaxial stretching. Med Eng Phys 2013;35:601–7. https://doi.org/10.1016/j.medengphy.2012.07.004.
[57] Slomka N, Or-Tzadikario S, Sassun D, Gefen A. Membrane-stretch-induced cell death in deep tissue injury: computer model studies. Cell Mol Bioeng 2009;2:118–32. https://doi.org/10.1007/s12195-009-0046-x.
[58] Zhang H, Landmann F, Zahreddine H, Rodriguez D, Koch M, Labouesse M. A tension-induced mechanotransduction pathway promotes epithelial morphogenesis. Nature 2011;471:99–103. https://doi.org/10.1038/nature09765.
[59] Tokuyama E, Nagai Y, Takahashi K, Kimata Y, Naruse K. Mechanical stretch on human skin equivalents increases the epidermal thickness and develops the basement membrane. PLoS One 2015;10:e0141989. https://doi.org/10.1371/journal.pone.0141989.
[60] Nagai Y, Yokoi H, Kaihara K, Naruse K. The mechanical stimulation of cells in 3D culture within a self-assembling peptide hydrogel. Biomaterials 2012;33:1044–51. https://doi.org/10.1016/j.biomaterials.2011.10.049.
[61] Naruse K, Yamada T, Sokabe M. Involvement of SA channels in orienting response of cultured endothelial cells to cyclic stretch. Am J Physiol 1998;274:H1532–8. https://doi.org/10.1152/ajpheart.1998.274.5.H1532.
[62] Undyala VV, Dembo M, Cembrola K, Perrin BJ, Huttenlocher A, Elce JS, et al. The calpain small subunit regulates cell-substrate mechanical interactions during fibroblast migration. J Cell Sci 2008;121:3581–8. https://doi.org/10.1242/jcs.036152.
[63] Angelini TE, Hannezo E, Trepat X, Fredberg JJ, Weitz DA. Cell migration driven by cooperative substrate deformation patterns. Phys Rev Lett 2010;104:168104. https://doi.org/10.1103/PhysRevLett.104.168104.
[64] Gefen A. The future of pressure ulcer prevention is here: detecting and targeting inflammation early. EWMA J 2018;19:7–13.
[65] Toume S, Gefen A, Weihs D. Printable low-cost, sustained and dynamic cell stretching apparatus. J Biomech 2016;49:1336–9. https://doi.org/10.1016/j.jbiomech.2016.03.026.
[66] O'Donnell-Tormey J, Nathan CF, Lanks K, DeBoer CJ, de la Harpe J. Secretion of pyruvate. An antioxidant defense of mammalian cells. J Exp Med 1987;165:500–14. https://doi.org/10.1084/jem.165.2.500.
[67] Gefen A, Weihs D. Mechanical cytoprotection: a review of cytoskeleton-protection approaches for cells. J Biomech 2016;49:1321–9. https://doi.org/10.1016/j.jbiomech.2015.10.030.
[68] Hinoi E, Takarada T, Tsuchihashi Y, Fujimori S, Moriguchi N, Wang L, et al. A molecular mechanism of pyruvate protection against cytotoxicity of reactive oxygen species in osteoblasts. Mol Pharmacol 2006;70:925–35. https://doi.org/10.1124/mol.106.024398.

[69] Wang Q, van Hoecke M, Tang XN, Lee H, Zheng Z, Swanson RA, et al. Pyruvate protects against experimental stroke via an anti-inflammatory mechanism. Neurobiol Dis 2009;36:223–31. https://doi.org/10.1016/j.nbd.2009.07.018.

[70] Kao KK, Fink MP. The biochemical basis for the anti-inflammatory and cytoprotective actions of ethyl pyruvate and related compounds. Biochem Pharmacol 2010;80:151–9. https://doi.org/10.1016/j.bcp.2010.03.007.

[71] Moriguchi N, Hinoi E, Tsuchihashi Y, Fujimori S, Iemata M, Takarada T, et al. Cytoprotection by pyruvate through an anti-oxidative mechanism in cultured rat calvarial osteoblasts. Histol Histopathol 2006;21:969–77. https://doi.org/10.14670/HH-21.969.

[72] Poteet E, Winters A, Xie L, Ryou MG, Liu R, Yang SH. In vitro protection by pyruvate against cadmium-induced cytotoxicity in hippocampal HT-22 cells. J Appl Toxicol 2014;34:903–13. https://doi.org/10.1002/jat.2913.

[73] Kang YH, Chung SJ, Kang IJ, Park JH, Bünger R. Intramitochondrial pyruvate attenuates hydrogen peroxide-induced apoptosis in bovine pulmonary artery endothelium. Mol Cell Biochem 2001;216:37–46. https://doi.org/10.1023/A:1011040026620.

[74] Ghersetich I, Brazzini B, Peris K, Cotellessa C, Manunta T, Lotti T. Pyruvic acid peels for the treatment of photoaging. Dermatol Surg 2004;30:32–6. Discussion 36.

[75] Onakpoya I, Hunt K, Wider B, Ernst E. Pyruvate supplementation for weight loss: a systematic review and meta-analysis of randomized clinical trials. Crit Rev Food Sci Nutr 2014;54:17–23. https://doi.org/10.1080/10408398.2011.565890.

[76] Levy A, Kopplin K, Gefen A. Computer simulations of efficacy of air-cell-based cushions in protecting against reoccurrence of pressure ulcers. J Rehabil Res Dev 2014;51:1297–319. https://doi.org/10.1682/JRRD.2014.02.0048.

[77] Levy A, Kopplin K, Gefen A. An air-cell-based cushion for pressure ulcer protection remarkably reduces tissue stresses in the seated buttocks with respect to foams: finite element studies. J Tissue Viability 2014;23:13–23. https://doi.org/10.1016/j.jtv.2013.12.005.

[78] Levy A, Kopplin K, Gefen A. Simulations of skin and subcutaneous tissue loading in the buttocks while regaining weight-bearing after a push-up in wheelchair users. J Mech Behav Biomed Mater 2013;28:436–47.

[79] Peko Cohen L, Levy A, Shabshin N, Neeman Z, Gefen A. Sacral soft tissue deformations when using a prophylactic multilayer dressing and positioning system. J Wound, Ostomy Cont Nurs 2018;45:1. https://doi.org/10.1097/WON.0000000000000461.

[80] Schwartz D, Levy A, Gefen A. A computer modeling study to assess the durability of prophylactic dressings subjected to moisture in biomechanical pressure injury prevention. Ostomy Wound Manage 2018;64:18–26.

[81] Levy A, Frank MB, Gefen A. The biomechanical efficacy of dressings in preventing heel ulcers. J Tissue Viability 2015;24:1–11. https://doi.org/10.1016/j.jtv.2015.01.001.

[82] Weihs D, Gefen A. Methods and compositions for prevention and treatment of pressure ulcers. 2018. PCT/IL2018/051424.

[83] Liang C-C, Park AY, Guan J-L. In vitro scratch assay: a convenient and inexpensive method for analysis of cell migration in vitro. Nat Protoc 2007;2:329–33. https://doi.org/10.1038/nprot.2007.30.

[84] Rodriguez LG, Wu X, Guan J-L. Wound-healing assay. Methods Mol Biol 2005;294:23–9. https://doi.org/10.1385/1-59259-860-9:023.

[85] Levy A, Kottner J, Gefen A. Release of sodium pyruvate from sacral prophylactic dressings: a computational model. Int Wound J 2019. https://doi.org/10.1111/iwj.13137. iwj.13137.

[86] Shoham N, Gottlieb R, Sharabani-Yosef O, Zaretsky U, Benayahu D, Gefen A. Static mechanical stretching accelerates lipid production in 3T3-L1 adipocytes by activating the MEK signaling pathway. Am J Physiol Physiol 2012;302:C429–41. https://doi.org/10.1152/ajpcell.00167.2011.

CHAPTER 7

Heel ulcers: investigating injurious tissue load thresholds in humans, based on a patient-specific computational heel model

Rinat Friedman[1], Noga Shabshin[2,3], Yohan Payan[4], Amit Gefen[1]

[1]Department of Biomedical Engineering, Faculty of Engineering, Tel Aviv University, Tel Aviv, Israel; [2]HaEmek Medical Center, Israel; [3]Hospital of the University of Pennsylvania, Penn Medicine, United States; [4]Laboratoire TIMC-IMAG, CNRS & University Grenoble Alpes, France

Chapter Outline

List of abbreviations

BW	body weight
3D	three dimensional
DTI	deep tissue injury
FE	finite element
FW	foot weight
HU	heel ulcer
PU	pressure ulcer

Innovations and Emerging Technologies in Wound Care. https://doi.org/10.1016/B978-0-12-815028-3.00007-9

1. Introduction

A pressure ulcer (PU) is an injury to the skin and/or underlying tissues pressed against a bony prominence, due to continuous pressure and shear forces [1]. Deep tissue injuries (DTIs) might become life-threatening and can ultimately result in death [2,3]. Mortality likelihood is increased almost by two in bed-bound patients with PUs, compared to bed-bound patients without them [4]. PUs prolong healing processes of other principal conditions, lengthen hospitalization [3,5], and carry immense health-care costs. It was previously reported that approximately 1.6 million PUs occurred yearly in US hospitals alone, with a total cost of 2.2–3.6 billion dollars. On average, DTIs increased health-care costs per patient by 14,000–23,000 dollars [5].

The heel is the second most common location for heel ulcers (HUs), occurring in 26% of all ulcer cases, and has the second highest percentage (38.5%) of DTIs [6]. HUs are formed when the soft tissues of the posterior heel (thin layers of skin and subcutaneous fat) are subjected to sustained deformations while the foot is weight bearing.

When in a supine position due to lengthy surgical procedures, long-term hospitalization, paralysis, or spinal cord injury, the soft tissues of the posterior heel are deformed by the weight of the foot when pressed between the rigid surface of the posterior calcaneus and the support surface [3,7–9]. Consequently, ischemia is formulated, rapidly leading to tissue deterioration and ultimately resulting in an ulceration of the area [2,10].

Compressive and tensile stresses and strains are the primary mechanical factors for the formation of ulcers of all severities. Friction is considered a secondary contributing factor [8,9,11]. Shear and tensile stresses will occur around the pressure point even for a completely perpendicular pressure [11]. This reaction is heightened around a bony prominence that acts as a peg around which the tissue is stretched and distorted [9,11].

Tissue distortion is formulated by shear and pressure stresses that entrap the tissue between an external support (e.g., a mattress) and an internal reaction surface (e.g., bone), which causes stretching or compression of blood vessels in the tissue, leading to vessels ischemia that results in necrosis of the tissue [8,9,12].

Friction promotes occurrence of shear stresses and thus stimulates the formation of ulcers. As the outer surface skin is kept immobilized against the support while the body keeps moving, a relative motion is produced between the skin and the rest of the body. This is a form of the "hammock effect" with the outer layer of the skin acting as the "tight cover." This relative motion of tissues leads to intertissue shear stresses and lateral strains, which result in ulcer formation [9,11].

The primary objectives of this study are follow: (1) Develop a three-dimensional (3D) computational, finite element (FE) modeling of a heel affected by an HU, based on a real case of HU, scanned by MRI. (2) Use the 3D model to investigate the case retrospectively and determine the thresholds of internal mechanical loads in skin and fat, which led to the onset and development of the HU in this case. (3) Effective (von Mises) stress, effective (Lagrangian) strain, and strain energy density are scalar parameters that can be computed by FE analysis and that are commonly used to evaluate the influence of external loads on biological tissues [13–17]. Our third goal was to evaluate the best parameter out of the three for one-to-one indication of injury.

2. Methods

2.1 Geometry

In this study, we used an MRI scan of a 72-year-old male subject (bodyweight = 95 kg) with a set of 41 T2-weighted 3 mm-spaced images, portraying an axial cut of the right foot. The calcaneus, fat, skin, and Achilles tendon tissues are demonstrated in the scan. A DTI with ulcerated skin and subcutaneous fat tissue is clearly visible in the scan (Fig. 7.1),

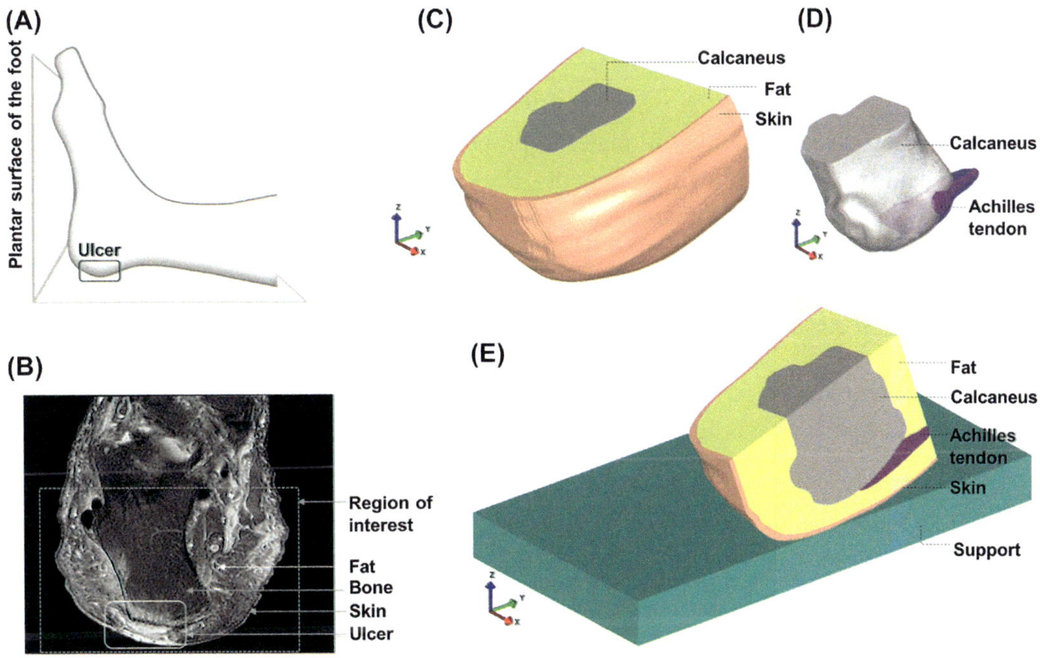

Figure 7.1: Heel modeling configuration.
(A) Ulcer location in relation to foot posture while in supine position. (B) Part of the MRI image set used for this project. Heel deep tissue injury (DTI) is clearly visible. (C) Finite element computational model of the heel. (D) Unobscured view of the inner components of the model: the calcaneus and the Achilles tendon (fat and skin layers were made transparent). (E) Midsagittal cross section of the heel model, rested on a mattress.

located on the posterior side of the heel, above the Achilles tendon insertion, which is typical for HUs that develop following a prolonged supine position (Fig. 7.1A,B). The analysis of the MRI data was conducted in close collaboration with an expert radiologist with specialization in detecting soft tissue damage by means of MRI, Dr. Nogah Shabshin from HaEmek Medical Center in Israel and the Hospital of the University of Pennsylvania in the United States.

2.2 Finite element modeling

The MRI images were segmented into masks, representing each of the tissues included in the scan. Simpleware ScanIP (Version 6) was used for the creation and meshing of the 3D model. All masks closely followed the MRI scan and were given physiological geometry (size and shape), while the ulcer region was replaced with healthy tissue structures (Fig. 7.1C–E). Number and type of mesh elements for each tissue are descried in Table 7.1. The heel was rested on a support with different stiffness levels and angles. The stress and strain levels that formed because of the foot weight were then calculated in the original ulcer site. Von Mises (effective) stress, shear stress, and Lagrangian strains were used for calculating the loading applied on the soft tissues of the heel in the subsequent injured area.

The model was solved using the FEBio Software Suite (University of Utah, US). PreView (ver. 1.18.2) was used for assigning material properties, boundary conditions, and model forces, as detailed in the following segments. PreView was also used for building and meshing the support surface ("mattress"). FEbio (ver 2.3.1) was used for numerical calculations and PostView (ver. 1.9.1) for force, stress, and strain analysis.

2.3 Mechanical properties

All tissues were assigned physiological mechanical properties according to the literature (Table 7.1). The skin was assigned "aged" mechanical properties according to Ref. [18].

Table 7.1: Tissue mechanical properties and number of element that were used in the model.

Tissue	Number of mesh elements	Type of mesh elements	Shear Modulus, G_{ins} (MPa)	Bulk Modulus, K (MPa)	Poisson's Ratio, υ	Elastic modulus, E (MPa)	Source
Skin	83,387	4-node	0.3247	32.357	0.495	0.970853[a]	[18]
Fat	213,170	linear	0.000286	0.0285	0.495	0.000855	[18]
Bone	65,971	tetrahedron	–	–	0.3	7000	[18,19]
Achilles	31,646		–	–	0.495	0.1945	[20,21]

[a]A rather high value for the elastic modulus of the skin was chosen according to Ref. [18]. Values in the range of E = 1 MPa for old skin tissue were also reported in other sources [22–24].

The skin, fat, and Achilles tendon were assigned nearly incompressible biophysical properties, due to their high water content [18,19]. Mechanical behavior of skin and fat was described using a Neo-Hookean model for isotropic hyperelastic materials [18,19]:

$$W = \frac{G_{ins}}{2}\left(\lambda_1^2 + \lambda_2^2 + \lambda_3^2 - 3\right) + \frac{1}{2}K(lnJ) \tag{1}$$

where W is the strain energy function, G_{ins} is the shear modulus, λ_i are the principal stretch ratios, K is the bulk modulus of the tissue, and J the determinant of the deformation gradient tensor. The calcaneus was considered an isotropic linear elastic solid [19].

For the intensions of this study and based on previous relevant works [18,19], elastic component of skin and fat was considered to be isotropic. The Achilles tendon retains transversely isotropic and linearly elastic properties when compressed perpendicularly to the main fiber axis [20], which was the case in this work, and so it was treated as an uncompressible isotropic elastic material [3]. Tendon's elastic modulus was taken as a mean between E_{11} and E_{22} for median strain condition from Ref. [20] and maximal E for moderate compression from Ref. [21]. Tissue properties used in the model are summarized in Table 7.1.

Mechanical properties of the support surface were chosen to be lineal elastic with elastic moduli of 40, 60, 80, and 100 kPa, based on our previous work which described elastic moduli and stiffness behavior for hospital mattresses [19,25,26]. The support was meshed with 8-node linear hexahedrons.

2.4 Boundary conditions

The bottom plane of the support was immobilized for translation and rotation motions in all directions. Static friction coefficient between heel and support was adopted from the literature [19]. As patient's precise foot weight was not known, anthropometric data (foot weight as % of total body weight) were used [27], indicating that foot weight is approximately $1.458 \pm 0.126\%$ of the total body weight. Two foot weights were modeled, accounting for a "light" and a "heavy" foot: 9.3N (1% of body weight) and 20.6N (2.2% of body weight), accordingly. The superior surface of the calcaneus was displaced downward and horizontally to simulate the load of the foot weight. As contact force is equal to the foot weight while resting on a support, the load levels were confirmed by verifying the contract force between the heel and the mattress for each case. Light and heavy foot weights were simulated by adding additional displacement of the calcaneus, without changing the width of the fat or skin layers, so the original geometry portrayed by the MRI reminded intact.

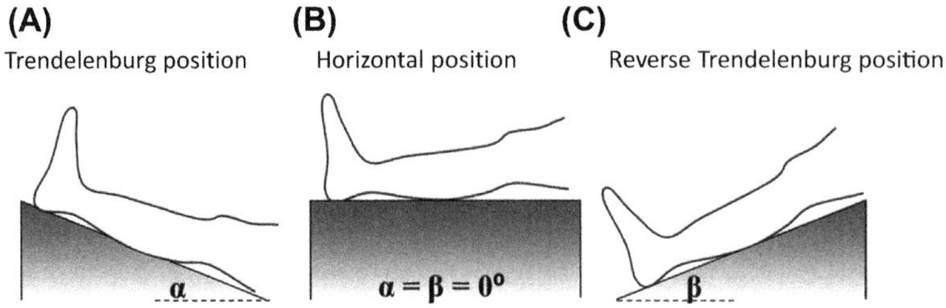

Figure 7.2: Three mattresses positions used in our work.
(A) Mattress in the Trendelenburg position, angle marked by α. $\alpha = 5$, 10, 20, and 30 degrees. (B) A horizontal mattress: $\alpha = \beta = 0$ degree. (C) Mattress in the reverse Trendelenburg position, angle marked by β. $\beta = 5$, 10, 20, and 30 degrees.

The foot was set in three mattress positions according to common surgical bed positions: Trendelenburg (Fig. 7.2A), horizontal (Fig. 7.2B), and revere Trendelenburg (Fig. 7.2C).

Several angles were simulated for the Trendelenburg and reverse Trendelenburg positions: 0 (horizontal), 5, 10, 20, and 30 degrees, chosen according to standard surgical bed angle range and common surgical practices for procedures requiring a nonhorizontal patient position [28–35].

2.5 Outcome measures

The skin and fat in original HU area were analyzed for effective stresses and Lagrangian strains. Outcome measures included maximal effective (von Mises) stress, maximal shear stress, and maximal Lagrangian strains. Outcome measures were compared between the various support angles and between fat and skin for each angle. In several cases (partial angles/mattress stiffness), additional measures included maximal compressive and tensile strains and distribution of strain energy density.

3. Results

An example of the FE model of the heel is presented in Fig. 7.3, rested on a horizontal 80 kPa support. Colors indicate the effective stress, Lagrangian strain, and strain energy density distribution. The original wound area is marked, and clearly indicating the subsequent DTI was a site for stress and strain concentration. A close-up of the loaded tissues in the wound area is also presented (Fig. 7.3D).

Computational data were cross-examined by category using graphs and tables so that large amount of data could be inspected simultaneously. In certain categories, some data were not considered for efficacy purposes, when no further conclusions could be drawn from it. For convenience, the angle range of the Trendelenburg position is marked by α, and the angle range of reverse Trendelenburg position is marked by β. The abbreviation "FW,"

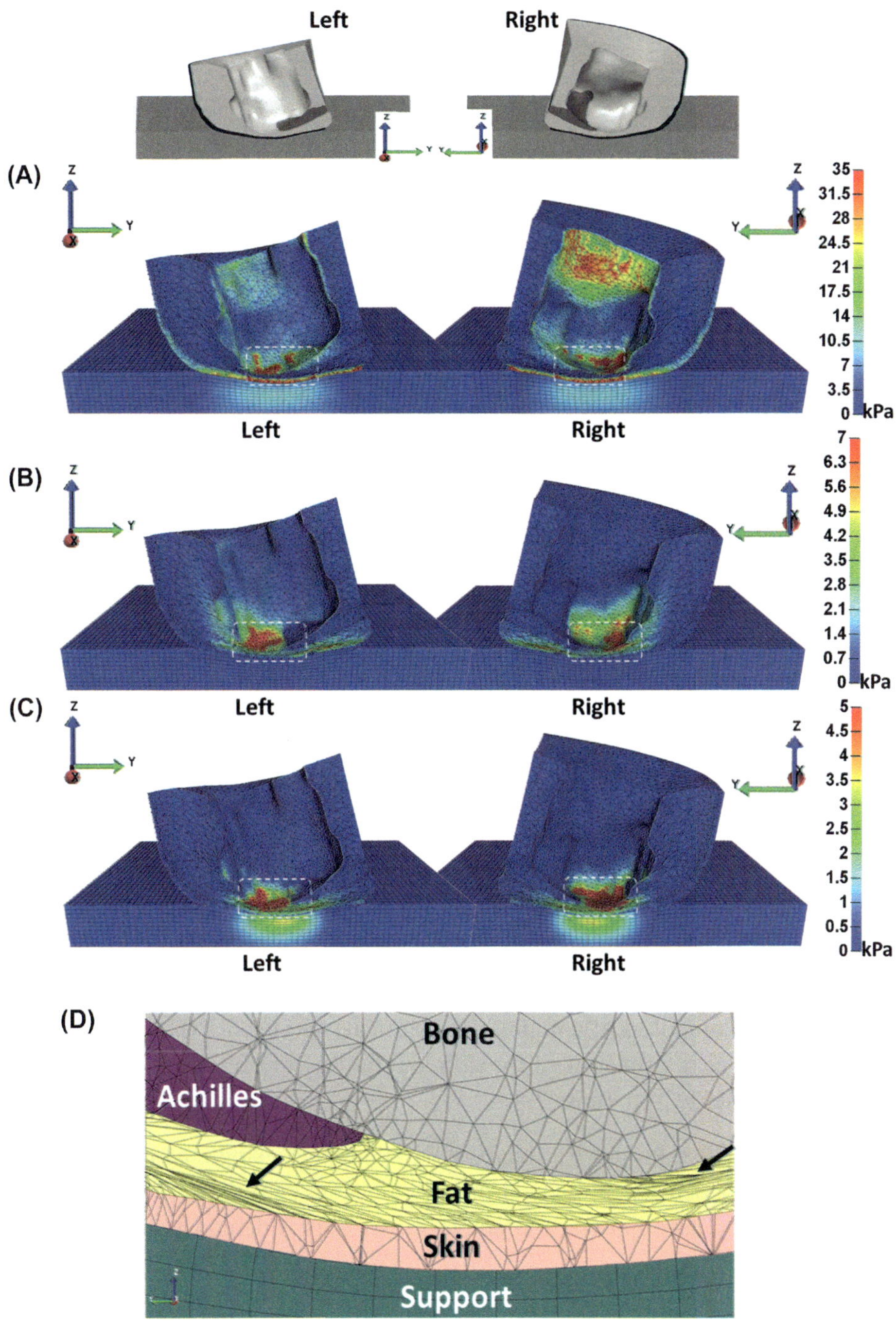

Figure 7.3: Example of stress and strain distribution in the soft tissues of the heel rested on a horizontal 80 kPa support and loaded by the natural weight of the foot.

The original location of the ulcer is marked by a white rectangle. Both sides of the heel are presented. The calcaneus was made transparent for presentation purpose only. (A) Effective stress (von Mises) distribution, (B) Lagrangian strain distribution (C) strain energy density distribution, and (D) a zoom in on the area on the original ulcer location. Extremely high deformations in the fat are clearly visible and indicated by black arrows.

used in the following figures, stands for "foot weight." For example, FW = 2.2%BW signifies that the foot weight is 2.2% of the total body weight.

Main results for effective and shear stress as function of tissue type (fat/skin), support angles, and mattress stiffness levels are presented in Fig. 7.4. A clear lineal connection between support stiffness stress levels is evident (Fig. 7.4A,B), in addition to higher stress levels in the skin as opposed to the fat (Fig. 7.4E–H).

Going from a 40 kPa mattress to a 100 kPa support, both the maximal effective and shear stresses were increased by more than 30% in the skin and more than doubled their values in the fat. Maximal shear stress was about 55% of the total maximal effective stress in both skin and fat, regardless of the angle. Results for α and β angles are summarized in Table 7.2.

We introduce what we call "injury thresholds," calculated for skin and fat tissue for low (FW = 1%BW) and high (FW = 2.2%BW) foot weight. Effective and shear stress injury thresholds were calculated using the average maximal effective and shear stresses and vary as function of mattress stiffness. Complete injury thresholds in kPa are described in Table 7.3.

The main results for Lagrangian strain injury thresholds for fat and skin, as function of support stiffness levels (Fig. 7.5) and angles (Fig. 7.6), are presented next. A clear lineal increase of strain in the skin as function of mattress stiffness is evident (Fig. 7.5A,B). However, strain levels in the fat show a far lesser correlation with mattress angles and stiffness (Figs. 7.5C,D and 7.6).

Lagrangian strain injury thresholds in skin and fat are presented in Table 7.4 for α and β angles, in addition to threshold levels averaged for all the angles for each mattress stiffness. It is evident that strain levels in the skin are much lower than in the fat tissue, with the difference escalating as the weight of the foot increases (Table 7.5).

Lagrangian strain is composed of compressive (vertical) and tensile (lateral) strains, the division between which was examined for skin and fat for several select support configurations, as described in Table 7.6.

Strain energy density distribution was analyzed graphically: a single and distinct zone of maximal strain energy density in the heel model is located directly in the middle of the eventual ulcer location, as opposed to effective stress. This is demonstrated in Fig. 7.7.

4. Discussion

An MRI scan of the right heel of a 72-year-old male subject, with a preexisting DTI at the time of the scan, was used to create patient-specific 3D FE model of the heel that included

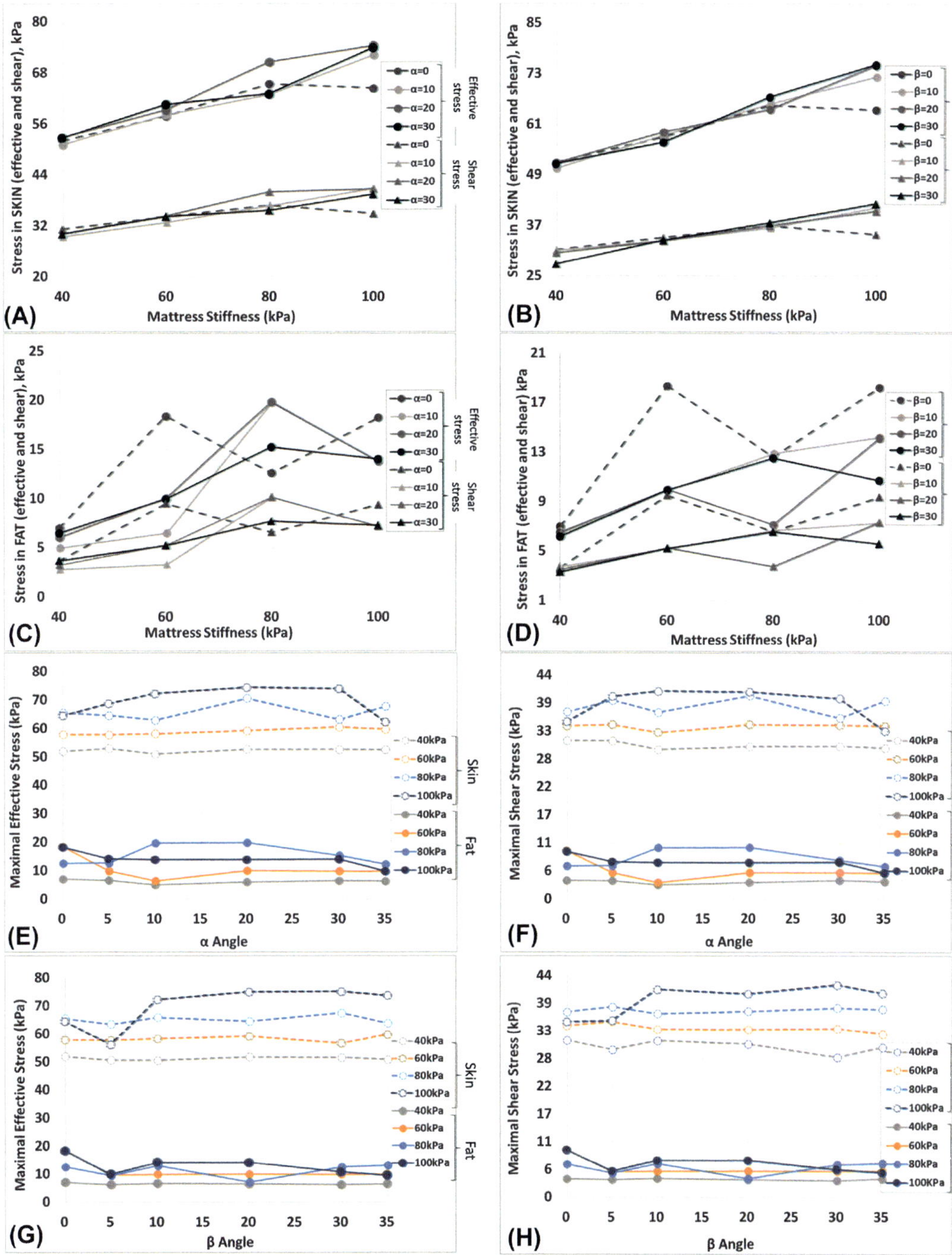

Figure 7.4: Maximal effective stress (kPa) versus maximal shear stress (kPa) as a function of mattress stiffness, FW = 2.2%BW: (A) α—skin, (B) β—skin, (C) α—fat, (D) β—fat. Maximal effective stress in skin versus fat, as function of the mattress angle: (E) α angle, (G) β angle. Maximal shear stress in skin versus fat, as function of the mattress angle: (F) α angle, (H) β angle.

Table 7.2: Maximal shear and effective stresses (averaged values for all α values and all β values) increase percentage, in skin and in fat, when going from a 40 kPa mattress to a 100 kPa mattress. Maximal shear stress as percentage of the total maximal effective stress, for α and β angles, for skin and fat.

	Maximal effective stress increase in %		Maximal shear stress increase in %		Maximal shear stress as % of total maximal effective stress	
Tissue	(α)	(β)	(α)	(β)	(α)	(β)
Skin	35%	34%	29%	29%	57.2%	57.6%
Fat	139%	108%	128%	97%	52.3%	52.5%

Table 7.3: Injury thresholds (kPa) in the skin (I) and fat (II): (FW1%BW − FW2.2%BW). Thresholds vary for each mattress stiffness. Effective stress injury thresholds for α and β angles are indicated by "effective stress (α)" and "effective stress (β)", respectively. Shear stress injury thresholds for α and β angles are indicated by "shear stress (α)" and "shear stress (β)", respectively.

(I) Skin injury thresholds (FW1%BW − FW2.2%BW), kPa				
Mattress stiffness (kPa)	Effective stress (α)	Effective stress (β)	Shear stress (α)	Shear stress (β)
40	48−52	48−51	27−30	27−30
60	55−59	55−58	32−34	32−34
80	48−66	61−65	35−38	35−37
100	62−69	62−69	36−38	35−37
(II) Fat injury thresholds (FW1%BW − FW2.2%BW), kPa				
Mattress stiffness (kPa)	Effective stress (α)	Effective stress (β)	Shear stress (α)	Shear stress (β)
40	3−6	3−6	2−3	2−4
60	4−11	4−11	2−6	2−6
80	4−15	4−11	2−8	2−6
100	4−15	4−13	2−7	2−7

the calcaneus, Achilles tendon, fat, and skin. The original DTI area was given properties of healthy tissues. Foot weight−related tissue loads that acted in the original wound area and which instigated HU formation were calculated and used to determine suggested injury thresholds. As foot weight was not expected to drastically change during the injury time frame (hour to days) of the patient's hospitalization, the load levels acting in the DTI area were also expected to remain mostly unvarying. We concluded that the computed stress and strain levels in the simulation (during the "healthy" state) would have been the same later on, when the ulcer was formed. This allowed us to estimate injury-causing stress and strain thresholds, as described in Tables 7.2 to 7.6. Over time, these load levels are expected to result in a DTI.

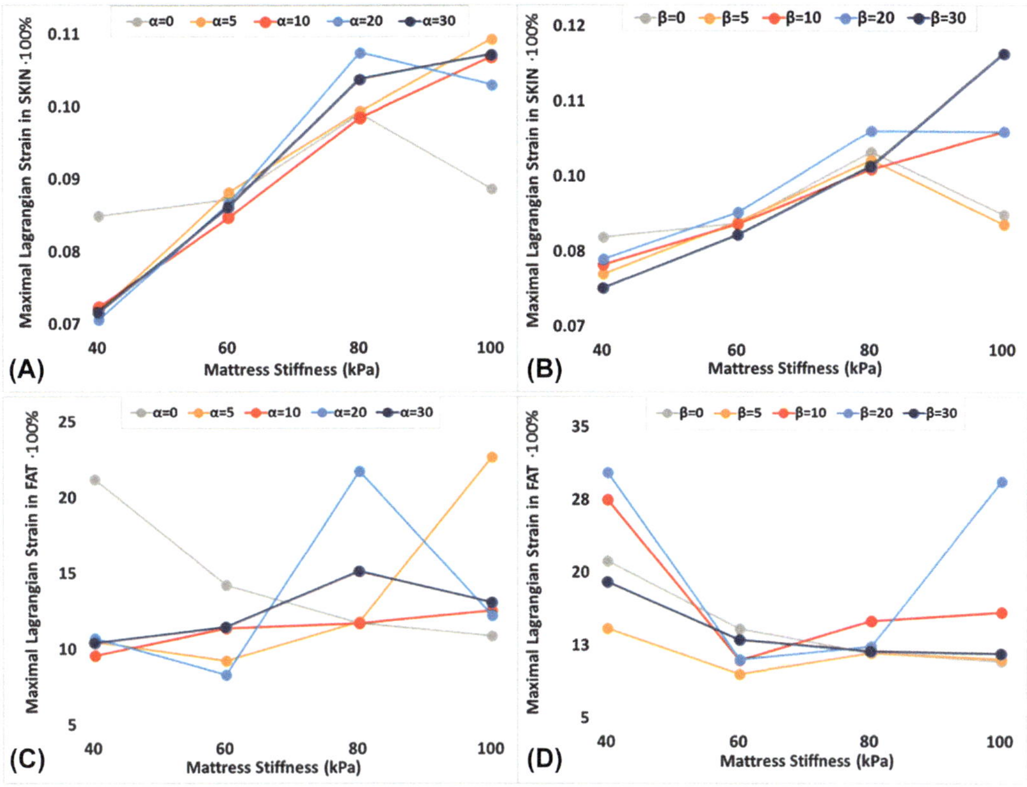

Figure 7.5: Maximal Lagrangian strain as function of mattress stiffness, FW = 2.2%BW, for both skin and fat.

Angle range for both α and β angles is = 0, 5, 10, 20, and 30 degrees. (A) Skin, α; (B) skin, β; (C) fat, α; (D) fat, β.

The fat tissue has exhibited effective and shear stress injury thresholds 14 to 6 and 16 to 6 times lower (for high and low foot weights), accordingly, compared to the skin, regardless of mattress stiffness and angle. This suggests the fat is much more susceptible to stress than the skin, as a lower stress is needed for injuring this tissue, possibly clarifying why ulcer onset is often initiated in the fat.

These results also exhibit the pivotal role shear stress has in ulcer formation, with the shear component of the total effective stress being larger than 50% for skin and fat for both angles, as demonstrated in Table 7.2.

The results indicate that fat tissue has a much greater tolerance to strain than the skin, as strain levels in the fat were 80–220 times higher than in the skin, for all foot weights, angles, and support stiffnesses.

Effective stress injury thresholds were significantly influenced by mattress stiffness. When going from a 40 kPa to a 100 kPa mattress, effective stress and shear stress injury

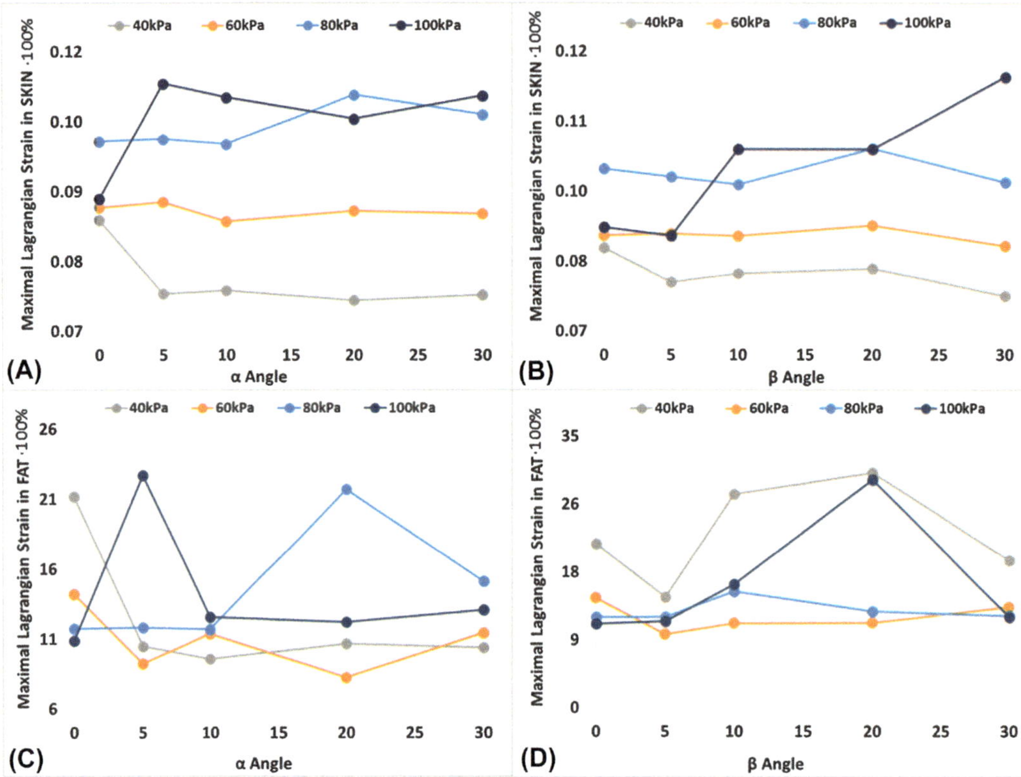

Figure 7.6: Maximal Lagrangian strain as function of mattress angle, FW = 2.2%BW, for both skin and fat.

Angle range for both α and β angles is 0, 5, 10, 20, and 30 degrees. (A) Skin, α; (B) skin, β; (C) fat, α; (D) fat, β.

Table 7.4: (A) Injury thresholds for Lagrangian strain is skin and fat. (B) Threshold levels of strain, averaged for all the angles for each mattress stiffness.

Mattress stiffness (kPa)	Skin—injury thresholds Strain (·100%) (FW1% BW – FW2.2%BW)		Fat—injury thresholds Strain (·100%) (FW1%BW – FW2.2%BW)		Skin—injury thresholds average of all angles Strain (·100%) (FW1% BW – FW2.2%BW)	Fat—injury thresholds average of all angles Strain (·100%) (FW1% BW – FW2.2% BW)
	(α)	(β)	(α)	(β)		
40	0.08–0.09	0.08–0.09	7–30	7–21	0.08–0.09	7–26
60	0.08–0.09	0.08–0.09	6–14	6–14	0.08–0.09	6–14
80	0.09–0.10	0.09–0.09	6–15	6–22	0.09–0.10	6–18
100	0.10–0.12	0.11–0.11	11–29	9–23	0.10–0.11	10–26

Table 7.5: Number of times maximal Lagrangian strain in fat is larger than in skin, as function of foot weight, low (BW = 1%) and high (BW = 2.2%), and as function of support stiffness (e.g., for 1%BW and a 40 kPa mattress, the maximal Lagrangian strain in fat is 88 times larger than in skin, and for 60 kPa mattress, maximal Lagrangian strain in fat is 70 times larger than in skin).

	Mattress stiffness (kPa)				
Foot weight as % of body weight	40	60	80	100	Average for all stiffnesses
FW is 1%BW	88	70	60	97	79
FW is 2.2%BW	303	161	193	232	222

Table 7.6: Transverse stretch as % of total strain versus vertical compression as % of total strain. Only two angles were examined: β = 0 and β = 30.

			Mattress stiffness (kPa) Stretch% of total strain – Compression% of total strain		
Tissue	Angle	Foot Weight	40	60	80
Skin	β = 0	1%BW	48%–52%	49%–51%	51%–49%
		2.2%BW	51%–49%	49%–51%	50%–50%
	β = 30	1%BW	50%–50%	49%–51%	50%–50%
		2.2%BW	52%–48%	54%–46%	50%–50%
Fat	β = 30	1%BW	89%–11%	88%–12%	85%–15%
		2.2%BW	95%–5%	91%–9%	93%–7%
	β = 30	1%BW	100%–0%	87%–13%	93%–7%
		2.2%BW	99%–1%	92%–8%	95%–5%

thresholds in the skin increased by 30% for both the "light" and "heavy" foot and by 30% and 110% in the fat for the "light" and "heavy" foot, accordingly.

Strain injury thresholds in the skin were significantly influenced by mattress stiffness, on average rising by 25% when going from a 40 kPa support to a 100 kPa support, with the highest strain levels resulting for the softest mattress (40 kPa). Foot weight increase from 1%BW to 2.2%BW caused strain injury thresholds of fat to triple for 80 and 100 kPa supports, to double for a 60 kPa support and to quadruple for a 40 kPa support. Strain injury thresholds of skin were less influenced by the weight increase, increasing by 11.5% on average when going from 1%BW to 2.2%BW. These results indicate that a mattress that is too soft might have a disadvantageous influence on HU formation, by causing greater deformations of soft tissues due to the sinking of the foot inside the mattress, while the friction between the foot and the mattress prevents relative movement of the outer skin layer, thus creating a drag effect of the inner soft tissues and resulting in a "hammock effect" with the skin acting as the "tight cover" [9,11]. The increase in foot weight had

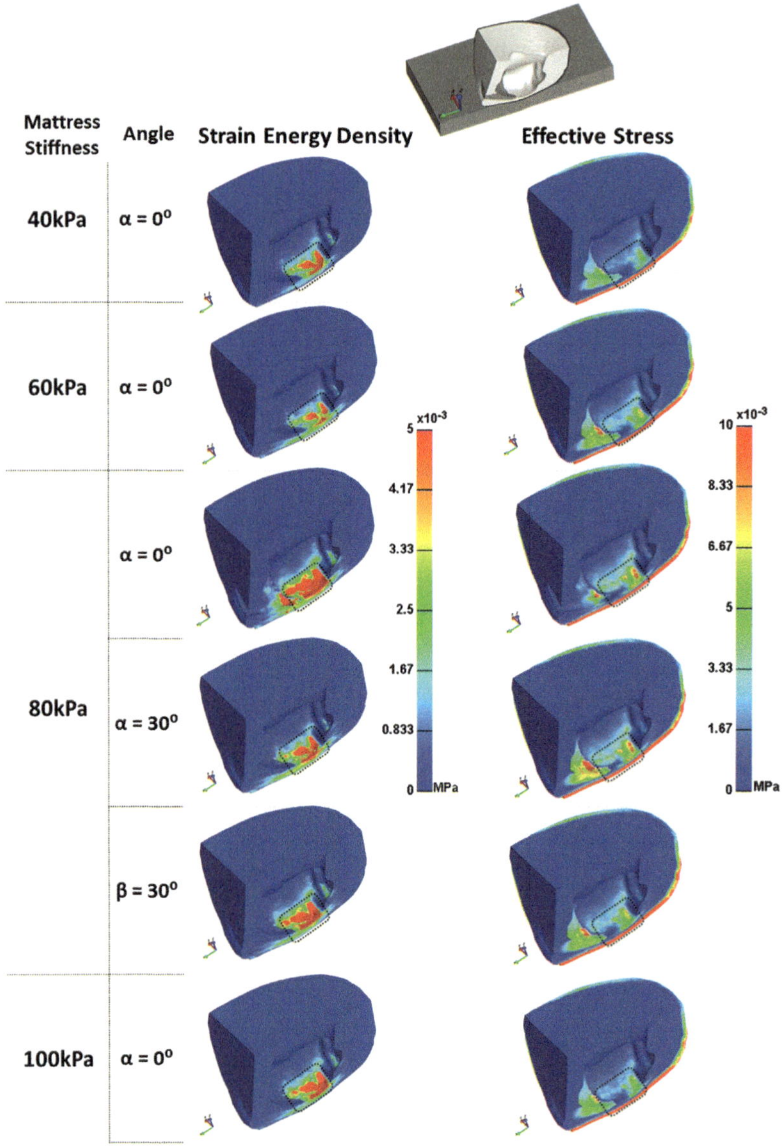

Figure 7.7: Strain energy density distribution in fat and skin versus effective stress distribution, for various mattress stiffnesses and angles.

The calcaneus was made transparent for presentation purpose only. Ulcer location is marked by a dotted square. It is evident that maximal levels of strain energy density are concentrated at the eventual ulcer location.

also a significant effect on stress levels, doubling the effective and shear stress levels in the fat for a 40 kPa mattress, while only increasing by 10% in the skin. Increasing the foot weight for 60, 80, and 100 kPa supports caused the stress injury thresholds to rise by 210% on average in the fat, while growing by 10% in the skin.

For all mattresses and angles, distribution of compressive tensile strains in the skin was equally 50%–50%. However, the fat had mostly experienced tensile strains, with compressive tensile distribution of 8%–92%. Following the downward compression due to the weight of the foot, the fat repositioned laterally, rather than actually compressing in place, due to its soft (compared with the skin) but uncompressible nature.

A small angle of 5–10 degree reduced effective and shear stress levels in the fat tissue by 25%–35% compared with the horizontal position, for both Trendelenburg and reverse Trendelenburg positions and all mattress stiffness levels but the 80 kPa mattress. An angle of 5 degree had also reduced strain levels for both positions with a 29% drop in the fat and a 4% in the skin, for all support stiffnesses but 100 kPa. A 10 degree angle had not improved strain levels, and in some cases even increased them, leading to the conclusion that a 5 degree-tilted bed (lifted or lowered) can greatly reduce strain and stress levels in both skin and fat, thus lessening the chances for ulcer formation.

Lastly, a correlation was observed between the eventual ulcer area and areas with distinct concentration of strain energy density, indicating that a high level of strain energy density is a good predicator for ulcer formation, which is also coherent with Ref. [17] and confirms the hypothesis suggested by Ref. [36], proposing to look at strains values (as opposed to stress values) as a predicator for a PU.

The 3D FE modeling of the heel presented in this study has allowed us to gain knowledge on the stress and strain levels involved in the formation of DTI of the heel.

We examined the influence of bed angle and foot weight on the resultant stress levels on the skin and fat tissues and determined injurious effective stress and shear stress thresholds. Although this work is based on a single MRI scan set and cannot provide universal DTI thresholds, the data offer an initial reference point to the size and range of ulcer causing stress and strain levels.

Acknowledgments

This research was supported by a grant from the Ministry of Science Technology and Space, Israel, and the CNRS, PICS program, France (YP, AG). The authors wish to extend their gratitude to the National Pressure Ulcer Advisory Panel (NPUAP) for funding this research with a minigrant.

References

[1] Black J, Baharestani M, Dorner B, Edsberg L, Langemo D, Posthauer ME, Ratliff C, Taler G, National Pressure Ulcer Advisory Panel. National pressure ulcer advisory panel's updated pressure ulcer staging system. Dermatol Nurs 2007;19(4):343–9.

[2] Fowler E, Scott-Williams S, McGuire JB. Practice recommendations for preventing heel pressure ulcers. Ostomy/Wound Manag October 2008;54(10):42–57.

[3] Russo A, Steiner C, Spector W. Hospitalizations related to pressure ulcers among adults 18 years and older, 2006 – statistical brief #64," healthcare cost and utilization project (HCUP) statistical briefs. December 2008.
[4] Brem H, Lyder C. Protocol for the successful treatment of pressure ulcers. Am J Surg July 2004;188(1, Suppl. 1):9–17.
[5] Aronovitch SA, Bechrich K. Hospital-acquired pressure ulcers: a comparison of costs in medical vs. surgical patients. Nurs Econ 1999;17(5):263–71.
[6] Vanderwee K, Clark M, Dealey C, Gunningberg L, Defloor T. Pressure ulcer prevalence in Europe: a pilot study. J Eval Clin Pract 2007;13(2):227–35.
[7] VanGilder C, Amlung S, Harrison P, Meyer S. Results of the 2008 – 2009 international pressure ulcer Prevalence™ survey and a 3-year, acute Care, unit-specific analysis. Ostomy/Wound Manag 2009;55(11):39–55.
[8] Gefen A, Farid KJ, Shaywitz I. A review of deep tissue injury development, detection, and prevention: shear savvy. Ostomy/Wound Manag 2013;59(2):26–35.
[9] Krzysztof SG. Pressure ulcer prevention. Part 1. Causes of pressure ulcers. Nurs Times 2002;98(11):41–4.
[10] Bauer JD, Mancoli JS, Phillips LG. Chapter 74: pressure sores. In: Thorne CH, Bartlett SP, Beasley RW, Aston SJ, Gurtner GC, Spear SL, editors. Grabb and Smith's plastic Surgery. 6th ed. Lippincott Williams & Wilkins; 2007. p. 722–9.
[11] International review. Pressure ulcer prevention: pressure, shear, friction and microclimate in context. A consensus document. London: Wounds International; 2010.
[12] Gefen A. The biomechanics of heel ulcers. J Tissue Viability 2010;19:124–31.
[13] Levy A, Frank MB, Gefen A. The biomechanical efficacy of dressings in preventing heel ulcers. J Tissue Viability 2015;24(1):1–11.
[14] Katzengolda R, Topazb M, Gefen A. Tissue loads applied by a novel medical device for closing large wounds. J Tissue Viability 2016;25(1):32–40.
[15] Mao H, Jin X, Zhang L, Yang KH, Igarashi T, Noble-Haeusslein L, King AI. Finite element analysis of controlled cortical impact-induced cell loss. J Neurotrauma 2010;27(5):877–88.
[16] Snedeker JG, Barbezat M, Niederer P, Schmidlin FR. Strain energy density as a rupture criterion for the kidney: impact tests on porcine organs, finite element simulation, and a baseline comparison between human and porcine tissues. J Biomech 2005;38(5):993–1001.
[17] Luboz V, Perrier A, Bucki M, Diot B, Cannard F, Vuillerme N, Payan Y. Influence of the calcaneus shape on the risk of posterior heel ulcer using 3D patient-specific biomechanical modeling. Ann Biomed Eng 2015;43(2):325–35.
[18] Levy A, Kopplin K, Gefen A. Simulations of skin and subcutaneous tissue loading in the buttocks while regaining weight-bearing after a push-up in wheelchair users. J Mech Behav Biomed Mater 2013;28:436–47.
[19] Sopher R, Nixon J, McGinnis E, Gefen A. The influence of foot posture, support stiffness, heel pad loading and tissue mechanical properties on biomechanical factors associated with a risk of heel ulceration. J Mech Behav Biomed Mater May 2011;4(4):572–82.
[20] Kuo P, Li P, Li M. Elastic properties of tendon measured by two different approaches. Ultrasound Med Biol September 2001;27(9):1275–84.
[21] Kot BCW, Zhang ZJ, Lee AWC, et al. Elastic modulus of muscle and tendon with shear wave ultrasound elastography: variations with different technical settings. PLoS One August 2012;7(8).
[22] Yarnitzky G, Yizhar Z, Gefen A. Real-time subject-specific monitoring of internal deformations and stresses in the soft tissues of the foot: a new approach in gait analysis. J Biomech 2006;39:2673–89.
[23] Spears IR, Miller-Young JE, Waters M, Rome K. The effect of loading conditions on stress in the arefooted heel pad. Med Sci Sport Exerc 2005;37:1030–6.
[24] Escoffier C, de Rigal J, Rochefort A, Vasselet R, Lévêque JL, Agache PG. Age-related mechanical properties of human skin: an in vivo study. J Investig Dermatol 1989;93:353–7.

[25] Tenenbaum S, Shabshin N, Levy A, Herman A, Gefen A. Effects of foot posture and heel padding devices on soft tissue deformations under the heel in supine position in males: MRI studies. JRRD – J Rehabil Res Dev 2013;50(8):1149–56.
[26] Linder-Ganz E, Gefen A. Stress analyses coupled with damage laws to determine biomechanical risk factors for deep tissue injury during sitting. J Biomech Eng 2008;131(1). 011003-011003–13.
[27] Clauser CE. Weight, volume, and center of mass of segments of the human body," Air Force Systems Command. Ohio: Wright- Patterson Air Force Base; 1969.
[28] Carey TW, Shaw A, Weber ML, DeVine JG. Effect of the degree of reverse Trendelenburg position on intraocular pressure during prone spine surgery: a randomized controlled trial. Spine J 2014;14(9):2118–26.
[29] Boyce JR, Ness T, Castroman P, Gleysteen JJ. A preliminary study of the optimal anesthesia positioning for the morbidly obese patient. Obes Surg 2003;13(1):4–9.
[30] Haure P, Cold GE, Hansen TM, Larsen JR. The ICP-lowering effect of 10 degrees reverse Trendelenburg position during craniotomy is stable during a 10-minute period. J Neurosurg Anesthesiol 2003;15(4):297–301.
[31] Gould C, Cull T, Wu YX, Osmundsen B. Blinded measure of Trendelenburg angle in pelvic robotic surgery. J Minim Invasive Gynecol 2012;19(4):465–8.
[32] Geerts BF, van den Bergh L, Stijnen T, Aarts LP, Jansen JR. Comprehensive review: is it better to use the Trendelenburg position or passive leg raising for the initial treatment of hypovolemia? J Clin Anesth 2012;24(8):668–74.
[33] Borahay MA, Patel PR, Walsh TM, Tarnal V, Koutrouvelis A, Vizzeri G, Jennings K, Jerig S, Kilic GS. Intraocular pressure and steep trendelenburg during minimally invasive gynecologic surgery: is there a risk? J Minim Invasive Gynecol 2013;20(6):819–24.
[34] Standard hospital bed, [Online]. Available: http://www.thecountrycaregroup.com.au/Capri-Bed-4-section-floor-auto-adjust-bed-with-trendelenburg-tilt-P110922.aspx; 2015.
[35] Standard hospital bed, [Online]. Available: http://www.hill-rom.com/usa/Products/Category/Hospital-Beds/Hill-Rom-1000-medical-surgical-bed/; 2015.
[36] Loerakker S, Manders E, Strijkers GJ, Nicolay K, Baaijens FT, Bader DL, Oomens CWJ. He effects of deformation, ischaemia and reperfusion on the development of muscle damage during prolonged loading. J Appl Physiol 2011;111(4):1168–77.

Further reading

[1] Pieper B, Langemo D, Cuddigan J. Pressure ulcer pain: a systematic literature review and national pressure ulcer advisory Panel white paper. Ostomy/Wound Manag February 2009;55(2):16–31.
[2] US Center of Disease Control and Prevention, National Center for Health Statistics. The national nursing home survey: 2004 overview. Series June 2009;13. Number 167.
[3] Gefen A, Cornelissen LH, Gawlitta D, Bader DL, Oomens CW. The free diffusion of macromolecules in tissue-engineered skeletal muscle subjected to large compression strains. J Biomech 2008;41:845–53.

CHAPTER 8

Shape memory polymers for design of smart stocking

Priyanka Gupta[1], Hairshkumar Narayana[2], Pramod Shankr[1], Bipin Kumar[1], Ning Pan[3]

[1]Department of Textile Technology, IIT Delhi, Hauz Khas, Delhi, India; [2]Department of Materials Engineering, University of British Columbia, Vancouver, British Columbia, Canada; [3]Biological & Agricultural Engineering, UC Davis, Davis, CA, United States

Chapter Outline

1. Introduction

The demand for textiles is rapidly changing in this 21st century with the unprecedented growth of science and technology, resulting in more ambitious requirement from a textile product. One of the major breakthroughs in the field of smart textiles is the introduction of stimuli-responsive polymers. Stimuli-responsive polymers can show significantly noticeable change in their configuration, shape, color, etc., on exposure to an external stimulus such as heat [1−3]. A number of commercial products of such polymers could be found today in the field of textiles, civil, biomedical, consumer products, aerospace, robotics, filtration, sensors, and actuators, etc (Fig. 8.1).

Innovations and Emerging Technologies in Wound Care. https://doi.org/10.1016/B978-0-12-815028-3.00008-0

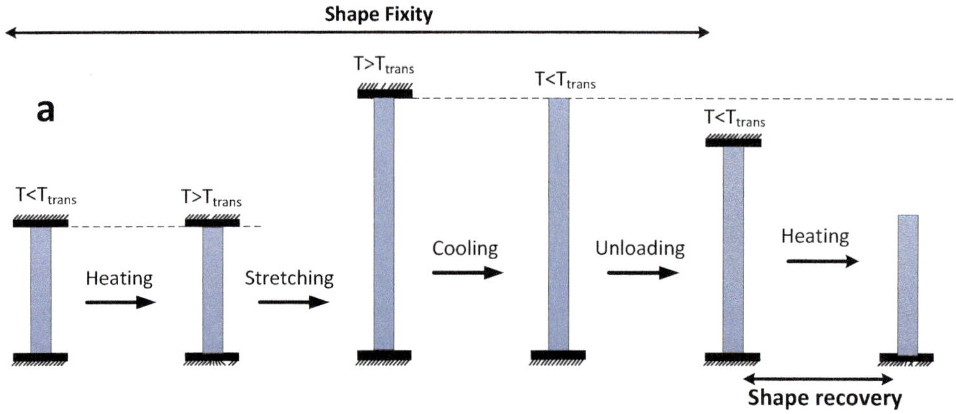

Figure 8.1
Shape memory effect in a shape memory polymer (SMP).

Among stimuli-responsive polymers, shape memory polymers (SMP) have gained scientific significance for past many decades. It has been attracting and steering the research direction of scientific community right from its inception toward numerous applications due to their compelling novel behaviors [4–6]. They demonstrate shape memory ability that permits programming or fixing of different deformed shapes from an original (permanent) shape and simultaneously allows the shape recovery to the original shape (Fig. 8.1) on an external stimulus such as heat, light, moisture, and magnetic field [7–11]. SMPs usually are composed of thermodynamically immiscible soft (reversible phase) and hard segments with different thermal transition temperature ranges attributed to shape fixation and shape recovery [12–14]. There are enormous possibilities of different molecular and structural designs for an SMP network including addition of foreign particles and nanomaterials to offer a broad spectrum of different properties. In addition, suitable and optimized polymer composition can also be developed specifically for the applications in, e.g., biomedical areas. Today, it is possible to make highly biodegradable, biocompatible, and comfortable memory polymer (MP)-based devices and bioactuators, which are compatible to human senses/tissues.

The most recent and unique application of SMP is found in the field of compression stockings [15]. Compression stockings are used in compression therapy, which is a very important health-care and medical treatment for patients suffering from venous and lymphatic disorders (Fig. 8.2) including venous ulcers, lymphedema, varicose veins, and deep vein thrombosis [16]. The main principle of compression therapy is to apply certain level of pressure around the affected tissue on the limb to reduce the venous hypertension (Fig. 8.2). Several researches have confirmed that the healing outcomes are better for patients receiving compression treatment compared with no compression [17]. Apart from stocking, other devices include bandages and intermittent pneumatic compression (IPC)

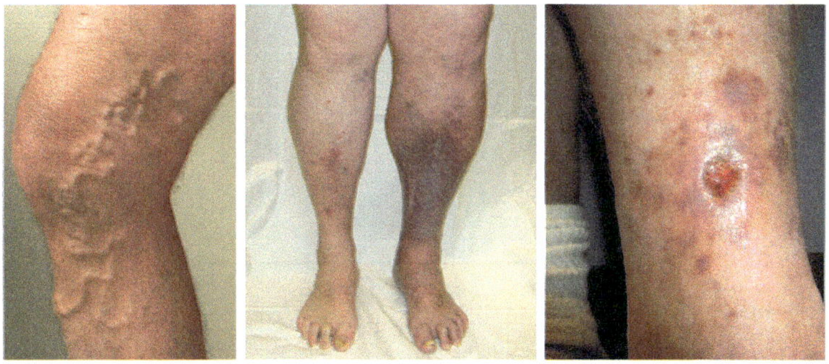

Figure 8.2
Different forms of chronic venous disorders.

systems [18]. Each of these devices holds its own significance for the treatment. Stocking is favored for low pressure (<50 mmHg) and allows easy application, whereas a bandage is size independent and permits high pressure. IPC is primarily recommended for the patients with poor calf muscle function or limited ankle mobility. More than two modes (stocking and IPC; bandage and IPC) are also used to provide benefits of both static and dynamic compression leading to healing of venous ulcers and alleviate symptoms in patients with chronic venous insufficiency [19].

Although compression products have been used for several decades, the conventional products have following shortcoming: (a) selection of stockings with proper sizing and fitting has always been a challenge for both health practitioners and manufacturers seeking better patient compliance and effective treatment; (b) for maximum treatment efficacy, a required pressure gradient from "toe to knee" is critical yet arduous to achieve in stocking mounting due to reasons including different leg attributes (shape or size) among patients and difference in stocking materials/designs, hence requiring trained personnel and raising the treatment cost to hospitals and patients; (c) for some patients, a dynamic (massaging) compression is recommended, but the required equipment such as the IPC is costly, noisy, bulky, and once attached, severely constraining the patients. Clearly, these inadequacies of conventional compression approaches demonstrate a compelling demand for multifunctional stocking in terms of controlled static pressure and dynamic (massage effect) pressure and thermal therapy (Fig. 8.3).

The use of SMP in the stocking has the potential of overcoming the aforementioned challenges and providing controlled static and dynamic compression benefits for the treatment whenever needed. This chapter primarily focuses on the research and development of shape memory stocking for smart compression management for chronic venous disorders. It is also aimed to cover the basic fundamentals of SMP, shape memory mechanisms and mechanics, related materials, and production methods. This could help readers to understand the topic and stimulate them to discover other medical application areas for the further growth of SMP field.

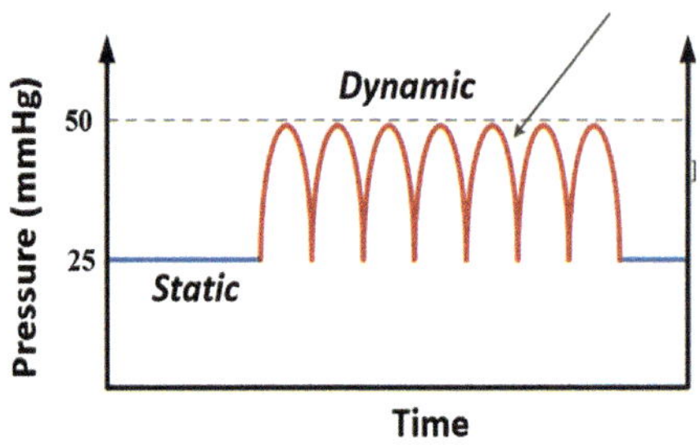

Figure 8.3
Possibility of static and dynamic compression using smart stocking.

2. Compression therapy and the physics of interface pressure

In healthy condition, the presence of veins and one-way valves allow the blood to move in one direction only (Fig. 8.4A). However, the action of foot and calf muscle movement often does not function properly, especially for older people due to their limited movement, resulting in increase in venous pressure. If not managed properly, venous hypertension causes excessive pooling of fluid (Fig. 8.4B), generates high pressure, and deprives oxygen and nourishment in the surrounding muscles. This cause skin deterioration, and eventually, the venous leg ulcers occur. Compression therapy provides a mean to treat such chronic venous disorders. The objectives here is to reduce the venous hypertension in the affected area, to reduce the swollen limb to minimize the size, and also to maintain a uniform pressure gradient in the leg from toe to knee (Fig. 8.4C) to improve the venous channeling to the heart. Compression also restores the valve function by bringing the walls of the veins closer together (Fig. 8.4D). External compression is given to the affected leg portion during compression therapy using compression materials such as bandage, stockings, or pressure garments, etc.

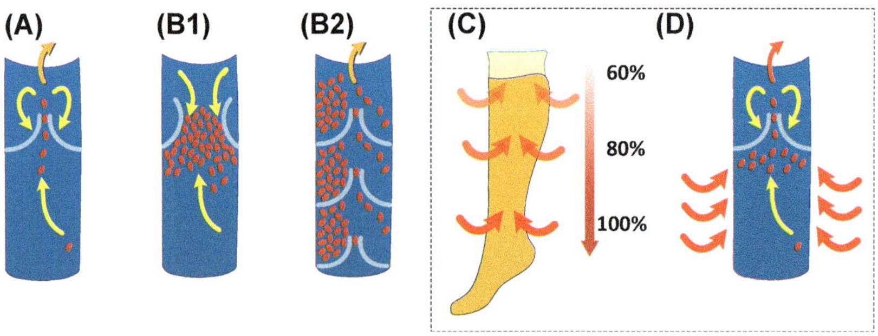

Figure 8.4
(A) Functioning of normal vein; (B) poor venous return due to damaged vein; (C) principle of compression therapy to apply gradient pressure using stocking; and (D) normalized vein function on therapy. B1 - Backflow of the venous return through a damaged vein; B2 - polling of venous return due to damaged veins

The efficacy of the compression treatment by a medical compression bandage is undoubtedly dependent on the pressure generated at the interfacial pressure between the bandage and the skin. The interfacial pressure depends on several factors such as limb shape and size, application technique, physical and structural properties of the bandage, physical activities taken by the patient, etc. The instantaneous pressure generated by a bandage just after its application is a function of the bandage tension, the number of layers applied, the bandage width, and the limb circumference, and this relation is based on the Laplace's law [20] for measuring pressure across a cylindrical thin film (Fig. 8.5).

To use Laplace's law in clinical practice, Thomas [21] proposed a simple equation according to which the pressure is found to be directly proportional to bandage tension and number of layers wrapped and inversely related to width and circumference of the limb to which the bandage is applied.

$$P = \frac{T \times n}{R \times W} \tag{1}$$

where P is the interface pressure (N/m^2), T is the longitudinal force applied to the bandage while wrapping (N), n is the number of layers of the bandage wrapped, R is the radius of the limb (m), and W is the bandage width (m). Because of this, the bandage applied with constant force to a leg of normal proportions will automatically produce pressure gradient from knee to ankle with the highest pressure at the ankle. The pressure generated by the bandage will gradually reduce upward the leg as the radius of limb increases.

3. Pressure requirement and challenges in conventional treatment

This interface pressure has to be quite accurate within certain limits and should not be below or above the prescribed level, otherwise it can lead to certain complications during treatment. Different class of stockings are required to provide light (Class I, 14–17 mmHg), medium (Class II, 18–24 mmHg), and strong (Class III, 25–35 mmHg) levels of

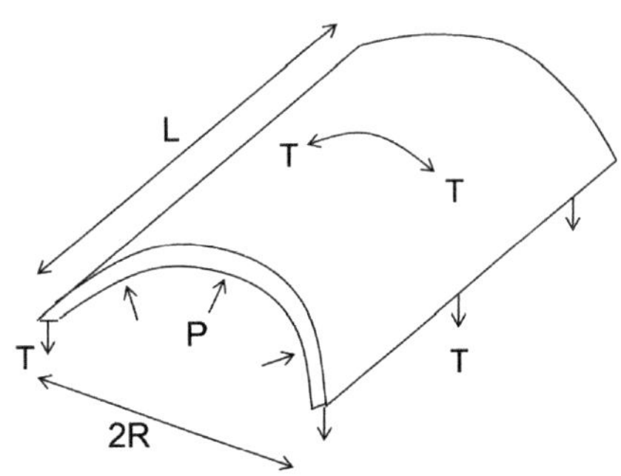

Figure 8.5
Laplace's law to relate the pressure and cylinder wall tension.

compression depending on the severity of the disease [22]. However, in practice, it is difficult to achieve the targeted pressure level due to various reasons, including mainly the different leg attributes (shape or size) among patients and difference in material (including both stockings and the legs) properties (time and temperature dependence). Moreover, pressure drop over time is also a major concern due to the time dependence of the system behaviors [23]. Several researchers showed that the interface pressure decreases over a period of time and hence reduces the effectiveness of the treatment of the compression system [24]. Kumar et al. [25,26] studied different fibrous material and fabric structure on the pressure performance and found that the pressure drop is more for nonelastomeric polymers (cotton, viscose, PET, nylon, etc.; >30% relaxation) as compared with elastomeric samples (polyurethane; < 10% relaxation). The decrease in the interface pressure beneath the wrapped bandage may occur because of relaxation of the internal stress in bandage and also due to reduction in swollen limb size. The relaxation of internal stress in the fabric structure is dependent on viscoelastic characteristics of fibers or yarns present in bandage and also on whether the wrapped bandage is in static or dynamic state.

As pressure drop is inevitable for almost all available stockings or bandages, and reinstalment or replacement of the stocking is needed once the pressure falls below the targeted level. Another reason for changing the stocking is for patient comfort, as there is often the need during night to remove the stocking so that it does not interfere with the sleep [27]. Clearly, the aforesaid inadequacies of the conventional approaches demonstrate a compelling demand for advanced compression system in compression management that allows modulating the pressure level via external control to easily change or readjust pressure whenever needed.

4. Research on smart actuator for compression management

Some researchers have made attempts to use smart material that can be actuated to allow external pressure control. As an example, Munch-Fals et al. [28] and Pourazadi et al. [29] developed electroactive compression bandage systems using smart sensors and dielectric elastomers, respectively, to control pressure. However, these systems are bulky and nonporous and therefore lead to noncompliance. Moein et al. [30] proposed shape memory alloy—based bandage, but the fabric structure having alloy wires is rigid and not easily confirmable compare with textile yarns. Also, the actuating temperature is beyond 100°C, which is not convenient to put on legs for practical usage. Bauerfeind [31] proposed modified polyurethane threads having acid functional filler that can be actuated by human skin's pH, but the system has to be rinsed in alkaline solution before application, leads to complication, and has no added pressure control. Other inventions relate to wearable systems in bandages/stockings for monitoring pressure variation over time [32], which has no function to manage pressure on the leg. SMP-based film actuator has been proposed by

Ahmad et al. (2012) for pressure bandage application. They have used temperature-responsive SMP strips attached to fabrics to control the compression by an external heat source. The use of SMP film actuators may significantly obstruct the moisture transmission and the permeability of the compression system and therefore may not be a viable solution for providing improved comfort to the patients.

5. Designing of smart memory stocking

Giving the complexities of pressure requirement discussed above, it is difficult if possible to achieve and maintain a desirable pressure level in treating chronic venous disease using the existing stockings. If looking at Eq. (1) more closely, however, the key factor causing the diminishing pressure P is the relaxed tension T in the stocking. If we can find a material whose internal tension T can be easily adjusted in such that it can compensate the deviations of the pressure from the initially designated level, we can then sustain the desired pressure on the leg. Furthermore, we can even modulate the internal stress so as to eliminate the need of multiple stocking sizes in achieving targeted pressures, i.e., a smart stocking!

5.1 Unique properties of shape memory polymer for smart compression

The challenges in compression products (i.e., easy application at constant extension and external pressure control) could be resolved if there is a possibility of shape and stress control in the fabric materials. SMPs could be useful here as they demonstrate shape memory ability that permits programming or fixing of different deformed shapes from an original (permanent) shape (Fig. 8.6A) and simultaneously allow the shape recovery to the original shape by external stimulus such as heat [33,34]. This shape control ability of SMP could be utilized in compression stocking to provide easy and alternative method of application; memory effect can be achieved by incorporating MP filaments in the fabric structure (Fig. 8.6B). To achieve easy application, the deformed shape (constant strain) of an SMP-based stocking could be programmed first using shape fixity potential of the SMP and then the stocking could be applied over the limb in loose state. The stocking could be then activated directly on the limb to initiate the recovery process that would result in the uniform shrinkage throughout the wrapping area, hence achieving nearly packed or tight in the entire length. This process could be easily achieved using external stimulus such as heat.

More recently, another associated phenomenon in SMP, i.e., stress memory, is also discovered [35]. Similar to shape memory, stress memory is a phenomenon whereby the stress in a polymer can be programmed, stored at temperature below its thermal transition temperature T_{trans}, and then retrieved reversibly, when needed, with an external thermal stimulus at $T > T_{trans}$ (Fig. 8.6C,D), thus to realize the in situ material stress control.

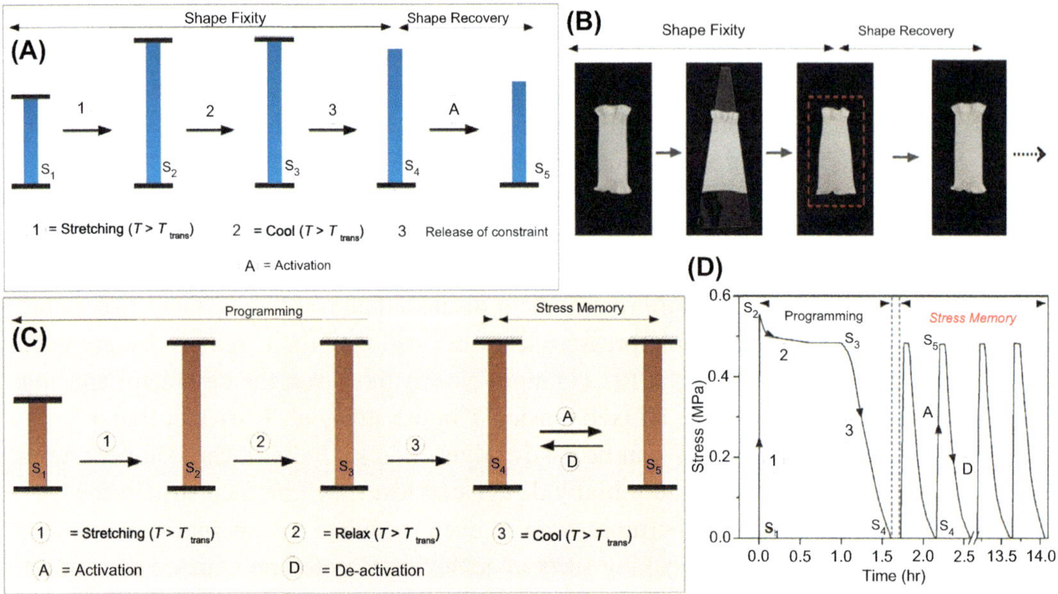

Figure 8.6

(A) Thermomechanical process to observe shape memory (fixity and recovery) in a memory polymer; (B) demonstration of shape memory using a memory fabric; (C) stress memory programming in the shape memory polymer (SMP); and (D) results of stress memory response of an SMP.

In other words, the integration of SMP fibers into stocking fabric structures could allow in situ stress control [36–38]. As the interface pressure depends on the internal stress in the fabric, so the possibility of stress alteration could help in achieving recommended pressure level in the wrapped state of the bandage. Also, this will allow the bandage to provide dynamic massage effect via alternating stress modulation and so would eliminate the need for IPC.

5.2 Memory polymer synthesis

The synthesis of semicrystalline MP has been carried out usually using two types of macroglycol (soft segment/reversible phase) to prepare two different MPs, and the materials used for the hard segment (fixed phase) are same. The macroglycols are (1) poly(ε-caprolactone) diols having molecular weight of 4000 g mol^{-1} (PCL-4000) and (2) poly(1,6-hexamethylene adipate) diol having molecular weight of 3000 g mol^{-1} (PHA-3000). 4,4′-methylene diphenyl diisocyanate (MDI) and chain extender 1,4-butanediol (BDO) are used as a fixed phase (hard segments). MP was prepared by using the bulk prepolymerization method, where PCL-4000 or PHA-3000 and MDI are kept for reaction at 80°C around 2–3 h. After completing the step formation of prepolymer, BDO was added as a chain extender and stirred using mechanical stirrer for 5 min to accomplish the reaction. The ratio (weighted ratio) of hard segment to (MDI + BDO) to soft segment was 28/72 for MP with PCL-4000 and 28.5/71.5 for MP with PHA-3000. All the reactions were performed under the nitrogen environment, and the temperature of less than 90°C was maintained. The synthesized prepolymer was casted into a preheated (100°C) polytetrafluoroethylene mold. The resultant polyurethane sheet with thickness ~3.0 mm was obtained. The prepolymer mold was kept in the vacuum oven for a period of 24 h at a temperature of 100°C. The casted MP sheets were crushed and extruded using a single screw extruder followed by pelletizer to get the chips for melt spinning as a raw material. MP chips was dried for 24 h at 100°C under vacuum condition to ensure the moisture level to be less than 100 ppm (parts per million) [39].

5.3 Memory fiber and fabric development

Polymeric molds were crushed and converted into pieces having smaller size. The crushed material was then used to get the polymer ribbon using single screw extruder and used as raw material for melt spinning. Polymer chips were stored overnight in a vacuum oven at 60°C temperature and pressure of 0.08 MPa to ensure moisture residual less than 100 ppm. The fibers were spun by melt spinning method using a single screw extruder (20 mm) with protected nitrogen environment. The spinneret was composed of 18 holes each of 0.4 mm in diameter, and the filter dimensions of the spinneret was 600 mazes/square inch. The speed of 475 m/min at winder and 15 m/min at overfeed was maintained. The linear density of the obtained filament was 150 denier (weight in grams of 9000 m of length) [40,41]. All the stocking fabrics were produced on MERZ (Model number—CC4-MED) medical compression stocking weft knitting machine. The machine cylinder diameter was 4.75 inches with 360 needles and 24 gages (24 needles/inch). The machine was anchored with four types of yarn input devices and each of nine yarn feeders and three furnishers for elastic yarn feeding. As the SMP filaments to be in a relaxed state, fleecy knit structure was used to insert as an inlay yarn (Fig. 8.7).

5.4 Pressure study

The interfacial pressure caused by the stocking was measured using a Kikuhime pressure sensor, and the experimental setup is shown in Fig. 8.8. Cylindrical tubes of different circumferences were used to act as a human leg. Before the application, the stocking was heated for 10 min using a hair drier and then applied over the cylindrical tube. The system was then allowed to cool down at room temperature for 1 h. The net increase in the interface pressure P above the baseline pressure was then recorded at three levels each for the two factors, temperature (30, 40, and 50°C) and prestrain ε (5.85%, 13.06%,

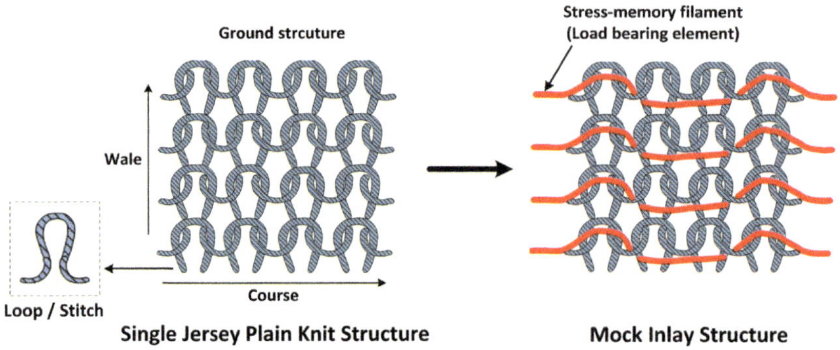

Figure 8.7
Scheme of stress memory filaments integration and structural details of the memory stocking.

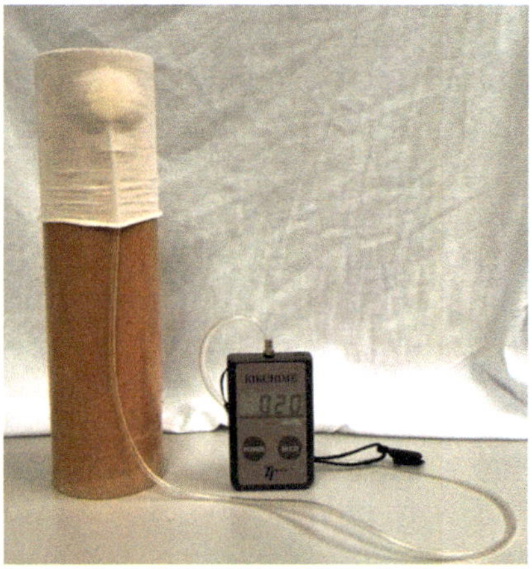

Figure 8.8
Setup for pressure measurement.

and 20.27%). The pressure variation over time was also examined to evaluate its long-term performance.

The pressure at different level of temperature was showed in Fig. 8.9. For a particular temperature, the pressure increases for the initial period (10 min) and finally reaches to a fixed value. The rate and amount of extra pressure generated is more at larger temperature ($P < .01$). SMP filaments had shown a range of transition temperature due to the fact that not all the switching segments would activate altogether at a fixed temperature point. It is expected that more switching segments are activated at upper extreme of T_{trans} ($\sim 50°C$) as compared with its lower extreme ($\sim 30°C$). Therefore, more amount of recovery stress will be generated in the SMP filament at higher temperature.

Other tests were also done to check the ability of the stocking in reproducing or readjusting the pressure when used multiple times. For the cyclic tests, the stocking was

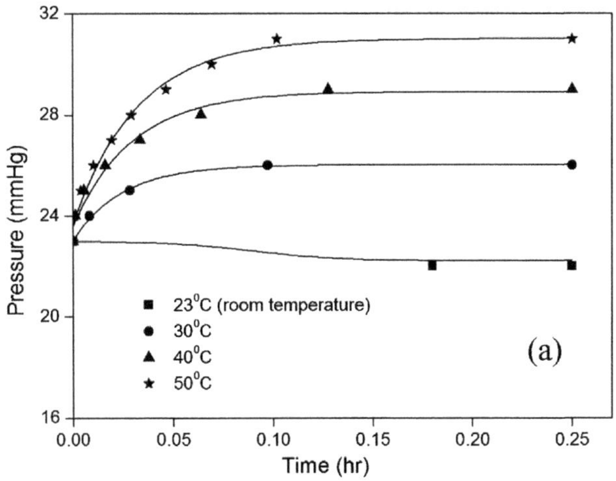

Figure 8.9
Pressure variation at different levels of temperature.

activated for 30 min in heated chamber and then placed at room temperature for 30 min. The procedure was repeated for 5 times, and the pressure variation was recorded. The nature of pressure variation over five cycles is summarized in Fig. 8.10. There is initial readjustment of pressure between the first two cycles. In the first cycle, the amount of pressure increase during heating is less than the pressure decrease during cooling. This resulted in lowering of baseline pressure (initial pressure at room temperature) at the end of the first cycle. This effect continued up to first two cycles, and the pressure variations become alike for subsequent cycles. Similar results were obtained at different strain levels. However, the decrease of baseline pressures was more at high strain level compared with low strain. The pressure drops of 10% and 20% were obtained, respectively, at 5.87% and

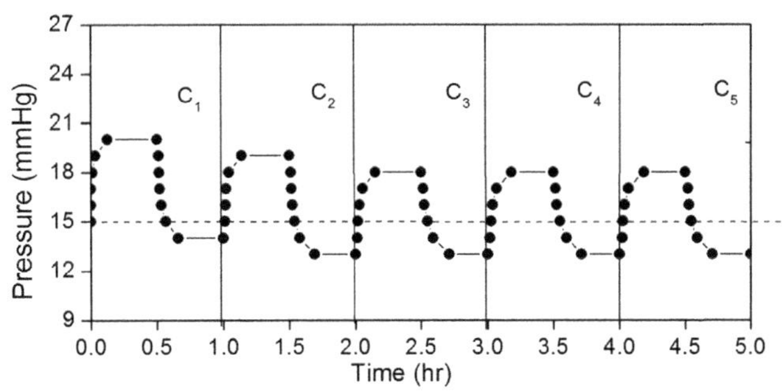

Figure 8.10
Pressure variation during sequential heating and cooling of the stocking up to five cycles (C_1–C_5: individual cycles). Each cycle includes heating of the stocking for 30 min in heated chamber ($T = 40°C$) and then cooling down for another 30 min at room temperature ($T = 23°C$).

20.27% strain levels. No significant variation was obtained from third cycle onward. The stocking showed the potential of reproducing the same amount of pressure even after multiple uses.

6. Summary and challenges ahead

MP-based stocking shows the potential to overcome the limitation of conventional stockings as it could provide the freedom to adjust pressure level externally whenever needed, i.e., a smart wound care product, during the course of compression therapy. Naturally, at this initial stage, there remain a few technical challenges before to ensure the stocking as a credible compression product. The transition temperature of the SMP filament needs to be controlled further to allow easy activation at still lower temperature range. Also in real practice, heating a stocking on the leg to a particular temperature is a challenge, and simple and efficient heating arrangement need to be assured to have precise control on pressure. Note that although pressure drop in the smart stocking can be compensated under heat stimuli, it can also be minimized by increasing the proportion of elastomeric SMP fibers in the structure. However, increasing the elastomeric fibers makes the structure tighter and less permeable and hence diminishes the comfort, and thus a proper blending with cotton or other fibers is preferred for comfort. Optimization should be done to obtain both desirable compression and sufficient comfort. All recent work limits to study of interfacial pressure on hard surface using cylindrical tubes. Further, it can be extended to soft surface to mimic the human limb and study the pressure-related parameters to deeply understand the influence of skin deformations on the applying pressure. Future work in SMP could include examining several other factors, e.g., different SMPs, hard segment content, etc.

Acknowledgments

The author acknowledges the financial support received from the Department of Science and Technology, India (project Codes: RP03454G/MI01567/MI01695).

References

[1] Lendlein A, Kelch S. Shape-memory polymers. Angew Chem Int Ed 2002;41(12):2034–57.
[2] Lendlein A, et al. Light-induced shape-memory polymers. Nature 2005;434(7035):879–82.
[3] Lendlein A, Behl M. Shape-memory polymers for biomedical applications. Adv Sci Technol 2009;54:96–102.
[4] Ratna D, Karger-Kocsis J. Recent advances in shape memory polymers and composites: a review. J Mater Sci 2008;43(1):254–69.
[5] Behl M, Lendlein A. Actively moving polymers. Soft Matter 2007;3(1):58–67.
[6] Kohlmeyer RR, et al. Shape-reprogrammable polymers: encoding, erasing, and re-encoding. Adv Mater 2014;26(48):8114–9.

[7] Choi W, et al. Synthesis of graphene and its applications: a review. Crit Rev Solid State Mater Sci 2010;35(1):52–71.
[8] Kim H, Abdala AA, Macosko CW. Graphene/polymer nanocomposites. Macromolecules 2010;43(16):6515–30.
[9] Koerner H, et al. Remotely actuated polymer nanocomposites – stress-recovery of carbon-nanotube-filled thermoplastic elastomers. Nat Mater 2004;3(2):115–20.
[10] Weigel T, Mohr R, Lendlein A. Investigation of parameters to achieve temperatures required to initiate the shape-memory effect of magnetic nanocomposites by inductive heating. Smart Mater Struct 2009;18(2).
[11] Zhao Y. Light-responsive block copolymer micelles. Macromolecules 2012;45(9):3647–57.
[12] Hu JL, Ji FL, Wong YW. Dependency of the shape memory properties of a polyurethane upon thermomechanical cyclic conditions. Polym Int 2005;54(3):600–5.
[13] Hu JL, et al. Crosslinked polyurethanes with shape memory properties. Polym Int 2005;54(5):854–9.
[14] Ji FL, et al. Effects of hard-segment content on the shape memory properties of segmented polyurethanes. In: Proceedings of 2005 International Conference on Advanced Fibers and Polymer Materials (Icafpm 2005), vol. 1 and 2; 2005. p. 64–7.
[15] Kumar B, Hu JL, Pan N. Smart medical stocking using memory polymer for chronic venous disorders. Biomaterials; 2015.
[16] Ramelet AA. Compression therapy. Dermatol Surg 2002;28(1):6–10.
[17] O'Meara S, et al. Compression for venous leg ulcers. Cochrane Database Syst Rev 2012;(11).
[18] Partsch H. Compression for the management of venous leg ulcers: which material do we have? Phlebology 2014;29:140–5.
[19] Feldman JL, et al. Intermittent pneumatic compression therapy: a systematic review. Lymphology 2012;45(1):13–25.
[20] Wertheim D, et al. Measurement of forces associated with compression therapy. Med Biol Eng Comput 1999;37(1):31–4.
[21] Thomas S. The production and measurement of sub-bandage pressure: Laplace's Law revisited. J Wound Care 2014;23(5):234–46.
[22] Sue J. Compression hosiery in the prevention and treatment of venous leg ulcers. J Tissue Viability 2002;12(2):67–74.
[23] Kumar B, Das A, Alagirusamy R. Study on interface pressure generated by a bandage using in vitro pressure measurement system. J Text Inst 2013;104(12):1374–83.
[24] Kumar B, Das A, Alagirusamy R. Science of compression bandages. Elsevier Science Limited; 2013.
[25] Kumar B, Das A, Alagirusamy R. Effect of material and structure of compression bandage on interface pressure variation over time. Phlebology 2014;29(6):376–85.
[26] Kumar B, Das A, Alagirusamy R. Study of the effect of composition and construction of material on sub-bandage pressure during dynamic loading of a limb in vitro. Biorheology 2012;50(1–2):83–94.
[27] Ziaja D, et al. Compliance with compression stockings in patients with chronic venous disorders. Phlebology 2011;26(8):353–60.
[28] Munch-Fals J, et al. Electro active elastic compression bandage. US: Google Patents; 2011.
[29] Pourazadi S, Ahmadi S, Menon C. Towards the development of active compression bandages using dielectric elastomer actuators. Smart Mater Struct 2014;23(6).
[30] Moein H, Menon C. An active compression bandage based on shape memory alloys: a preliminary investigation. Biomed Eng Online 2014;13.
[31] Bauerfeind HB. Shape memory element for medical aids. Google Patents; 2013.
[32] Taylor M. Bandage pressure sensor. US: Google Patents; 2008.
[33] Leng JS, et al. Shape-memory polymers and their composites: stimulus methods and applications. Prog Mater Sci 2011;56(7):1077–135.
[34] Xie T. Tunable polymer multi-shape memory effect. Nature 2010;464(7286):267–70.
[35] Hu JL, Kumar B, Narayan HK. Stress memory polymers. J Polym Sci B Polym Phys 2015;53(13):893–8.

[36] Kumar B, Hu JL, Pan N. Smart medical stocking using memory polymer for chronic venous disorders. Biomaterials 2016;75:174–81.
[37] Kumar B, et al. A smart orthopedic compression device based on a polymeric stress memory actuator. Mater Des 2016;97:222–9.
[38] Kumar B, Hu J, Pan N. Memory bandage for functional compression management for venous ulcers. Fibers 2016;4(1).
[39] Hu JL, Kumar B, Narayana H. Stress memory polymers. J Polym Sci B Polym Phys 2015;53(13):893–8.
[40] Narayana H, et al. Stress-memory polymeric filaments for advanced compression therapy. J Mater Chem B 2017;5(10):1905–16.
[41] Meng QH, Hu JL. Study on poly(epsilon-caprolactone)-based shape memory copolymer fiber prepared by bulk polymerization and melt spinning. Polym Adv Technol 2008;19(2):131–6.

CHAPTER 9

Emerging electroporation-based technologies for wound care

Alexander Golberg
Porter School of Environment and Earth Sciences, Tel Aviv University, Tel Aviv, Israel

Chapter Outline

1. What is electroporation?

Using electric fields for medical purposes dates to antiquity, Greeks and Romans treated headaches and pain with electric rays and torpedo fish [1]. In the 18th century, Galvani demonstrated that electric pulses act on nerves and muscles, and Nollet performed the first systematic investigation of the disruptive effects of electric discharges on the skin, such as the occurrence of "red spots" that we now identify primarily as damage to the capillaries because of the electroporation of the endothelial cells membranes [2]. What we know now as nerve cell electroporation was observed as the electrical conductivity changes in nerves exposed to the electric fields, reported by Ritter in 1802 [3], Frankenhaeuser and Widén in 1954 [4], and Stampfli and Willi in 1957 [5]. Fuller, in 1898, investigated sterilization of river by high-voltage discharges [6] and reported what was likely the first evidence on nonthermal electric-pulse inactivation of microbes, nowadays understood as cell membrane irreversible electroporation (IRE). The first systematic studies of microbe inactivation by electroporation were published by Sale and Hamilton in the series of three papers in 1967–68 [7–9].

Innovations and Emerging Technologies in Wound Care. https://doi.org/10.1016/B978-0-12-815028-3.00009-2

All these century-long observations that pulsed electric field change conductivity of tissues and can kill cells started to get new explanations with the development of electroporation technology, which probably took a root in 1982 with the seminal work of Neumann and colleagues. In this work, the authors used pulsed electric fields to deliver a DNA molecule to a cell by temporal permeabilization of cell lipid membrane [10]. A month later, the same group coined this technology "electroporation" [11].

Several theoretical descriptions of the change of cell membrane permeability to large molecules have been proposed over years. Some assumed a certain type of deformation of the membrane lipids [12—14], their phase transition [15], and the breakdown of interfaces between the domains with different lipid compositions [16], while others suggested denaturation of membrane structural proteins [17]. However, all these theories did not fit the experimental observations and eventually were rejected [18]. Current consensus describes electroporation as the formation of aqueous pores in the lipid bilayer that enable molecular transport [18—20]. The theory of aqueous pore formation, largely based on thermodynamical considerations, describes the formation of aqueous pores as started by penetration of water molecules into the lipid bilayer of the membrane, which leads to reorientation of the adjacent lipids with their polar headgroups toward these water molecules [18] Unstable pores with nanosecond lifetimes can form even in the absence of an external electric field, but presence of such a field induces a voltage across the lipid bilayer (i.e., a difference between the electric potentials on the inner and the outer side of the membrane), thus reducing the energy required for penetration of water into the bilayer [18]. This increases both the probability of pore formation and the pores' average lifetime, resulting in a larger number of pores formed in the bilayer per unit of area and per unit of time, with the pores being more stable than in the absence of the electric field.

There are four general ranges of electric pulse amplitude and duration, each characterized by different effects on the cells exposed to electric fields. For weak and short pulses, the pores, even if formed, are too small and short-lived for measurable transport; hence, there is no detectable electroporation. For stronger and/or longer pulses, the pores are formed to provide a temporary pathway for transport, but after the pulses they reseal, the transport ceases, and most cells retain their viability. This effect is called reversible electroporation. The correlation between the induced transmembrane voltage and the electroporation-mediated transport is demonstrated particularly clearly by combining potentiometric measurements and monitoring of transmembrane transport on the same cell [21]. As the amplitude and/or duration of the pulses increases, an increasing fraction of pores reseals too slowly to preserve cell viability, leading to cell death, but without thermal damage—this is nonthermal IRE [22]. Finally, for even stronger and/or longer pulses, IRE becomes accompanied by thermal damage. These four ranges partly overlap both because pore formation is stochastic [18] and because the exposed cells generally vary in size and shape [21]. The range boundaries also depend on the cell type and the properties of the cells' surrounding medium.

2. An overview of electroporation-based technologies in medicine

The medical applications of pulsed electric fields technology probably started in 1982 with the seminal work of Neumann and colleagues [11]. The authors used pulsed electric fields to temporary permeabilize a cell membrane to deliver foreign DNA molecules to cells. This method, coined "reversible electroporation," emerged into novel clinical applications: electrochemotherapy (ECT), first demonstrated by Okino and Mohri in 1987 [23], and gene electrotherapy or DNA vaccination that followed [24]. Lee and colleagues suggested that the tissue electric trauma has a nonthermal, IRE nature of the damage, in addition to thermal effects [25]. Davalos, Mir, and Rubinsky in 2005 proposed that pulsed electric fields can nonthermally ablate clinically relevant volumes of tissue, for example, solid tumors [26]. Miklavcic et al. showed that electric field distribution correlates with the effectiveness of ECT [27]—this serves as the basis for careful treatment planning of electroporation technologies in tissues (Fig. 9.1). Treatment planning is a critical part of the successful electroporation treatment protocol as it enables to predict precise electrode location for maximum electric field strengths focusing on the target tissue and minimum damage to nontarget tissues [28,29] (Fig. 9.1).

3. Normal skin regeneration after irreversible electroporation

Skin wound healing and scar formation is a complex process that results in the formation of the dense extracellular matrix without normal skin appendages such as hair and glands. The absence of a scarless healing model in adult mammals prevents the development of successful therapies. To address this problem and provide a model for scarless wound healing studies, a rat model was developed, which showed complete scarless skin regeneration with all accessory organs following IRE [31,32]. Pulsed electric fields at 500 V, with 70 μs pulse duration, and 1000 pulses delivered at 3 Hz, applied through two electrodes separated by 2 mm (Fig. 9.2 shows the procedure and a model of the electric field distribution in the normal skin), lead to massive cell death. However, the extracellular matrix architecture of the skin was preserved. Six months after the ablation, the epidermis, sebaceous glands, panniculus carnosus, hair follicles, microvasculature, and arrector pili muscle were altogether reformed in the entire ablated area (Fig. 9.3). These results suggest a key role of the extracellular matrix architecture in the differentiation, migration, and signaling of cells during scarless wound healing and show that IRE is a useful tool for scarless tissue regeneration [31].

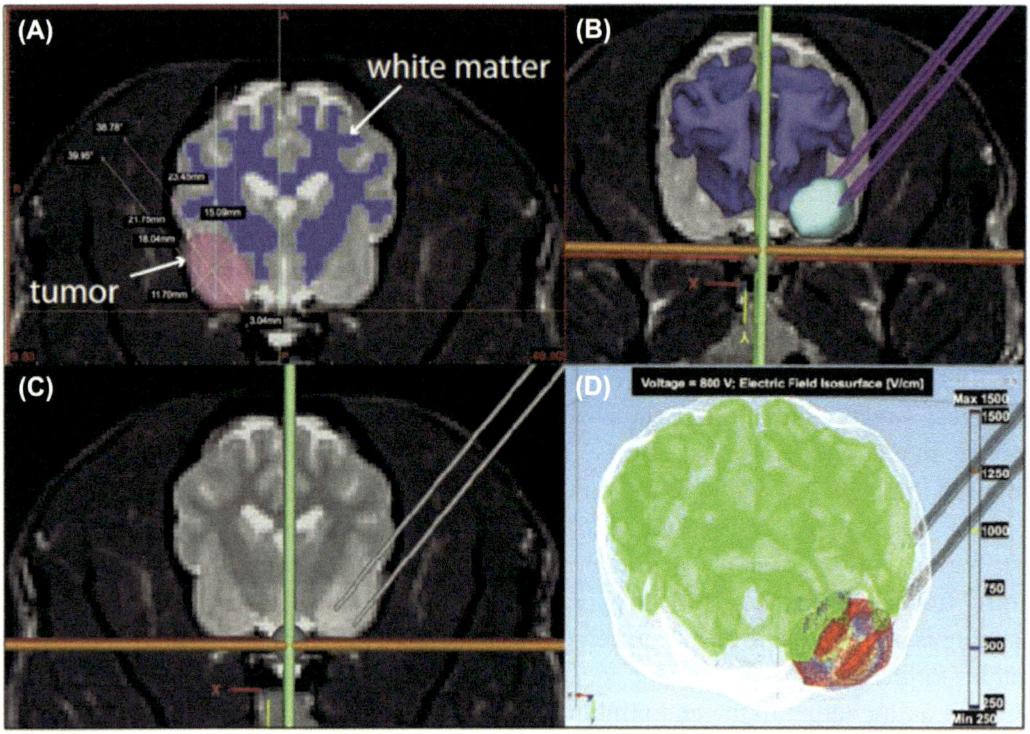

Figure 9.1

MRI-based planning of irreversible electroporation (IRE) treatment of malignant glioma in canine patients. (A). White matter and tumor segmentation and corresponding measurements (e.g., distance and angles) for electrode placement. (B) Three-dimensional volumetric reconstruction of the gray matter, white matter, and tumor tissue from MR images and the corresponding electrode geometry used during surgical procedures. (C) Planned electrode insertion depth and trajectory for delivering intratumoral pulsed electric fields. (D) Computational simulation of the electric field distribution surrounding the active electrodes during pulse delivery to compute tumor coverage and cell kill probabilities. *Figure adapted from Garcia PA, et al. Predictive therapeutic planning for irreversible electroporation treatment of spontaneous malignant glioma. Med Phys 2017;44:4968–4980 with permission under the terms of the Creative Commons Attribution Non-Commercial License CC BY-NC.*

4. Skin rejuvenation with partially irreversible electroporation

Genetic background and environmental stresses such as aging, trauma, and chronic metabolic diseases, including diabetes, often lead to alterations in skin color, texture, and barrier function. Loss of skin function leads to atrophy, infection, chronic wounds, laxity, and rhytides. Epidermal atrophy—characterized by a thinning of the epidermis and an increase in fragility—is observed in ~32% of individuals over 60 years of age [33]. As electroporation affects precisely cell membrane and does not affect extracellular matrix, it was suggested that it could lead to cell-based therapy for skin rejuvenation. The cell-based

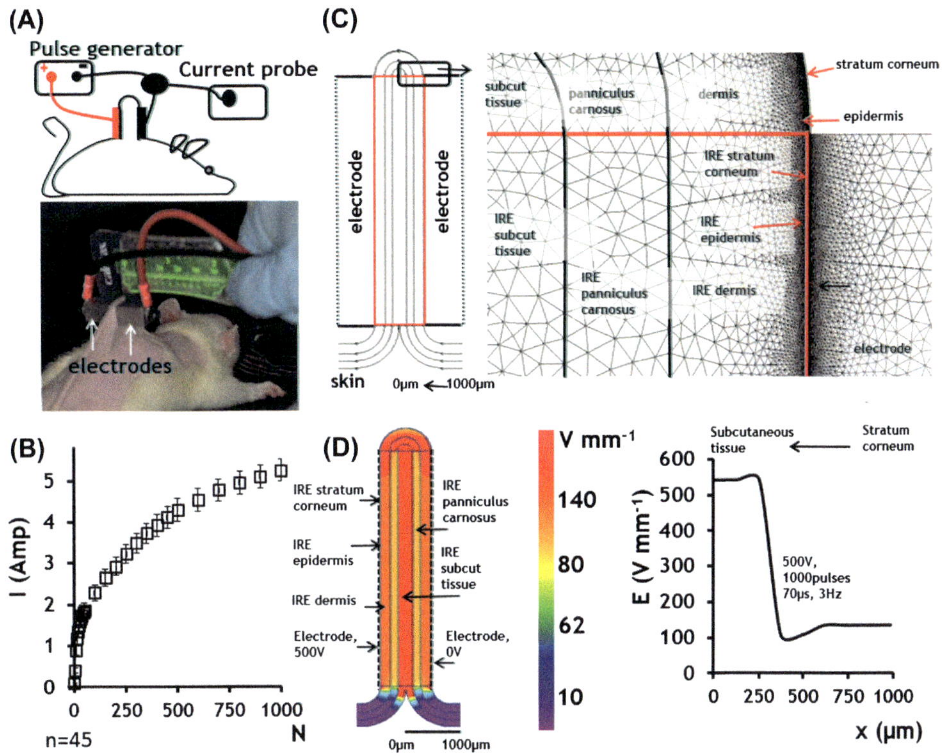

Figure 9.2: Normal skin irreversible electroporation (IRE).
(A) Electrode positioning. (B) Current measured during pulse delivery (n = 45). (C) Geometry and mesh of the finite element numerical model of rat skin, with all major composing layers. The skin is divided into two regions: (1) control and (2) irreversibly electroporated skin (IRES). IRES layers are tagged with IRE sign. (D) The electric field distribution in different skin layers under various boundary conditions. The panel on the left shows the two-dimensional map of the electric field distribution. The plot of the right shows the numerical values of the electric field strength in different skin layers. *Reprinted from Golberg A, et al. Skin regeneration with all accessory organs following ablation with irreversible electroporation. J Tissue Eng Regenerat Med 2017. https://doi.org/10.1002/term.2374 with a permit.*

therapy concept is that the energy is delivered to the skin cells and not to the bulk tissues as done by other physical methods.

Partial IRE protocol consisting of applied voltage: 500 V, pulse duration: 70 μs, number of pulses: 200, frequency of pulses delivery: 3 Hz led to the increased epidermis thickness (up to 3 weeks after treatment)—important for infection prevention, angiogenesis, nutrient supply, and waste removal, increased new capillary production and metabolism, and measured by the rate of glucose uptake (Fig. 9.4A) [34]. Application of pulsed electric field in partially irreversible electroporation (pIRE) regime on the normal rat skin led to >72% increase of total collagen content 3 weeks after the treatment in comparison with untreated

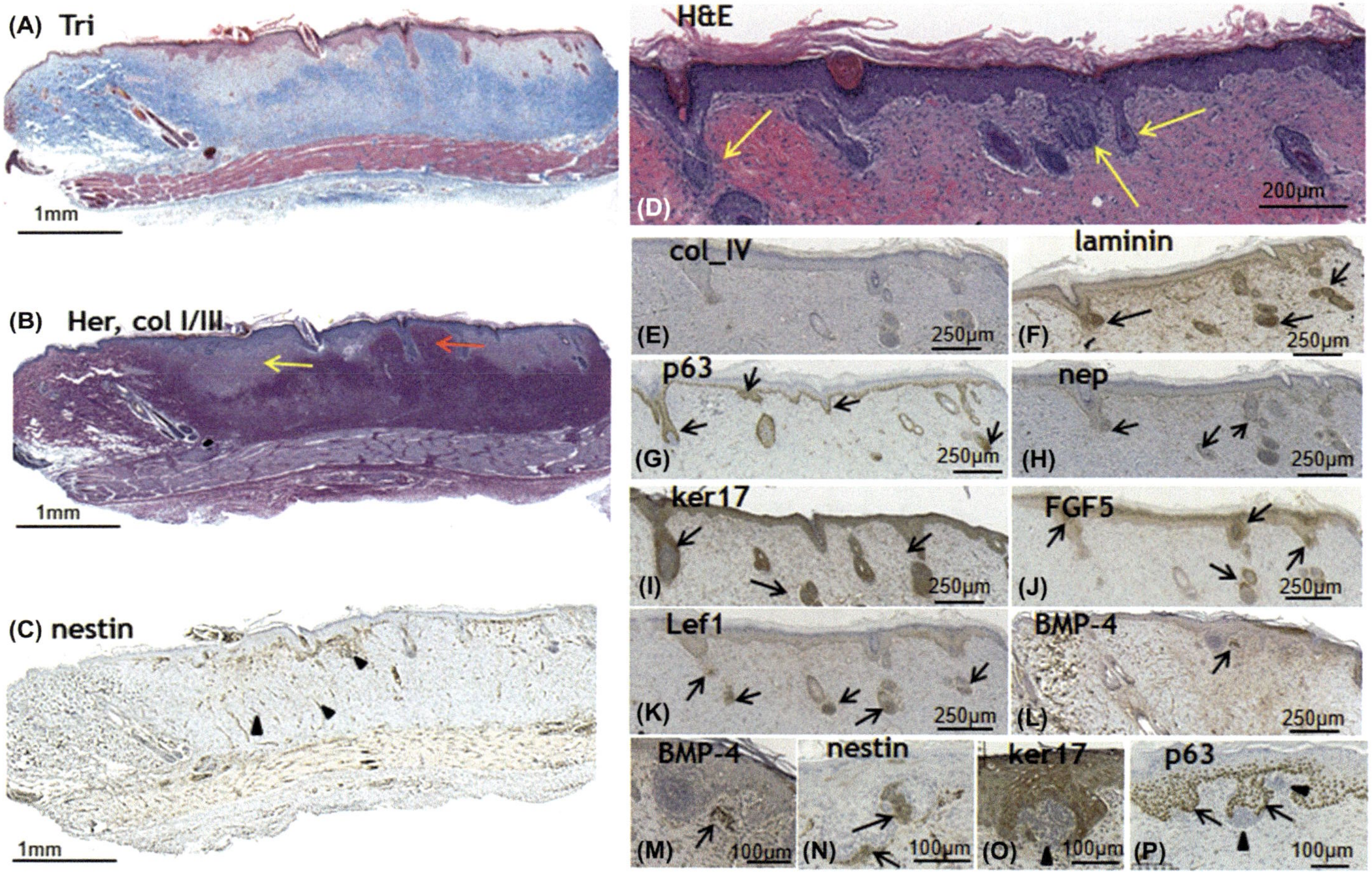

Figure 9.3: Skin regeneration 3 weeks after irreversible electroporation (IRE).
(A) Extracellular matrix structure (Masson's trichrome stain). (B) The area is rich with fresh type III collagen (*blue* in the shown Herovici staining). *Yellow arrow* indicates the areas with new uncrosslinked collagen. *Red arrow* shows the formation of a new hair in the middle of the ablated zone. (C) Angiogenesis detected with nestin staining. Reduced formation of new vessels in comparison with 3 days and 1 weeks time points. Hair folliculogenesis at the IRE-ablated site 3 weeks after the injury (D–P). (D) Pathology of the site reveals the formation of new hair follicles at multiple locations inside the injured area (H&E staining). *Arrows* show the newly formed follicles. (E) Increased expression of collagen IV in the basement membrane. (F) Increased expression of laminin in the basement membrane. (E) Increased expression of nephronectin in the basement membrane. (G, P) Increased expression of p63 in the external part of the forming hair follicles, absence in the hair papilla (*arrowheads*). (I, O) Increased expression of ker17 in the external part of the forming hair follicles, absence in the hair papilla (*arrowheads*). (J) Increased expression of FGF-5 in the hair bulging area. (K) Increased expression of Lef-1 in the hair papilla. (L) Increased expression of BMP-4 in the hair papilla area. (N) Increased expression of nestin in the hair papilla area. Combined, the expression of the markers displayed in (D–P) demonstrates the generation of new hair follicles at the ablated site. Five animals were used for this time point. *Reprinted from Golberg A, et al. Skin regeneration with all accessory organs following ablation with*

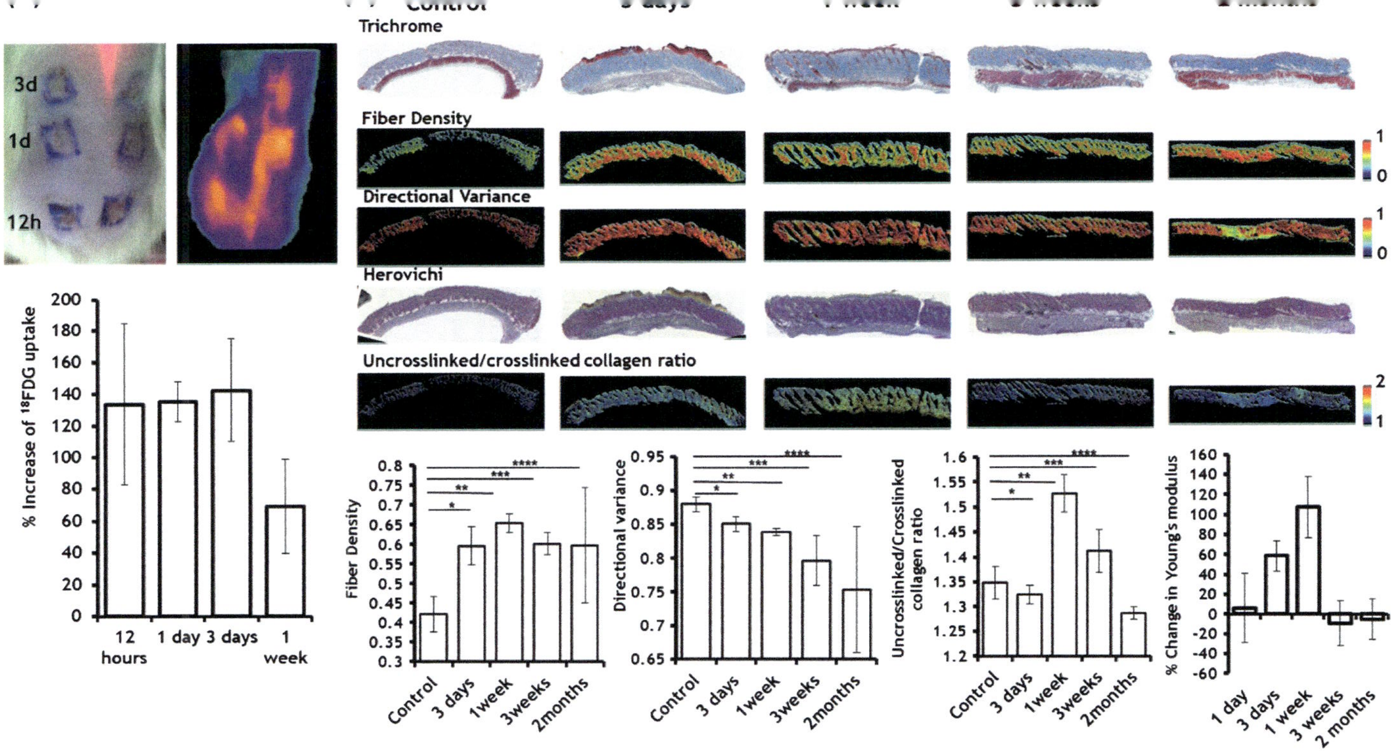

Figure 9.4: Skin rejuvenation with partially irreversible electroporations.

(A) The increased metabolism of skin areas treated by PEF as detected using ^{18}FDG uptake. (B) The effect of PEF on the extracellular matrix. The *upper panel* (Masson's trichrome staining) enabled the identification of collagen-containing pixels and demonstrated significantly higher density within the PEF-treated areas of the skin. The *lower panel* shows the pixel-wise fiber directional analysis and enabled quantification of the local strength of collagen fiber alignment, which demonstrated reduced variance in fibers up to 3 weeks after PEF but no significant difference in the directional variation within the PEF-treated area 2 months after the treatment. Secretion of new fibers to the PEF-treated area was detected by Herovici staining. Quantitative analyses of fiber density in the center of PEF-treated area were significantly elevated, starting with day 3 after PEF treatment and continuing up to 3 weeks after the treatment. Two months after treatment, the fiber density of the PEF-treated area was not significantly different from the control, untreated skin (*) p-val = 0.03; (**) p-val = 0.009; (***) p-val = 0.02. Quantitative analyses showed no significant difference in fiber alignment, indicating no scar formation at the PEF area. Quantification of *blue/red* in the digitized Herovici stained slides shows significant increase of the *blue* staining (collagen type III) 1 week after the PEF, (*) p-val = 0.01. For each experimental point, we analyzed at least four slides, each obtained from a different animal. Young's modulus change of the PEF-treated skin in comparison with untreated control. An increase in Young's modulus was observed 3 days and 1 week after PEF treatment of the skin, (*) p-val = 0.0048; (**) p-val = 0.0028. Five animals were used for each time point. In all plots, error bars show ± SEM. Statistical analysis was performed by first using one-way ANOVA for multivariate analysis and post hoc using *Dunnett's* tests to assess significance between individual groups. Significance was set at $P < 0.05$. *Figure adapted with permission from Golberg A, et al. Skin rejuvenation with non-invasive pulsed electric fields. Sci Rep 2015;5:10187.*

skin. This amount reduced to ~50% increase of total collagen content 2 months after the treatment in comparison with untreated skin [34]. Electroporation increased fiber density and Young's modulus but did not change fiber directional variance (which marks scars) (Fig. 9.4B).

5. Third-degree burn scar area reduction with partially irreversible electroporation

Hypertrophic scars (HTSs) after trauma and burn injury remain a major clinical problem that leads to physical, esthetic, functional, psychological, and social stresses in thousands of surviving injured patients [35]. In adults, the formation of HTSs depends on the burn depth and individual response; in children, however, HTSs are formed even in superficial wounds, burns, and donor areas. Current data show that alterations in coagulation, inflammation, angiogenesis, fibroplasia, contraction, remodeling, and mechanical tension correlate with the formation of HTS [36–39], but thus far, the mechanisms that induce HTS are not well understood. Various procedures such as surgical excision, intralesional steroid injection, cryotherapy, laser therapy, irradiation, mechanical compression dressing, silicone sheet applications, intralesional interferon injection, or combinations of techniques have been tested to treat the scars over the years [40–44].

Encouraging results for tissue regeneration have been observed recently in animal studies that have used pulsed electric fields for partial tissue ablation, a procedure we coined as pIRE [45]. Irreversible electroporation is a tissue ablation method where specific pulsed electric fields cause irreversible damage to cells by affecting their cell membranes while sparing the neighboring tissue scaffold, large blood vessels, and other accessory structures [29,32,46–48]. It was found that pIRE using 200 pulses of 250 V and 70 μs duration, delivered at 3 Hz every 20 days during a total of five therapy sessions after the initial burn injury, resulted in a 57.9% reduction of the scar area in comparison with untreated scars and structural features approaching those of normal skin as described histologically and quantified by polarization-sensitive optical coherence tomography (PS-OCT) (Fig. 9.5) [45]. The pIRE model, which could generate scars with different sizes, also enabled the development of a linear regression model that predicts the size of a burn scar 6 months after third-degree burn injuries in rats based on early postinjury PS-OFDI and measurements of scar area [49]. This model opens new possibilities for quantitative and objective assessment of scar severity that could help to determine the optimal course of therapy for burn injuries.

6. Burn wound disinfection with partially irreversible electroporation

Pathogens, *Acinetobacter baumannii*, methicillin-resistant *Staphylococcus aureus, Pseudomonas, and Klebsiella*, which are increasingly resistant to various antimicrobial

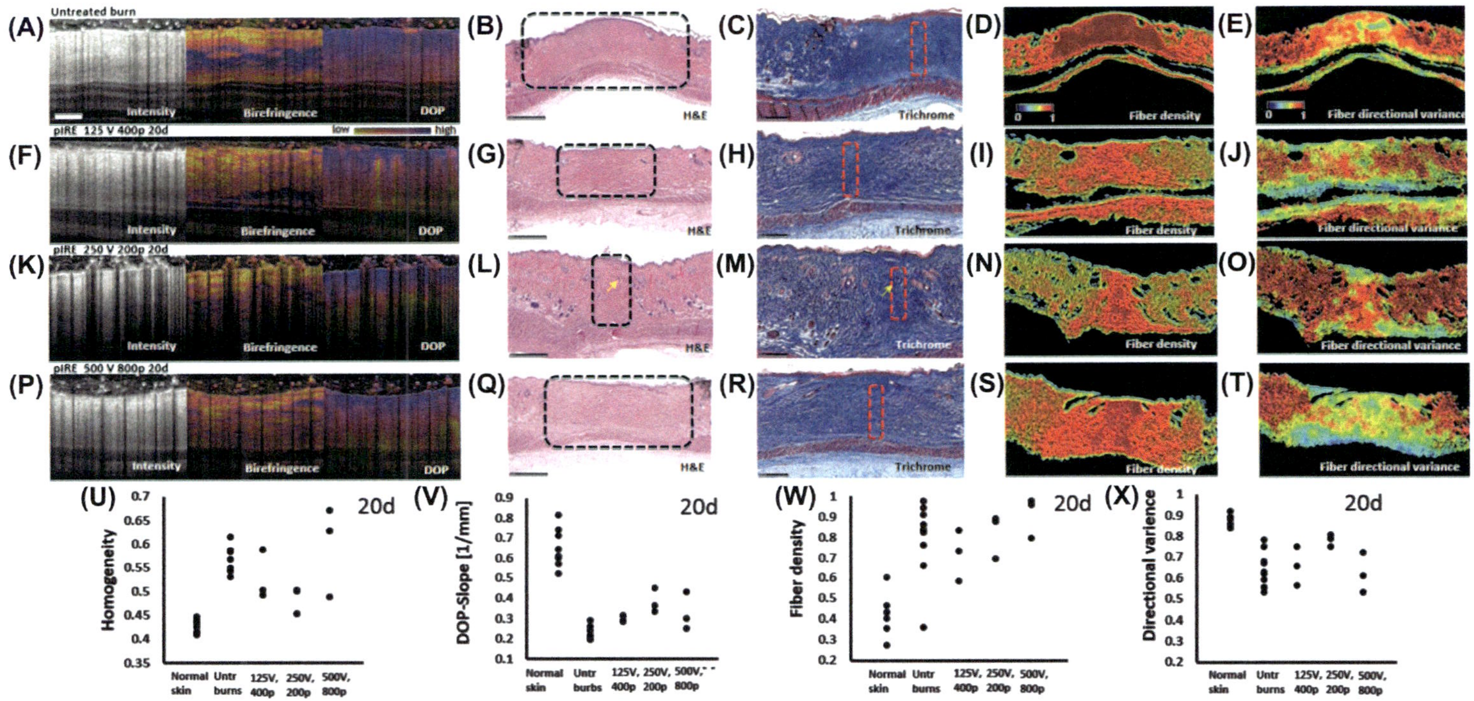

Figure 9.5

Skin regeneration following partially irreversible electroporation of third-degree burns 6 months after the injury. Untreated burns (n = 9): (A) polarization-sensitive optical coherence tomography (PS-OCT); (B) histology H&E; (C) histology Masson's trichrome; (D) fiber density; and (E) fiber directional variance. Burns treated with 125 V, 400 pulses every 20 days (n = 3): (F) PS-OCT; (G) histology H&E; (H) histology Masson's trichrome; (I) fiber density; and (J) fiber directional variance. Burns treated with 250 V, 200 pulses every 20 days (n = 3): (K) PS-OCT; (L) histology H&E; (M) histology Masson's trichrome; (N) fiber density; and (O) fiber directional variance. Burns treated with 500 V, 800 pulses every 20 days (n = 3): (P) PS-OCT; (Q) histology H&E; (R) histology Masson's trichrome; (S) fiber density; and (T) fiber directional variance. In all partially irreversible electroporation (pIRE) protocols, pulse duration was 70 μs and pulse delivery frequency was 3 Hz. (U) Quantification of PS-OCT homogeneity (n = 3 for normal skin and pIRE, n = 9 for untreated burns). (V) Quantification of PS-OCT DOP slope (n = 3 for normal skin and pIRE, n = 9 for untreated burns). (W) Quantification of fiber density (n = 3 for pIRE and n = 9 for untreated controls). (W) Quantification of fiber directional variance (n = 3 for pIRE and n = 9 for untreated controls). All individual measurements are reported. Scale bar: A, C, F, H, K, M, P, R = 500 μm, B, G, L, Q = 1 mm. *Figure adapted from Golberg A, et al. Preventing scars after injury with partial irreversible electroporation. J Invest Dermatol 2016with permission.*

agents, are a major problem in burn patient's recovery [50]. Recently, IRE was also applied to treat the third-degree wound infections (Fig. 9.6). Using 80 pulses of 5000 V cm^{-1}, disinfection with 4.91 ± 0.71 Log10 reduction of *A. baumannii* 3 hours after treatment was achieved (Fig. 9.6) [51,52]. The efficacy of disinfection was correlated with both the electric field strength and the number of delivered pulses. Increasing the pulse number from 40 to 80 led to a 255% increase in the reduction of bacterial load in the wound, from 1.49 ± 0.07 Log10 to 5.30 ± 0.85 Log10 when all other parameters were set constant. Increasing the field, however, from 250 to 500 V mm^{-1}, while keeping the number of pulses at 40, led to only a ~37% increase in the log reduction of bacterial load in the burn, from 1.49 ± 0.07 Log10 to 2.04 ± 0.29 Log10.

Surprisingly, the efficiency capability did not correlate with the increase of the invested energy. For delivery of 80 pulses with 250 V mm^{-1} at 1 Hz, ~9 J are needed. For delivery of 40 pulses with 500 V mm^{-1} at 1Hz, ~18 J are needed. These energy consumption findings are interesting: they show that increasing the number of pulses from 40 to 80 leads to a significantly larger bacterial load reduction than increasing the electric field strength from 250 to 500 V mm^{-1}. Therefore, electroporation-based technologies does not necessarily depend on delivered energy. These findings are strikingly different from heat/radiation-based disinfection, where the bacterial load reduction directly correlates with consumed energy. Nevertheless, these findings are consistent with the current electroporation theory. According to the current theory, increasing the field strength increases the total electroporated surface of the cell membrane. Increasing the number of pulses, after the electroporation threshold potential is reached, increases the number and size of the aqueous pores of the membranes at the electroporated site [53–55].

7. Genetic engineering of chronic wounds with reversible electroporation

Diabetes, infection or venous diseases, and nonhealing chronic wounds require therapeutic strategies to reinforce or replace topically applying recombinant growth factors, which are expensive and can jeopardize wound care [56,57]. Transfection of cells in the wound area with foreign DNA that leads to increased local secretion of growth factors is an emerging and promising approach for chronic wound care (Fig. 9.7) [58]. Plasmid expression is reduced in weeks after transfection [59,60], which is sufficient to enhance wound closure kinetics without leading to uncontrolled growth [61].

Vascular endothelial growth factor (VEGF) has been a target for wound cell transfection with electroporation. Successful transfection in the rat model was reported with 3 pulses lasting 50 ms, 160 V/cm [62], 1 Hz; 8 pulses lasting 20 ms, 200 V/cm [63], 1 Hz; and 72 pulses lasting 150 ms, 500 V/cm, 6.6 Hz [64]. Fibroblast growth factor-basic was transfected in the rat models using three pulses lasting 50 ms, 50 V/cm, 1 Hz, followed by

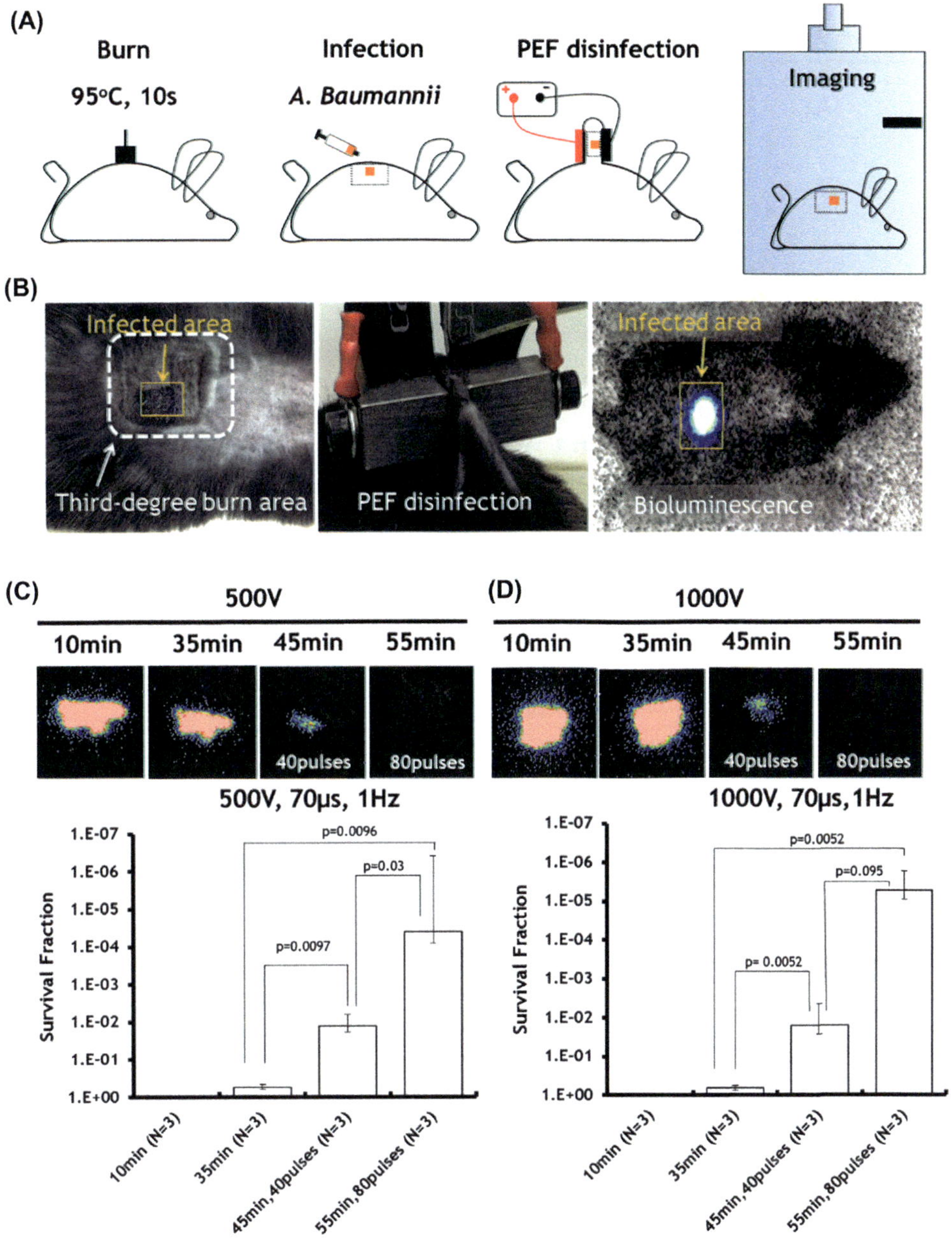

Figure 9.6: Pulsed electric field disinfection in vivo: C57BL/6 black mice model.
(A) Schematic illustration of procedures we used in this study. The 1 cm [2] burn injury was followed by dispersion of *Acinetobacter baumannii* on part of the wound. Next pulsed electric field was applied using two plate electrodes. *A. baumannii* infection load was quantified using bioluminescent imaging. (B) The left panel shows digital photography of the burned (*white frame*) and infected (*orange frame*) areas of the skin. The central panel shows digital photography of the applied electrodes. The right panel shows the images of the mice as observed inside the dark imaging box. Orange frame shows the infected area as detected by a strong bioluminescent signal emitted from bacteria. (C) Electric field strength 500 V. (D) Electric field strength 1000 V. (C, D)The top panel shows the postburn time when the images were taken. The bottom panel shows the survival fraction of microorganisms as detected by the top panel images (logarithmic, inversed scale). N shows the number of animals per group. Error bar ± standard deviation of the mean. *Figure adapted from Golberg A, et al. Eradication of multidrug-resistant* A. baumannii *in burn wounds by antiseptic pulsed electric field. Technology 2014;2:153–160 with permission.*

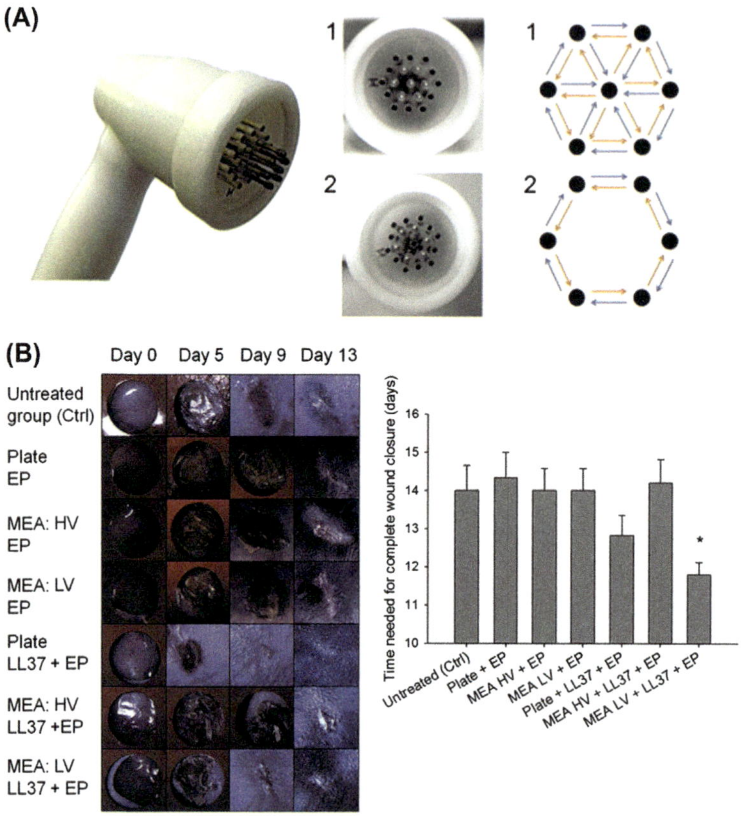

Figure 9.7

(A) Noninvasive multielectrode array developed for skin reversible electroporation applications. (B) Time needed for the complete wound closure after the intradermal delivery of the plasmid encoding LL37 around the wounded area. Left: Images of the wounds taken under the stereomicroscope at the designated times. EP = the group was treated by application of electric pulses; LL37 = hCAP-18/LL-37 encoding plasmid. Right: The number of days needed for the complete wound repair. Error bars indicate SEM. A*P value < 0.05 compared with the control group. N = 6 mice per group. *Images reprinted with permission from Kos S, et al. Gene electrotransfer into skin using noninvasive multi-electrode array for vaccination and wound healing. Bioelectrochemistry 2017;114:33–41.*

three pulses of the opposite polarity [65] and pulses (number not reported) with duration of 150 ms, 250 V/cm [66]. Keratinocytes growth factor 1 was overexpressed in the diabetic mouse [59] and septic wound rat models [60] using six pulses lasting 100 μs, 1800 V/cm, 8 Hz. Hypoxia-inducible factor 1-alpha was transfected by electroporation in the diabetic mouse model using 10 pulses lasting 20 ms, 800 V/cm, 8 Hz [67]. hCAP-718/LL-3 was transfected in normal and diabetic mouse skin using a combination of short high-voltage pulses with long, low voltage pulses as follows: one pulse lasting 100 μs, 700 V/cm, immediately followed by one pulse lasting 400 ms at 200 V/cm [68].

Importantly, already at the early studies, it was demonstrated that electric field alone significantly improves cutaneous wound closure [69]. This observation suggests the electrostimulation potential of electrical parameters used in electroporation context. Accumulated experimental data and clinical trials support the benefit role of electrostimulation for promoting wound healing process, especially in a context of pathologic chronic wounds [70–72].

8. Summary and future technology development needs

Recent in vivo data showed the potential of various electroporation application in the wound healing ranging from burn scar treatment and disinfection to gene therapy of chronic wounds. However, most of the studies were done with rodents and in less extent on pigs. Murine skin, though, is both anatomically and biologically different from the human skin. Therefore, future work should focus on the skin with a structure similar to humans. This could be achieved either by using animal models with transplanted human skin or using tissue-engineered approaches [73] that allow producing in vitro human skin substitutes for basic electroporation studies [74]. Both approaches will still have significant differences with human skin in terms of inflammation responses, which plays a critical role in wound healing. Yet, they could provide essential information, needed first in human trials, on electric field distributions and thermal effects in environmental similar to human skin.

References

[1] Bullock TH, Hopkins CD, Popper AN. Electroreception. New York: Springer; 2005.

[2] Nollet J. Recherches sur les causes particulieres des phénoménes électriques. In: Guerin CHL, Delatour LF, editors; 1754. Paris.

[3] Noad H. Lectures on electricity:comprosing galvnism, magnetism,electromagentism, maneto-and thermo-electricity, and electro-physiology. George Knoght and Sons; 1849.

[4] Frankenhaeuser B, Widén L. Anode break excitation in desheathed frog nerve. J Physiol 1956;131:243–7.

[5] Stampfli R, Willi M. Membrane potential of a Ranvier node measured after electrical destruction of its membrane. Experientia 1957;13:297–8.

[6] Fuller G. Report on the investigations into the purification of the Ohio river water at Lousville Kentuki. Company DVN; 1898.

[7] Hamilton W, Sale JH. Effects of high electric fields on microorganisms II. Mechanism of action of the lethal effect. Biochim Biophys Acta – Gen Subj 1967;148:789–800.

[8] Sale J, Hamilton W. Effects of high electric fields on micro-organisms. 3. Lysis of erythrocytes and protoplasts. Biochim Biophys Acta 1968;163:37–43.

[9] Sale JH, Hamilton W. Effects of high electric fields on microorganisms: I. Killing of bacteria and yeasts. Biochim Biophys Acta – Gen Subj 1967;148:781–8.

[10] Wong T-K, Neumann E. Electric field mediated gene transfer. Biochem Biophys Res Commun 1982;107:584–7.

[11] Neumann E, Schaefer-Ridder M, Wang Y, Hofschneider PH. Gene transfer into mouse lyoma cells by electroporation in high electric fields. EMBO J 1982;1:841–5.

[12] Michael D, O'Neill M. Electrohydrodynamic instability in plane layers of fluid. J Fluid Mech 1970;41:571–80.
[13] Crowley JM. Electrical breakdown of bimolecular lipid membranes as an electromechanical instability. Biophys J 1973;13:711–24.
[14] Steinchen A, Gallez D, Sanfeld A. A viscoelastic approach to the hydrodynamic stability of membranes. J Colloid Interface Sci 1982;85:5–15.
[15] Sugár I. A theory of the electric field-induced phase transition of phospholipid bilayers. Biochim Biophys Acta 1979;556:72–85.
[16] Cruzeiro-Hansson L, Mouritsen OG. Passive ion permeability of lipid membranes modelled via lipid-domain interfacial area. BBA – Biomembr 1988;944:63–72.
[17] Tsong TY. Electroporation of cell membranes. Biophys J 1991;60:297–306.
[18] Weaver JC, Chizmadzhev YA. Theory of electroporation: a review. Bioelectrochem Bioenerg 1996;41:135–60.
[19] Spugnini EP, et al. Ultrastructural modifications of cell membranes induced by 'electroporation' on melanoma xenografts. Microsc Res Tech 2007;70:1041–50.
[20] Kotnik T, Kramar P, Pucihar G, Miklavčič D, Tarek M. Cell membrane electroporation – part 1: the phenomenon. IEEE Electr Insul Mag 2012;28:14–23.
[21] Kotnik T, Pucihar G, Miklavčič D. Induced transmembrane voltage and its correlation with electroporation- mediated molecular transport. J Membr Biol 2010;236:3–13.
[22] Golberg A, Yarmush ML. Nonthermal irreversible electroporation: fundamentals, applications, and challenges. IEEE Trans Biomed Eng 2013;60:707–14.
[23] Okino M, Mohri H. Effects of a high-voltage electrical impulse and an anticancer drug on in vivo growing tumors. Jn J Cancer Res Gann 1987;72:1319–21.
[24] Nomura M, et al. In vivo induction of cytotoxic T lymphocytes specific for a single epitope introduced into an unrelated molecule. J Immunol Methods 1996;193:41–9.
[25] Lee RC, Kolodney MS. Electrical injury mechanisms: electrical breakdown of cell membranes. Plast Reconstr Surg 1987;80:672–9.
[26] Davalos RV, Mir LM, Rubinsky B. Tissue ablation with irreversible electroporation. Ann Biomed Eng 2005;33:223–31.
[27] Miklavcic D, et al. The importance of electric field distribution for effective in vivo electroporation of tissues. Biophys J 1998;74:2152–8.
[28] Kos B, et al. Robustness of treatment planning for electrochemotherapy of deep-seated tumors. J Membr Biol 2010;236:147–53.
[29] Golberg A, Bruinsma BG, Uygun BE, Yarmush ML. Tissue heterogeneity in structure and conductivity contribute to cell survival during irreversible electroporation ablation by 'electric field sinks'. Sci Rep 2015;5:8485.
[30] Garcia PA, et al. Predictive therapeutic planning for irreversible electroporation treatment of spontaneous malignant glioma. Med Phys 2017;44:4968–80.
[31] Golberg A, et al. Non-thermal, pulsed electric field cell ablation: a novel tool for regenerative medicine and scarless skin regeneration. Technology 2013;1:1–8.
[32] Golberg A, et al. Skin regeneration with all accessory organs following ablation with irreversible electroporation. J Tissue Eng Regenerat Med 2017. https://doi.org/10.1002/term.2374.
[33] Mengeaud V, Dautezac-Vieu C, Josse G, Vellas B, Schmitt A-M. Prevalence of dermatoporosis in elderly French hospital in-patients: a cross-sectional study. Br J Dermatol 2012;166:442–3.
[34] Golberg A, et al. Skin rejuvenation with non-invasive pulsed electric fields. Sci Rep 2015;5:10187.
[35] Sen CK, et al. Human skin wounds: a major and snowballing threat to publich health and the economy. Wound Repair Regen 2009;17:763–71.
[36] DiPietro LA. Angiogenesis and scar formation in healing wounds. Curr Opin Rheumatol 2012;1. https://doi.org/10.1097/BOR.0b013e32835b13b6.

[37] Aarabi S, et al. Mechanical load initiates hypertrophic scar formation through decreased cellular apoptosis. FASEB J 2007;21:3250–61.
[38] Wong VW, et al. Focal adhesion kinase links mechanical force to skin fibrosis via inflammatory signaling. Nat Med 2011;18:148–52.
[39] Rees PA, Greaves NS, Baguneid M, Bayat A. Chemokines in wound healing and as potential therapeutic targets for reducing cutaneous scarring. Adv Wound Care 2015;4:687–703.
[40] Leventhal D, Furr M, Reiter D. Treatment of keloids and hypertrophic scars: a meta-analysis and review of the literature. Arch Facial Plast Surg 2006;8:362–8.
[41] Mofikoya BO, Adeyemo WL, Abdus-salam AA. Keloid and hypertrophic scars: a review of recent developments in pathogenesis and management. Niger Q J Hosp Med 2007;17:134–9.
[42] Rabello FB, Souza CD, Farina Júnior JA. Update on hypertrophic scar treatment. Clinics (Sao Paulo). 2014;69:565–73.
[43] Ault P, Plaza A, Paratz J. Scar massage for hypertrophic burns scarring-A systematic review. Burns 2017. https://doi.org/10.1016/j.burns.2017.05.006.
[44] Mari W, et al. Novel insights on understanding of keloid scar: article review. J Am Coll Clin Wound Spec 2015;7:1–7.
[45] Golberg A, et al. Preventing scars after injury with partial irreversible electroporation. J Invest Dermatol 2016.
[46] Charpentier KP, et al. Irreversible electroporation of the pancreas in swine: a pilot study. HPB (Oxford) 2010;12:348–51.
[47] Rubinsky B, Onik G, Mikus P. Irreversible electroporation: a new ablation modality—clinical implications. Technol Cancer Res Treat 2007;6:37–48.
[48] Phillips M, Maor E, Rubinsky B. Nonthermal irreversible electroporation for tissue decellularization. J Biomech Eng 2010;132:91003.
[49] Kravez E, et al. Prediction of scar size in rats six months after burns based on early post-injury polarization-sensitive optical frequency domain imaging. Front Physiol 2017;8.
[50] Keen EF, et al. Prevalence of multidrug-resistant organisms recovered at a military burn center. Burns 2010;36:819–25.
[51] Golberg A, et al. Pulsed electric fields for burn wound disinfection in a murine model. J Burn Care Res 2014. https://doi.org/10.1097/BCR.0000000000000157.
[52] Golberg A, et al. Eradication of multidrug-resistant *A. baumannii* in burn wounds by antiseptic pulsed electric field. Technology 2014;2:153–60.
[53] Escoffre J-M, et al. What is (still not) known of the mechanism by which electroporation mediates gene transfer and expression in cells and tissues. Mol Biotechnol 2009;41:286–95.
[54] Gabriel B, Teissié J. Direct observation in the millisecond time range of fluorescent molecule asymmetrical interaction with the electropermeabilized cell membrane. Biophys J 1997;73:2630–7.
[55] Gabriel B, Teissié J. Time courses of mammalian cell electropermeabilization observed by millisecond imaging of membrane property changes during the pulse. Biophys J 1999;76:2158–65.
[56] Mudge EJ. Recent accomplishments in wound healing. Int Wound J 2015;12:4–9.
[57] Smiell JM, et al. Efficacy and safety of becaplermin (recombinant human platelet-derived growth factor-BB) in patients with nonhealing, lower extremity diabetic ulcers: a combined analysis of four randomized studies. Wound Repair Regen 1999;7:335–46.
[58] Yao F, Eriksson E. Gene therapy in wound repair and regeneration. Wound Repair Regen 2000;8:443–51.
[59] Marti G, et al. Electroporative transfection with KGF-1 DNA improves wound healing in a diabetic mouse model. Gene Ther 2004;11:1780–5.
[60] Lin MP, et al. Delivery of plasmid DNA expression vector for keratinocyte growth factor-1 using electroporation to improve cutaneous wound healing in a septic rat model. Wound Repair Regen 2006;14:618–24.

[61] Kos S, et al. Gene electrotransfer into skin using noninvasive multi-electrode array for vaccination and wound healing. Bioelectrochemistry 2017;114:33–41.
[62] Rezende FC, et al. Electroporation of vascular endothelial growth factor gene in a unipedicle transverse rectus abdominis myocutaneous flap reduces necrosis. Ann Plast Surg 2010;64:242–6.
[63] Ferraro B, Cruz YL, Coppola D, Heller R. Intradermal delivery of plasmid VEGF(165) by electroporation promotes wound healing. Mol Ther 2009;17:651–7.
[64] Basu G, et al. Prevention of distal flap necrosis in a rat random skin flap model by gene electro transfer delivering VEGF(165) plasmid. J Gene Med 2014;16:55–65.
[65] Fujihara Y, Koyama H, Nishiyama N, Eguchi T, Takato T. Gene transfer of bFGF to recipient bed improves survival of ischemic skin flap. Br J Plast Surg 2005;58:511–7.
[66] Ferraro B, Cruz YL, Baldwin M, Coppola D, Heller R. Increased perfusion and angiogenesis in a hindlimb ischemia model with plasmid FGF-2 delivered by noninvasive electroporation. Gene Ther 2010;17:763–9.
[67] Liu L, et al. Age-dependent impairment of HIF-1alpha expression in diabetic mice: correction with electroporation-facilitated gene therapy increases wound healing, angiogenesis, and circulating angiogenic cells. J Cell Physiol 2008;217:319–27.
[68] Steinstraesser L, et al. Skin electroporation of a plasmid encoding hCAP-18/LL-37 host defense peptide promotes wound healing. Mol Ther 2014;22:734–42.
[69] Lee PY, Chesnoy S, Huang L. Electroporatic delivery of TGF-??1 gene works synergistically with electric therapy to enhance diabetic wound healing in db/db mice. J Investig Dermatol 2004;123:791–8.
[70] Zhao M, et al. Electrical signals control wound healing through phosphatidylinositol-3-OH kinase-gamma and PTEN. Nature 2006;442:457–60.
[71] Kloth LC. Electrical stimulation technologies for wound healing. Adv Wound Care 2014;3:81–90.
[72] Koel G, Houghton PE. Electrostimulation: current status, strength of evidence guidelines, and meta-analysis. Adv Wound Care (New Rochelle) 2014;3:118–26.
[73] Athanasiou KA, Eswaramoorthy R, Hadidi P, Hu JC. Self-organization and the self-assembling process in tissue engineering. Annu Rev Biomed Eng 2013;15:115–36.
[74] Madi M, Rols M-P, Gibot L. Efficient in vitro electropermeabilization of reconstructed human dermal tissue. J Membr Biol 2015;248:903–8.

CHAPTER 10

Deep tissue pressure injury: a clinical perspective regarding a condition that evolves under the skin

Joyce Marie Black[1], Christine Thies Berke[2]

[1]College of Nursing, University of Nebraska Medical Center, Omaha, NE, United States; [2]Wound Ostomy Services, Nebraska Medicine, Omaha, NE, United States

Chapter Outline

Pressure ulcer or pressure injury (PI) has plagued man for centuries. It was long assumed that the force of pressure applied to the skin led to death of the skin cells. Pressure ulcers were understood to be progressive, that is, once the skin was injured, deep tissues and structures also could be injured over time. Therefore, a stage 1 became a stage 2 and so on. However, this understanding did not explain skin and soft tissue wounds that looked like bruises and became full-thickness injury in a matter of days. Today, clinicians are aware of a type of pressure ulcer/injury that does not begin on the skin, but rather the muscle and soft tissues, that is, deep tissue pressure injury (DTPI). A qualitative review of early cases showed that the patient was confined to a surface (e.g., hard floor, operating room [OR] table) or had a significant change in physical condition (e.g., sepsis, hypotension) around 48 h before initial presentation of purple skin. The findings also showed that blistering of the purple skin occurred in another 24–48 h. The cases below show this pattern:

> *A 50-year-old male was involved in a car crash on March 1. He was initially taken to a local hospital for emergency care, intubated, sedated and send by air ambulance to a*

Innovations and Emerging Technologies in Wound Care. https://doi.org/10.1016/B978-0-12-815028-3.00010-9

major medical center due to hemorrhage. The day after admission, he was desaturating with turning and hypotensive requiring vasoactive drugs. Seventy-two hours after the accident, the skin of the buttocks was maroon and deep purple with superficial skin loss.

A 70-year-old female was admitted with septic shock from urosepsis and renal failure on June 15th. She was intubated the next day and remained hypotensive requiring 2 vasoactive medications. Stage I pressure ulcers were present on the buttocks on June 17th and 2 purple tissue/DTI were documented later the same day.

An obese 72-year-old female had a 4 hour operation on September 5th and had 2 hypotensive episodes during the case for which she was transfused. She continued to be hypotensive in the post anesthesia care unit and received a fluid bolus. She was turned twice during the evening and nighttime hours. Twenty four hours after surgery, erythematous areas of the right and left buttock were noted and "skin tears" were noted the next day.

Is DTPI a new condition? A review of the older literature offers proof that clinicians have known about these wounds for many years, but they were not included in the early PI staging systems. In the 19th century, Jean-Martin Charcot described a rosy, dark red, or violet color in the skin from which the epidermis tears off and leaves bright red open wounds. Eschar follows. When the tissue was examined after death, he noted that the muscle was involved and the wound was extensive. These ulcers were called "decubitus ominosus" and were found in persons with acute brain and spinal injury and preceded death [1]. During the same period, Paget wrote that ulcers could erupt from intact skin and they may be purple or yellow from the extravasation of blood. The deeper tissues die, muscle, and the sloughing that follows of the skin and fat leaves an empty place [2]. In 1975, Shea developed a staging system for pressure ulcers that included closed pressure ulcers. He defined these ulcers as ones with subcutaneous necrosis without skin ulceration. The most commonly used staging system initially did not include any language for DTPI. A four-stage system was developed by the International Association of EnteroStomal Therapists in 1988 and by the National Pressure Ulcer Advisory Panel in (NPUAP) 1989. Deep tissue injury (DTI) was added to the staging system in 2007 by NPUAP (Black) and revised in 2016 (Edsberg).

The current definition of DTPI is as follows: Intact or non-intact skin with localized area of persistent non-blanchable deep red, maroon, purple discoloration or epidermal separation revealing a dark wound bed or blood filled blister. Pain and temperature change often precedes skin color changes. Discoloration may appear differently in darkly pigmented skin. This injury results from intense and/or prolonged pressure and shear forces at the bone–muscle interface. The wound may evolve rapidly to reveal the actual extent of tissue injury or may resolve without tissue loss. If necrotic tissue, subcutaneous tissue, granulation tissue, fascia, muscle, or other underlying structures are visible, this indicates a full-thickness PI (unstageable, stage 3, or stage 4). Do not use DTPI to describe vascular, traumatic, neuropathic, or dermatologic conditions ([3]).

Mechanisms of DTPI studies on the formation and pathogenesis of DTI have been well explained by Oomens and colleagues. This section is a summary of their report; the reader is referred to the paper for more depth. To understand how DTPI develops quickly and evolves quickly, a series of experiments were designed to examine cells placed under strain. When tissue is deformed by slow and moderate pressure, blood vessels are occluded and the tissues become ischemic and acidotic and eventually die. When the intensity of the pressure is increased, very quickly direct damage to the cell membrane occurs.

A number of authors have focused on determining an injury threshold specific to muscle tissue in relation to DTPI development [4–6]. Linder-Ganz et al. reported that cell death occurred in animal model muscles subjected to 260–525 mmHg for 15 min–1 h or in muscles with 68 mmHg applied pressure for more than 2 h, indicating that magnitude of pressure is an important factor in muscle cell death. A sigmoid pressure–time curve is useful to outline the finite failure strength of muscle tissue at short times and load levels that muscle tissue can bear over long periods [7].

Shear within the layers of tissue also develops and contributes to tissue damage. Muscle cells are the most sensitive tissue to the effect of shear due to their arrangement in bundles and most sensitive to ischemia due to their high metabolic rate (Gefen, Gefen). Recently, the impact of DTI in adipose tissue has also been described [8].

The role of ongoing tissue damage from reperfusion injury following an initial ischemic injury is well known in the myocardium and brain. Reperfusion of blood causes cellular edema, tissue damage, and overproduction of reactive oxygen species, which may injure the endoplasmic reticulum (ER). An inflammatory response is incited, and a mass of inflammatory mediators is released in the ischemic tissue, which, in turn, results in the overproduction of cytotoxic oxygen–derived free radicals (Jiang). Reperfusion injury also occurs in DTI and should be considered a significant factor in the etiology. Cui and colleagues found that in an animal model, there is change in the pressure-injured cell's ER and an abnormal activation of Akt kinase pathway during injury and reperfusion.

Overall, this evidence suggests that DTI, as a result of mechanical loading, occurs because of a combination of ischemia, lymphatic blockage, reperfusion, and cell deformation [9]. It is important that clinicians are aware of the etiology of DTI, to understand that a visual assessment is not a reliable indicator of early detection [10]. Regular palpation of the skin may enable identification of early changes in turgor.

Further research focused on the sequence of events in the development of DTI to include specific patient comorbidities may assist clinicians to initiate earlier preventative strategies.

1. Natural history

The earliest reports of the natural history of DTPI were done by Black and Berke (2007). This retrospective review of medical records discovered a pattern of changes in skin appearance over time. The event of intense pressure occurred 48 h before the skin changed color to purple or maroon. Their work guided in the initial definitions of DTPI by NPUAP. A later paper by Honaker confirmed the time frame, noting an average of 2.14 days from the inciting pressure event to the change in skin color. In 2008, Baharestani examined 200 cases of DTPI and reported that 72 h after admission to the hospital, the purple area sloughed the epidermal like a blister but the wound bed remained purple. At 7 days, the wound was often necrotic and a full-thickness PI. Sullivan reported that no signs of resolution were evident before the fourth day.

The natural history of DTPI has changed as clinicians have become increasing aware of the condition. Early reports of DTPI include a rapid evolution of the wound to stable eschar on the heels or unstageable wounds on the coccyx, sacrum, buttocks, or ischial tuberosities. Osteomyelitis and infection were common. The most recent report of natural history, now influenced by more aggressive off-loading, silicone dressings, and noncontact low-frequency ultrasonic mist, has reduced the severity of the ulcers. Sullivan reported on 77 cases of DTPI and reported that only 9.3% of the DTPI became full thickness.

2. Prevention

Prevention of DTPI begins with a thorough understanding of the profile of risk in the patient population being served. Obviously, there is no prevention for patients who develop DTPI outside of the hospital. Spinal immobilization, poorly padded emergency department (ED) stretchers, and restricted body positioning on narrow stretchers are pressure ulcer risks unique to the ED. Backboards were designed as an extrication tool, yet reports indicate that patients remain on these boards for extensive periods of time while being transported to the hospital and while in the ED. General pressure ulcer prevention can be done in the ED and should include turning and pressure reduction surfaces. A high priority should be placed on removing patients from backboards as soon as possible. Spinal precautions can be maintained without a backboard by using a firm mattress, cervical collar, and logrolling. Applying multilayer foam dressings while in ED can also significantly reduce pressure ulcer rates (Santamaria). DTPI occurs in the OR also. The duration of surgery is the main predictor of PI in the OR; the location of the PI is dependent on the position for surgery. PI occurs on the buttocks, heels, and occiput in patients who are supine and on the face and chest in patients who are prone. Prevention in the OR centers on additional padding placed on high-risk body areas either with dressings or mattresses (Lupe, Engels, Brindle). Interface pressure and immersion have been measured on various OR table mattresses. Fluid immersion surfaces and air-inflated static seat cushions provided lower measured peak interface pressures in the sacral area. A surface with the properties of a lower measured peak interface pressure, while increasing the area of skin contact over a larger surface area, is required for minimizing risk for pressure ulcer development, which was best by the air-inflated seat cushion (Kirkland Walsh).

3. Risk for deep tissue injury

There is no predictive risk assessment tool for DTPI. However, there are unique aspects of their development that must be considered. First, the pressure has to be intense enough to injure the muscle cells. This intense pressure need not be of long duration; however, the exact parameters of time to injury have not been described. In the initial case series by Black and Berke, patients were seen to have had had a period of "confinement" to a hard surface such as the floor, a transport gurney/backboard, and the OR table. Patients at risk for DTPI will often have such exposure to pressure in the prior 48 h before the skin color changes. DTPI can begin before admission. Patients with a history of being "found down" often have no idea of how long they were on the ground before someone found them. Other patients self-medicate at home hoping to "feel better soon" but in the interim do nothing more than sit in a recliner chair. These long periods of immobility equal pressure and can create the DTI at the bone–muscle interface.

During an acute care stay, the risk of DTPI is highest when the patient is exposed to intense pressure, such as during surgery or interventional radiology. A study of DTPI in ICU concluded that for every hour in surgery, the odds of developing DTPI increased by about 20% [11].

When the surface is not hard, such as a standard hospital bed, ischemia likely plays a greater role in the development of DTPI, and the period of exposure to pressure is longer. DTPI can develop during critically ill periods when the patient is in shock or being dialyzed and is not moved or cannot be moved. A study of DTPI in ICU concluded that for every mmHg decrease in diastolic pressure, the odds of developing DTPI increased by about 7.5% [11]. There are many comorbid conditions that increase risk for DTPI and vary by the type of patient being studied. In acute care, they include peripheral artery disease, respiratory disease, and diabetes [12].

In long-term care (aged care) settings, residents who have mobility impairments and neurogenerative diseases such as multiple sclerosis were found to have increased risk. DTPI was also associated with malnutrition, dehydration, anemia, infection, and incontinence in this group [13]. There is also reported risk from sitting on unpadded toilet seats [14].

4. Early detection of deep tissue pressure injury

Pressure ulcer staging is primarily dependent on visible changes of the skin and soft tissues. Misclassification of PI is a potential risk until redness resolves or necrosis occurs. Attempts have been made to improve the assessment of deeper tissues suspected of injury from pressure and shear forces. Identifying the presence of deep tissue damage/injury more quickly helps the clinician use prevention and treatment resources quicker and more appropriately to potentially mitigate progression of the damage/injury as well as hasten recovery and healing.

Farid et al. [15] examined 85 individuals who presented with pressure-related intact discolored areas of skin (PRIDAS) with a handheld infrared thermographic device to identify temperature differences between the center portion of PRIDAS and adjacent tissues (5–10 cm). PRIDAS with cooler skin center temperatures and absence of capillary refill (negative blanching) were more likely to develop necrosis than those PRIDAS whose center temperatures were warmer and had positive capillary refill (positive blanching). Cox et al. conducted a similar study in seven skilled nursing facilities (long-term care units) and found similar results; patients with cooler skin temperatures at the center of discolored skin areas with negative blanching were more likely to progress to necrotic wounds. Early detection could help identify those PRIDAS at greater risk for progression/necrosis of the underlying discolored tissues and escalated early treatment.

Ultrasound (US) has also been proposed as an assessment tool to identify early signs of internal pressure–related tissue changes before visual external skin changes related to pressure/shear. Akins et al. [16] showed that US used to detect risk factors of patients with spinal cord injuries (tissue thickness measurements, flattening of the ischial tuberosities) were as effective as MRI-preferred detection of anatomical changes in persons with prolonged immobilization and decrease sensory perception. US is an excellent imaging model to evaluate the musculoskeletal system [17]. US can detect soft tissue changes and damage within 24 h; changes appear as hyperechoic (white dense areas, indicate early destruction, rigor mortis) rather than hypoechoic (translucent dark areas) changes seen later after the tissue decomposes and liquefies [18]. Gefen et al. [17] discuss the presence of rigor mortis in early tissue destruction, which causes palpable thickening that can last for up to 36–72 h after injury. He called this "stiffening" in comparison with the surrounding normal muscle tissue. Schenier et al. [18] found that

35% of their 23 study participants with a Braden score of 16 or less developed pressure-related skin necrosis that was detected by US (13 body sites per patient) on admission to the ED; they evaluated both deep (bone level) and shallow (subcutaneous) tissue levels with the US. They noted evolution of US + DTI at bone level took 6–7 days for skin changes and US + subcutaneous DTI took 2 days for skin manifestations to become visibly apparent.

DTI involving muscle destruction contributes to rhabdomyolysis that can be chemically detected at specific periods after the muscle destruction has occurred. Creatine kinase (CK) is an enzyme specific to muscle tissue. An elevation of CK in the absence of changes in troponins can be a sign of skeletal muscle damage (Gefen, 2016). CK blood levels start to rise within 12 h of muscle destruction and then peak within the next 12 h to 3 days, declining after the third day. Knowing the history of the patient (impaired mobility, sitting for prolonged periods of time or unable to stand, unwitnessed fall/found down/unconscious) may guide ordering laboratory tests to determine presence of deeper muscle tissue damage that is not visible on admission to the ER/acute care unit. Vigilance of skin exams can be ordered if CK tests are elevated watching for visible signs of DTI.

A scanner used to detect changes in subepidermal moisture (SEM) levels (a purported biophysical marker of damaged tissue) has recently been developed, evaluated, and launched in the United Kingdom. It has shown promising results in the United Kingdom for early evaluation and identification of at-risk skin areas for DTI. The scanner uses sensor technology to measure tissue electrical capacitance through the application of low amplitude signals from electrode structures placed on the patient's skin; these values indicate changes to SEM. This could potentially be another bed side diagnostic tool for detection and monitoring of chronic preventable conditions [19].

5. Differential diagnosis

There are many conditions that cause skin to become purple and rapidly develop gangrene. It is important to distinguish those problems from DTPI. DPTI will appear in a patient with a history of pressure and or shear in the body part with the wound. The DTPI is palpable and painful (if the patient is able to sense pain). The differential diagnoses for DTPI are presented in Fig. 10.1; it is beyond the scope of this chapter to discuss all of these conditions.

6. Treatment

The treatment of patients at high risk for development of DTPI must include off-loading the area. Patients who have been in the OR should be positioned in a different position than was used for surgery. Patients who have been in dialysis should be positioned on the

Differential Diagnoses of Deep Tissue Pressure Injury			
Arterial Inflow Diseases	*Venous Outflow Diseases*	*Inflammatory*	*Traumatic*
Arterial Insufficiency	Heart failure	Calciphylaxis	Bruises
Distal Embolization	Venous Congestion	Echthyma Gangrenosum	Blood blister
Gluteal compartment syndrome		Necrotizing fasciitis	Ecchymosis
Use of injected heroin or cocaine		Pyoderma Gangrenosum	Hematoma
Vasopressor agents for hypotension		Toxic epidermal necrosis	Morell Lavelle Lesion
		Vasculitis	Skin Tear
		Warfarin necrosis	

(A)

(B)

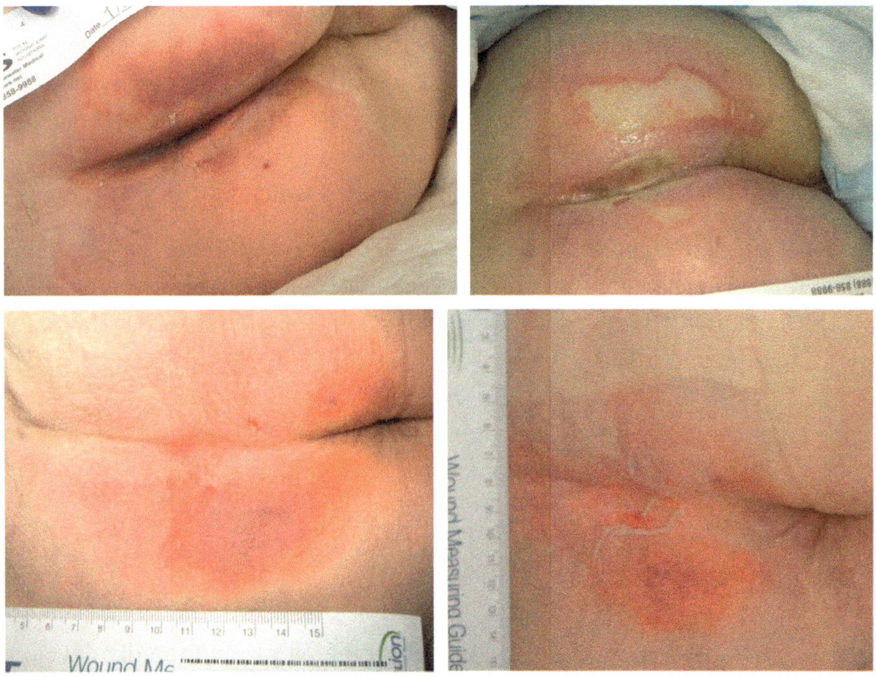

Figure 10.1
(A) Deep tissue pressure injury not treated with MIST. (B) Deep Tissue Pressure Injury Treated with MIST.

side for recovery. Removing pressure from the damaged tissue will aid in its recover through the restoration of blood flow.

Once the DPTI has been identified, soft multilayer silicone foam dressings (Sullivan) and air-fluidized beds (Allen) have been shown to reduce the extent of the DTPI. Both these therapies likely reduce the severity of the injury by reducing pressure and shear in the DTPI. In a longitudinal prospective historical case–control study (N = 60), Honaker showed that the addition of noncontact low-frequency US, within 5 days of onset, reduced the severity of DTPI, its total surface area, and its final PI stage. The surface area of the DTPI in the treatment group decreased by a mean of 8.8 cm^2 compared with a mean decrease in the control group of only 0.3 cm^2. In this study, 57% of the DPTI treated with standard care became unstageable compared with 3% of the group treated with Non Contact Low Frequency Ultrasound (NCLFUS).

7. Outcomes

Outcomes of DTPI can be devastating if the injury is not detected early and not treated aggressively. In the case series by Black and Berke, few of the DTIs healed (10%), and healing versus nonhealing was correlated to the age of the patient ($P < 0.05$) and relative immobility ($P < 0.05$). Debridements were done in 55% of the cases, but few (5%) underwent closure with a flap. Sepsis from the DTI occurred in 28% of patients, and 21% of the patients died as a result of the DTI. Diabetes and peripheral vascular disease were correlated with sepsis ($P < 0.007$ and 0.000 respectively). In a multivariate model, DTI of the sacrum increased the risk of sepsis ($P < 0.003$). DTI of the heel required amputation in 13%, and amputation was more likely in patients with diabetes ($P < 0.000$) and peripheral vascular disease ($P < 0.05$). Children appear to resolve more rapidly, and with less tissue loss than in adults, the reason for this difference is not understood.

References

[1] Levine J. Historical perspective on pressure ulcers: the decubitus ominosus of Jean_Martin Charcot. J Amer Geri Soc 2005;53(7):1248–51.

[2] Cox J, Kaes L, Martinez M, Moles D. A prospective, observational study to assess the use of thermography to predict progression of discolored intact skin to necrosis among patients in skilled nursing facilities. Ostomy Wound Manag 2016;62(10):14–33.

[3] Edsberg LE, Black JM, Goldberg M, McNichol L, Moore L, Sieggreen M. Revised national pressure ulcer advisory panel pressure injury staging system: revised pressure injury staging system. J Wound Ostomy Continence Nurs 2016 Nov/Dec;43(6):585–97.

[4] Gefen A, van Nierop B, Bader DL, Oomens CW. Strain-time cell death threshold for skeletal muscle in a tissue-engineered model system for deep tissue injury. J Biomech 2008;41(9):2003–12.

[5] Linder-Ganz E, Engelberg S, Scheinowitz M, Gefen A. Pressure-time cell death threshold for albino rat skeletal muscles as related to pressure sore biomechanics. J Biomech 2006;39(14):2725–32.

[6] Stekelenburg A, Strijkers GJ, Parusel H, Bader DL, Nicolay K, Oomens CW. Role of ischemia and deformation in the onset of compression-induced deep tissue injury: MRI-based studies in a rat model. J Appl Physiol 2007;102(5):2002–11.

[7] Leopold E, Gefen A. A simple stochastic model to explain the sigmoid nature of the strain-time cellular tolerance curve. J Tissue Viability 2012;21(1):27–36.

[8] Edsberg L, Call E, Black J. Histology of deep tissue pressure injury: what can be learned from cadavers. In: National pressure ulcer advisory panel conference, Las Vegas, NV; 2018.

[9] Peart J. The aetiology of deep tissue injury: a literature review. Brit J Nurs 2016;25(15):840–3.

[10] Berlowitz D, Brienza D. Are all pressure ulcers the result of deep tissue injury? A review of the literature. Ostomy Wound Manage 2007;53(10):34–8.
[11] Kirkland-Kyhn H, Teleten O, Wilson M. A retrospective, descriptive, comparative study to identify patient variables that contribute to the development of deep tissue injury among patients in intensive care units. Ostomy Wound Manage 2017;63(2):42–7.
[12] Richbourg L, Smith J, Dunzweiler S. Suspected deep tissue injury evaluated by North Carolina WOC nurses. J Wound Ostomy Continence Nurs 2011;38(6):644–60.
[13] Ahn H, Cowan L, Garvan C, Lyon D, Stechmiller J. Risk factors for pressure ulcers including suspected deep tissue injury in nursing home facility residents: analysis of national minimum data set 3.0. Adv Skin Wound Care 2016;29(4):178.
[14] Lustig M, Levy A, Kopplin K, Ovadia-Blechman Z, Gefen A. Beware of the toilet. The risk for deep tissue injury during toilet sittings. J Tissue Viability 2018;27:23–31.
[15] Farid KJ, Winkleman C, Rizkala A, Jones K. Using temperature of pressure-related intact discolored areas of skin to detect deep tissue injury; an observational, retrospective, correlational study. Ostomy Wound Manag 2012;58(8):20–1.
[16] Akins JS, Vallely JJ, Karg PE, Kopplin K, Gefen A, Poorjary-Mazzotta P, Brienza DM. Feasibility of freehand ultrasound to measure anatomical features associated with deep tissue injury risk. Med Eng Phys 2016;38:839–44.
[17] Gefen A, Farid KJ, Shaywitz I. A review of deep tissue injury development, detection, and prevention: shear savvy. Ostomy Wound Manag 2013;59(2):26–35.
[18] Schenier J, Farid K, Raden M, Demisse S. Ultrasound to detect pressure-related deep tissue injuries in adults admitted via the emergency department: a prospective, descriptive, pilot study. Ostomy/Wound Manag 2017;63(3):36–46.
[19] Lester R. Real world evidence evaluating a novel early-detection device for HAPU reduction. In: eAbstract K2.3 presented to 19th EPUAP 2017 annual meeting, Belfast, Norther Ireland; September 20–22, 2017.

Further reading

[1] Allen L, McGarrah B, Barrett D, Stenson B, Turpin PG, Vangilder C. Air-fluidized therapy in patients with suspected deep tissue injury: a case series. J Wound Ostomy Cont Nurs 2012;39:555–61.
[2] Baharestani M. Case series of deep tissue injury. In: Podium presentation. European pressure ulcer advisory meeting, Berlin, Germany; August 28, 2007.
[3] Berke CT, Black J. Retrospective study to identify a risk profile for pressure related deep tissue injury. In: Poster presentation. National Wound Ostomy Continence Nurse Conference, Salt Lake City, Utah; June 9–13, 2007.
[4] Black JM, Baharestani MM, Cuddigan J, Dorner B, Edsberg L, Langemo D, Posthauer ME, Ratliff C, Taler G. National pressure ulcer advisory panel. National Pressure Ulcer Advisory Panel's updated pressure ulcer staging system Adv Skin Wound Care May 2007;20(5):269–74.
[5] Brindle CT, Wegelin JA. Prophylactic dressing application to reduce pressure ulcer formation in cardiac surgery patients. J Wound Ostomy Continence Nurs March–April 2012;39(2):133–42.
[6] Cui F, Pan Y, Xie H, Wang X, Shi H, Xiao J, Zhang H, Chang H, Jiang L. Pressure combined with ischemia/reperfusion injury induces deep tissue injury via endoplasmic reticulum stress in a rat pressure ulcer model. Int J Mol Sci March 2016;17(3):284.
[7] Engels D, Austin M, McNichol L, Fencl J, Gupta S, Kazi H. Pressure ulcers: factors contributing to their development in the OR. AORN J 2016;103(3):271–81.
[8] Gefen A, Gefen N, Linder-Ganz R, Margulies S. In vivo muscle stiffing under bone compression promotes deep pressure sores. J Biomech Eng 2005;127(3):512–26.
[9] Honaker J, Brockopp D, Moe K. Suspected deep tissue injury profile: a pilot study. Adv Skin Wound Care 2014;27(3):133–40.

[10] Honaker JS, Forston MR, Davis EA, Weisner MM, Morgan JA, Sacca E. The effect of adjunctive noncontact low frequency ultrasound on deep tissue pressure injury. Wound Repair Regen November 2016;24(6):1081–8.

[11] Kirkland-Walsh H, Teleten O, Wilson M, Raingruger B. Pressure mapping comparison of four OR surfaces. AORN J 2015;102(1):61.e1–9.

[12] Lupe L, Zambrana D, Cooper L. Prevention of hospital-acquired pressure ulcers in the operating room and beyond: a successful monitoring and intervention strategy program. Inter Anes Clinics 2013;51(1):128–46.

[13] Jiang L, Tu Q, Wang Y, Zhang E. Ischemia-reperfusion injury-induced histological changes affecting early stage pressure ulcer development in a rat model. Ostomy Wound Manage 2011;57(2):55–60.

[14] Naccarato MK, Kelechi T. Pressure ulcer prevention in the emergency department. Adv Emer Nurs J 2011;33(2):155–62.

[15] Paget J. Clinical lecture on bedsores. Students J Hosp Gazette (London) 1873;1:144–6.

[16] Preston A, Rao A, Strauss R, Stamm R, Zalman D. Deep tissue pressure injury: a clinical review. Amer J Nurs 2016;117(5):50–7.

[17] Shea JD. Pressure sores: classification and management. Clin Orthop Relat Res 1975;Oct(112):89–100.

[18] Sullivan RA. Two-year retrospective review of suspected deep tissue injury evolution in adult acute care patients. Ostomy Wound Manage 2013;59(9):30–9.

[19] Sullivan R. Use of a soft silicone foam dressing to change the trajectory of destruction associated with suspected deep tissue pressure ulcers. Medsurg Nurs J 2015;24:237–42. 267.

[20] Wounds UK Consensus Document. Recognizing, managing and preventing deep tissue injury (DTI). London: Wounds UK; 2017. Available from: www.wounds-uk.com.

CHAPTER 11

Skin health and integrity

Jan Kottner[1,2], Dimitri Beeckman[2], Annika Vogt[1], Ulrike Blume-Peytavi[1]

[1]Charité-Universitätsmedizin Berlin, Department of Dermatology and Allergy, Clinical Research Centre for Hair and Skin Science, Berlin, Germany; [2]Ghent University, Faculty of Medicine and Health Sciences, Department of Public Health and Primary Care, University Centre for Nursing and Midwifery, Skin Integrity Research Group, Gent, Belgium

Chapter Outline

1. Introduction

The skin is widely regarded as the largest organ of the human body. Its main function is to protect from various external insults and to maintain the internal homeostasis and ultimately survival. This includes thermoregulation, innate and adaptive immune functions, sensory perception, vitamin D production, and many more [1]. In addition, the outer appearance and the skin sensations are important for well-being, self-esteem, attractiveness, and communication [2,3]. The skin is optimally designed to fulfill all these functions and tasks. However, there may be periods of enhanced skin vulnerability in life, which render individuals more prone to the development of "skin problems," e.g., very early in life when the skin is not fully mature, when having dermatological or other systemic diseases, or at advanced age and at the end of life. Personal hygiene behaviors such as frequent bathing, leisure activities including extended UV exposure, or unfavorable

Innovations and Emerging Technologies in Wound Care. https://doi.org/10.1016/B978-0-12-815028-3.00011-0

occupational environments such as wet work may compromise the skin as well [4]. Signs and symptoms of impaired skin integrity may include unpleasant sensory feelings, inflammation, cutaneous lesions, or loss of cutaneous substance (wounds). This may not only result in reduced quality of life, itching, pain, disfigurement, and disability but also pose secondary risks including allergies and secondary infection. Overall, the global burden of skin diseases is increasing with increasing age [5]. Thus, the older we get, the more important skin problems become. This statement is supported by the high prevalence of self-reported skin concerns in the elderly [6,7].

The overall aim of this chapter is to define the concept of skin integrity and skin health and to discuss general risks and ways to maintain or promote skin integrity, especially in the context of chronic diseases, old age, and care dependency.

2. The skin structure and function

The skin is a complex structure that hosts various subcompartments and cell populations that interact in many different ways to build and maintain the barrier. The two main layers, namely the epidermis and the dermis, are interconnected via the dermal–epidermal junction lay on the subcutaneous fat tissue.

2.1 The epidermis

The epidermis is interrupted by hair follicles and sweat glands, except at the palms and soles where no hair follicles are present. At most hairy skin areas, the epidermis is approximately 70 μm thick [8,9] and it is much thicker at the heels and palms. The lowermost part (stratum basale) consists of epidermal stem cells that constantly proliferate. Daughter cells (keratinocytes) are produced and differentiate via their journey to the upper epidermal layers. The result of this differentiation is so-called nucleus-free corneocytes, i.e., flat platelet-like protein structures embedded in a lipid matrix forming the stratum corneum. The turnover time of epidermal cells is estimated to be approximately 28 days, but it is much longer in aged skin [10]. The stratum corneum forms the outermost part of the skin and is in direct contact to the environment. It not only forms an extremely resistant and strong barrier against external insults but also limits the diffusion of water molecules from the highly hydrated keratinocytes to the outside [11]. Natural moisturizing factors such as amino acids, lactate, and urea within the stratum corneum are by-products of the differentiation process and help to keep water in this layer, rendering the skin pliable and allowing enzymatic processes.

Within the stratum corneum, there is a gradient of the pH from approximately 7 at the stratum granulosum to 5 at the skin surface. This "acid mantle" facilitates the growth of the residual skin flora, hereby inhibiting pathologic microbial colonization. Because many enzymes involved in epidermal differentiation are pH dependent, pH also exerts direct effects on the stratum corneum cohesion, desquamation, and permeability homeostasis [12].

2.2 The dermis

Depending on the skin area, the dermis is on average 1–4 mm thick [13]. The upper layer, the stratum papillare, is a loose structure of fine connective tissue fibers that fill the papillary dermis and supplies the epidermis by means of capillary loops The density of the dermal papillae decreases with age, which results in a flattening of the dermal epidermal junction [14]. The lower dermal layer, the stratum reticulare, is a dense fibrous structure containing coarse collagen fibers forming a three-dimensional network providing the tensile cutaneous strength.

The dermis is loosely connected to the hypodermis. It consists mostly of fatty tissue that provides thermal and mechanical protection. Depending on the skin area and the individual anatomy, the subcutaneous layer may be very thick (e.g., at waist) or it may be almost absent (e.g., bridge of the nose). This presence and thickness of this layer seems to be related to the intrinsic pressure ulcer risk [15].

3. What is skin integrity?

Although the concept of "skin integrity" is widely used in many different areas and health-care contexts [16,17], a formal definition is lacking so far. In medicine, a disease is often defined as the deviation from the normal, i.e., healthy state. Based on observed signs, symptoms, and/or the results of diagnostic tests, a medical diagnosis is made, which guides the treatment strategy [18]. Especially, in dermatology, the visual examination of the skin and the description of cutaneous lesions play a key role for diagnosing skin diseases [19]. Therefore, from a medical/dermatological perspective, one may define skin integrity with absence of any erythema, visible scaling, or distinct cutaneous lesions.

The NANDA International nursing diagnosis classification contains two skin integrity related diagnoses. "Impaired skin integrity" is defined as "altered epidermis and/or dermis" and "risk for impaired skin integrity" is defined as "susceptible to alteration in epidermis and/or dermis, which may compromise health" [20]. Similar to the medical perspective, skin integrity is here defined as an alteration from the normal.

However, this conceptual approach may be too simplistic. Just observing a macule, a mole, a mild erythema, or fine scales does not necessarily justify a dermatological or nursing diagnosis. A large variety of dermatological diagnoses do not have immediate clinical relevance for the skin integrity. For instance, androgenetic alopecia may be cosmetically disturbing and possibly affecting psychological well-being, but despite hair loss, the scalp skin is intact. Another example is skin aging: compared with young adults, the epidermis and dermis of aged subjects are "altered" in many ways [21,22]. Does this mean that all aged subjects have an "Impaired skin integrity" according to the NANDA International?

On the other hand, there are "skin conditions" that have no medical diagnoses but are practically relevant for subjects. For instance, the so-called "skin types" such as oily or dry are important for personal skin care behavior and the susceptibility to develop skin diseases [23]. "Sensitive skin" is another clinically relevant category. Subjects, who declare themselves to have sensitive skin, describe symptoms such as irritable skin leading to tightness, burning, or tingling sensations. There is a common agreement in the skin research community that such a concept exists, but despite sophisticated research methods, the actual underlying morphological and/physiological explanation in not entirely clear yet [24]. However, empirical evidence indicates that these symptoms occur in those subjects after exposure to irritants (such as cosmetic products), who have an underlying (subclinical) skin barrier impairment [25]. Thus, the concept of sensitive skin is rather a response pattern than a structural entity. A similar clinical phenomenon is irritant contact dermatitis (ICD). The pathophysiological pathways of ICD are described and a number of risk factors have been identified. However, the mere exposure to an irritant does not automatically lead to ICD. There seems to be a number of individual intrinsic characteristics that increase or decrease the susceptibility for ICD development [26].

These examples clearly indicate that morphological characteristics, functional features, and individual response patterns are tightly connected. Skin function is a key aspect of skin integrity. The skin constantly adapts to changing environmental conditions and stressors. For instance, after UV-induced damage, the melanogenesis is increased to protect the skin from further damage [27]. Disruption of the skin barrier, such as the exposure to irritants such as alkaline soap, damages and removes the protein and lipid structures of stratum corneum, resulting in an immediate release of preformed lipids and an increased synthesis of bilayers to repair the damage [28]. Repeated irritant exposure may also lead to a "hardening effect" that may be regarded as a protective response as well [29,30]. Similarly, repeated friction (e.g., due to walking barefoot) leads to increased proliferation of the stratum corneum protecting the foot. Inflammation per se (e.g., due to infection) is a basic physiological response of the skin to restore its integrity and function [31].

These examples indicate that the skin not only adapts to the environment but also constantly compensates risks. This characteristic, which may be referred to as "vitality" or "functional capacity," was originally conceptualized in the context of general aging [32], but it may also be applied to more specific areas such as the skin [33,34]. Vitality may be defined as the capacity to adapt to changing environmental conditions. The better the functional capacity of the skin, the better is its resistance to possible threats. Aging in general and skin aging in particular are associated with a decline in functional capacity [32,34,35]. An impressive example is the water barrier function of the skin barrier indicated by the transepidermal water loss (TEWL). By just comparing TEWL of adults and aged subjects, there are nearly no differences detectable. However, when the skin is stressed, differences in the functional response become obvious [36,37].

On the continuum from skin integrity to compromised skin, "skin failure" may be regarded as the endpoint of complete cutaneous dysfunction. This concept was first introduced into the literature in 1991 and defined as the inability to maintain skin functions and of the mechanical barrier to prevent the penetration of foreign materials [38]. Skin failure is the consequence of acute severe dermatological diseases (e.g., toxic epidermal necrolysis) leading to a number of life-threatening consequences such as hemodynamic instability, hypothermia, metabolic abnormalities, and critical fluid, electrolyte and nutrient loss [39,40]. Later, this concept was also used in the wound and pressure ulcer care community to describe tissue necrosis due to hypoperfusion [41,42].

Taking the aforementioned concepts and examples together, the following definition of skin integrity is proposed: skin integrity is the combination of an intact cutaneous structure and a functional capacity that is high enough to preserve it. This definition indicates that intact epidermal and dermal structures are only one aspect of skin integrity. In addition, the skin must be functionally competent to compensate internal and external hazards. At the same time, functional capacity is not a fixed property. It does change during the life course, during changing environmental conditions, or during diseases. The important question is whether the functional capacity is high enough to maintain the cutaneous structure.

4. Internal challenges to skin integrity

During the life course, there are two periods in which the functional capacity of the skin is low: very early and very late in life [33]. Because the skin of newborns and infants is not fully mature, it is more fragile and vulnerable to a wide range of internal and external insults [43]. Later in life, intrinsic skin aging causes a gradual accumulation of damage in the molecular structures that increases the risk for various skin conditions and diseases such as xerosis, infections, and injuries [44,45]. Skin growth, maturation, and aging are normal physiological processes.

In contrast, there are many pathological processes challenging the skin integrity. For instance, systemic diseases such as diabetes mellitus compromise the skin structure and function in many ways. Advanced glycation end products accumulate and dermal collagen is gradually being destroyed, which renders the skin stiffer, is less able to deform in response to mechanical loads, and is therefore more susceptible to deformation injury and wounding [46,47]. Because of the autonomic neuropathy, the skin becomes drier leading to xerosis, cracks, fissures, and diabetic foot ulcers [48]. Foot deformity and gait problems increase callus formation with subsequent risk for foot ulceration [49,50].

Other intrinsic characteristics such as obesity are associated with an impaired skin barrier and abnormal blood flow [51,52], leading to a wide spectrum of obesity-related skin diseases [53]. Systemic medication such as intake of diuretics is considered to cause skin dryness.

There are also regional risks for impaired skin integrity. For instance, venous stasis in the lower legs prevents diffusion of oxygen and nutrients, leading to stasis dermatitis, edema, hemosiderin deposits in the dermis, dryness, and stiffening and to a high risk for leg ulceration.

Finally, dermatological diseases per se directly affect the cutaneous structure and function in many ways, ranging from inflammatory skin diseases (e.g., atopic dermatitis, rosacea, acne) to tumors (e.g., actinic keratosis, melanoma). Associations between genetic and internal predisposition and the actual occurrence of skin conditions and diseases are emerging, but we are still at the beginning [43,54,55].

5. External challenges to skin integrity

External influences are well known to impair skin integrity. Chronic sun exposure initiates molecular processes leading to cutaneous damage and premature skin aging [56]. Clinical signs of photodamage vary considerably among individuals and include epidermal atrophy, epidermal thickening, irregular pigmentation, and deep wrinkles. Of particular clinical relevance are tumors such as actinic keratosis, basal cell carcinoma, and melanoma [34]. Chronic sun exposure is also considered a major contributor to the development of a process called dermatoporosis, which is characterized by extremely atrophic and fragile skin in advanced age with a high susceptibility to shear-type injuries and laceration [57]. Smoking and air pollution are additional well-recognized aggravation factors of impaired skin integrity and premature aging [56].

Repeated and prolonged exposure to water and other irritants such as surfactants, solvents, oils or urine and/stool damage the skin barrier in many ways: the skin surface pH is elevated; natural moisturizing factors are removed from the stratum corneum; and the lipid bilayers are disturbed [58]. If the functional (repair) capacity of the skin is exceeded, ICD may develop. This is not only caused by occupational exposure (e.g., hairdressers, construction workers, or health-care professionals) [43] but also due to excessive bathing and cleansing habits [4]. However, individual predisposition seems to be an important predictor as to whether ICD develops or not [26,59]. An impressive example is incontinence-associated dermatitis (IAD). Although the prevalence of incontinence and IAD in geriatric care is high, the majority of incontinent patients do not develop an IAD [60,61].

In the context of aging, illness, and care dependency, the skin is often exposed to sustained mechanical loads leading to skin and soft tissue deformation. This may damage cutaneous structures leading to pressure or diabetic foot ulcers [60,62].

6. How to promote skin integrity?

6.1 Internal measures

The internal vulnerability and functional capacity to maintain skin integrity seem to be highly variable and depend on not only known (e.g., chronological age, presence of a dermatological disease) but also a number of unknown factors. For instance, higher melanin content in the epidermis renders darker skin prototypes less susceptible to damages due to chronic sun exposure. Genetic, anatomical, hormonal variations of skin structure and function in general and even psychological factors and the daytime can increase or decrease the susceptibility to internal or external threads [34,43,63,64]. The risk for developing ICD is also highly individual [26,55]. Recently, the concept of "biomechanical risk" was introduced to characterize the intrinsic response of soft tissue deformation to extrinsic applied forces indicating pressure ulcer risk that varies substantially between individuals [65]. However, based on our current understanding it seems to be impossible to influence or treat any of these internal characteristics. Knowing and/or measuring internal predictors may help to estimate individual risks, but the overall level of therapeutic control is low.

On the other hand, there are interventions helping to promote skin integrity indirectly. This includes the maintenance of an adequate body weight, a balanced diet, nonsmoking, and the control of local (dermatological) and systemic diseases. The promotion of activity and mobility are important measures to maintain independency and to avoid confinement to bed. Medication-induced cutaneous side effects such as xerosis cutis caused by diuretics [66] must be reduced and other internal risk factors should be controlled to the greatest extend possible.

6.2 External measures

Compared with the intrinsic predisposition, there are much more opportunities to protect and to enhance skin integrity from the outside. The most important intervention is the avoidance of hazards. UV protection across the life course is one cornerstone [21]. This includes the limitation of exposure to UV radiation with avoidance of sunburns, wearing of UV-protective textiles, and the application of sunscreens [67,68]. Sunscreens are also recommended to protect against air pollution in combination with skin cleansing and caring [69]. Smoking should be avoided or stopped.

A major "modern" problem in western societies is the repeated and extensive exposure to various skin cleansing (e.g., water, soap) and care activities and products (e.g., hydrophilic emulsions, plant oils). Such an excessive exposure has a number of side effects such an increased skin surface pH and the destruction of corneocytes and lipid layers of the stratum corneum. This leads to irritation, dryness, inflammation, and eventually to ICD. Preservatives or fragrances may enhance irritation and increase the risk for allergic dermatitis especially when the skin integrity is impaired. The best method to avoid harm is the strict avoidance of any unnecessary skin care activities [17].

The avoidance of the irritant and/or the allergens is also crucial in occupational environments and care environments. For instance, the promotion of continence and/or effective ways to hold urine in continence products is the cornerstone of IAD prevention [60]. If the exposure to irritants is inevitable, the skin must be protected. This may include gloves or the application of the so-called barrier products to delay and/or to modify the irritant exposure. However, gloves and barrier products themselves may trigger ICD [43].

In case of (sub)clinical skin barrier impairments such as dry skin, mild and nonirritating lipophilic leave-on products should be used to increase the epidermal moisture content and to alleviate symptoms of dry skin. Although differences in product performance are well known and side effects occur [17,70], empirical evidence suggests that the regular application seems to be more important for dry skin treatment than the particular formulation [71,72].

If the skin is highly vulnerable as it occurs during severe illness or immobility, additional measures must be implemented to create a safe environment. This includes safe patient-handling techniques, special support surfaces, or low-friction textiles to maintain an optimal skin microclimate and to reduce pressure, shear, and friction [73].

7. Outlook

The concept of skin integrity is essential in nursing, medical and health-care practice, and research. It is not just an intact cutaneous structure but also the functional capacity that needs to be higher than the internal and external risks and hazards (Fig. 11.1). The lower the internal and external risks are, the better the self-repair capacity, and the more effective the preventive and corrective interventions are, the better the skin integrity. Thus, the framework of primary and secondary prevention is directly applicable to the promotion of skin health and integrity [21,74].

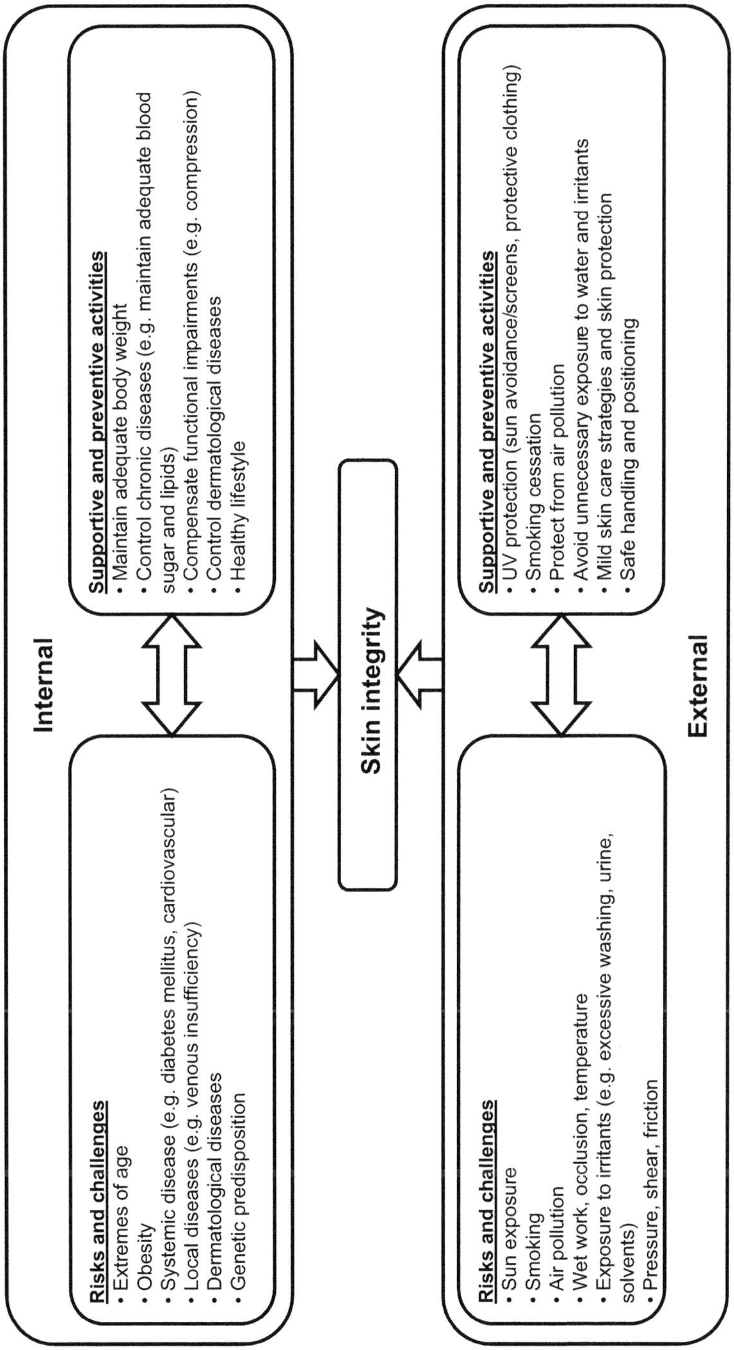

Figure 11.1
Conceptual framework of skin integrity.

During the last years, substantial advances have been made in many areas of dermatology, and skin research and innovative interventions to actively enhance skin integrity and/or to "reverse" skin aging have been proposed. For instance, topically applied vitamin C restructures the papillary dermis and increases the density of dermal papillae [14]. Topically applied retinoids and hyaluronate fragments have been shown to reverse dermatoporosis [57], and recently, the oral intake of bioactive collagen peptides to increase dermal collagen and elastin has been proposed [75]. External acidification of the skin seems to increase the cohesion of the stratum corneum and thus the barrier function [76,77]. However, whether these approaches will be actually implemented widely in clinical practice, whether patients will have a clinical relevant benefit, and whether this is cost-effective is unclear at the moment.

According to our current understanding, it seems that rather basic and "simple" strategies are very powerful to promote skin integrity [23,69,78]:

- Protect the skin from chronic sun exposure;
- Protect the skin from extreme temperatures and dryness (e.g., wear protective clothes when it is cold);
- Protect the skin from exposure to irritants and occupational hazards;
- Keep skin cleansing, showers, and baths short;
- Use mild and nonalkaline skin cleansers;
- Apply a lipophilic skin care leave-on product when the skin is dry;
- Don't smoke.

This chapter provided a broad and multiperspective overview about the concept of skin integrity. Therefore, many physiological and pathological details were omitted. However, the proposed definition and the relationships between the internal and external factors help to put this concept into a broad context. Obviously, a healthy life maintains and promotes not only health in general but also skin health in particular. In vulnerable populations, the maintenance of skin integrity is a major professional responsibility, and there seems to be room for improvement to provide better skin care to enhance patient quality of life, safety, and well-being.

References

[1] Chuong CM, Nickoloff BJ, Elias PM, Goldsmith LA, Macher E, Maderson PA, et al. What is the 'true' function of skin? Exp Dermatol 2002;11(2):159–87.
[2] Gupta MA. Aging skin: some psychsomatic aspects. In: Farage MA, editor. Textbook of aging skin. Heidelberg: Springer-Verlag; 2015.
[3] Stepanova EV, Strube MJ. Attractiveness as a function of skin tone and facial features: evidence from categorization studies. J Gen Psychol 2018;145(1):1–20.
[4] Iizaka S. Skin hydration and lifestyle-related factors in community-dwelling older people. Arch Gerontol Geriatr 2017;72:121–6.

[5] Hay RJ, Johns NE, Williams HC, Bolliger IW, Dellavalle RP, Margolis DJ, et al. The global burden of skin disease in 2010: an analysis of the prevalence and impact of skin conditions. J Invest Dermatol 2014;134(6):1527–34.

[6] Cowdell F, Dyson J, Long J, Macleod U. Self-reported skin concerns: an epidemiological study of community-dwelling older people. Int J Older People Nurs 2018:e12195.

[7] Gontijo Guerra S, Vasiliadis HM, Preville M, Berbiche D. Skin conditions in community-living older adults: prevalence and characteristics of medical care service use. J Cutan Med Surg 2014;18(3):186–94.

[8] Gambichler T, Matip R, Moussa G, Altmeyer P, Hoffmann K. In vivo data of epidermal thickness evaluated by optical coherence tomography: effects of age, gender, skin type, and anatomic site. J Dermatol Sci 2006;44(3):145–52.

[9] Trojahn C, Dobos G, Richter C, Blume-Peytavi U, Kottner J. Measuring skin aging using optical coherence tomography in vivo: a validation study. J Biomed Opt 2015;20(4):045003.

[10] Masaki H, Yamashita Y, Kyotani D, Honda T, Takano K, Tamura T, et al. Correlations between skin hydration parameters and corneocyte-derived parameters to characterize skin conditions. J Cosmet Dermatol 2019 Feb;18(1):308–14.

[11] Kottner J, Vogt A. Transepidermal water loss. Textbook of cosmetic dermatology. Boca Raton: CRC Press; 2017. p. 28–31.

[12] Ali SM, Yosipovitch G. Skin pH: from basic science to basic skin care. Acta Derm Venereol 2013;93(3):261–7.

[13] Ropke MA, Alonso C, Jung S, Norsgaard H, Richter C, Darvin ME, et al. Effects of glucocorticoids on stratum corneum lipids and function in human skin-A detailed lipidomic analysis. J Dermatol Sci 2017;88(3):330–8.

[14] Sauermann K, Jaspers S, Koop U, Wenck H. Topically applied vitamin C increases the density of dermal papillae in aged human skin. BMC Dermatol 2004;4(1):13.

[15] Kottner J, Gefen A, Lahmann N. Weight and pressure ulcer occurrence: a secondary data analysis. Int J Nurs Stud 2011;48(11):1339–48.

[16] Campbell JL, Coyer FM, Osborne SR. The skin safety model: reconceptualizing skin vulnerability in older patients. J Nurs Scholarsh 2016;48(1):14–22.

[17] Kottner J, Surber C. Skin care in nursing: a critical discussion of nursing practice and research. Int J Nurs Stud 2016;61:20–8.

[18] McKenna H. Nursing theories and models. London: Routledge; 1997.

[19] Nast A, Griffiths CE, Hay R, Sterry W, Bolognia JL. The 2016 International League of Dermatological Societies' revised glossary for the description of cutaneous lesions. Br J Dermatol 2016;174(6):1351–8.

[20] Herdman TH, Kamitsuru S. NANDA international nursing diagnoses definitions and classification 2018–2020. New York: Thieme; 2018.

[21] Kottner J, Lichterfeld A, Blume-Peytavi U, Kuhlmey A. [Skin health promotion in the elderly]. Z Gerontol Geriatr 2015;48(3):231–6.

[22] Chang AL, Wong JW, Endo JO, Norman RA. Geriatric dermatology review: major changes in skin function in older patients and their contribution to common clinical challenges. J Am Med Dir Assoc 2013;14(10):724–30.

[23] Baumann L. Understanding and treating various skin types: the Baumann skin type indicator. Dermatol Clin 2008;26(3):359–73. vi.

[24] Raj N, Voegeli R, Rawlings AV, Doppler S, Imfeld D, Munday MR, et al. A fundamental investigation into aspects of the physiology and biochemistry of the stratum corneum in subjects with sensitive skin. Int J Cosmet Sci 2017;39(1):2–10.

[25] Li Z, Hu L, Elias PM, Man MQ. Skin care products can aggravate epidermal function: studies in a murine model suggest a pathogenic role in sensitive skin. Contact Dermatitis 2018;78(2):151–8.

[26] Ale IS, Maibach HI. Irritant contact dermatitis. Rev Environ Health 2014;29(3):195–206.

[27] Brenner M, Hearing VJ. The protective role of melanin against UV damage in human skin. Photochem Photobiol 2008;84(3):539–49.

[28] Feingold KR, Schmuth M, Elias PM. The regulation of permeability barrier homeostasis. J Investig Dermatol 2007;127(7):1574–6.

[29] Fluhr JW, Darlenski R, Angelova-Fischer I, Tsankov N, Basketter D. Skin irritation and sensitization: mechanisms and new approaches for risk assessment. 1. Skin irritation. Skin Pharmacol Physiol 2008;21(3):124–35.

[30] Park SY, Kim JH, Cho SI, Kim KI, Cho HJ, Park CW, et al. Induction of a hardening phenomenon and quantitative changes of ceramides in stratum corneum. Ann Dermatol 2014;26(1):35–42.

[31] Mogensen TH. Pathogen recognition and inflammatory signaling in innate immune defenses. Clin Microbiol Rev 2009;22(2):240–73. Table of Contents.

[32] Beier W. [The concept of vitality and the Ries biological index]. Z Gerontol 1985;18(6):353–7.

[33] Blume-Peytavi U, Kottner J, Sterry W, Hodin MW, Griffiths TW, Watson RE, et al. Age-associated skin conditions and diseases: current perspectives and future options. Gerontol 2016;56(Suppl. 2):S230–42.

[34] Yaar M, Gilchrest BA. Photoageing: mechanism, prevention and therapy. Br J Dermatol 2007;157(5):874–87.

[35] Ruiz-Torres A, Agudo A, Vicent D, Beier W. Measuring human aging using a two-compartmental mathematical model and the vitality concept. Arch Gerontol Geriatr 1990;10(1):69–76.

[36] Kottner J, Lichterfeld A, Blume-Peytavi U. Transepidermal water loss in young and aged healthy humans: a systematic review and meta-analysis. Arch Dermatol Res 2013;305(4):315–23.

[37] Ghadially R, Brown BE, Sequeira-Martin SM, Feingold KR, Elias PM. The aged epidermal permeability barrier. Structural, functional, and lipid biochemical abnormalities in humans and a senescent murine model. J Clin Investig 1995;95(5):2281–90.

[38] Irvine C. 'Skin failure'—a real entity: discussion paper. J R Soc Med 1991;84(7):412–3.

[39] Inamadar AC, Palit A. Acute skin failure: concept, causes, consequences and care. Indian J Dermatol Venereol Leprol 2005;71(6):379–85.

[40] Vaishampayan SS, Sharma YK, Das AL, Verma R. Emergencies in dermatology : acute skin failure. Med J Armed Forces India 2006;62(1):56–9.

[41] Delmore B, Cox J, Rolnitzky L, Chu A, Stolfi A. Differentiating a pressure ulcer from acute skin failure in the adult critical care patient. Adv Skin Wound Care 2015;28(11):514–24. quiz 25-6.

[42] Olshansky K. "Skin failure" as a cause of pressure ulcers: accurate terminology, misnomer, or cop out? Ostomy Wound Manag 2012;58(8):6.

[43] Blume-Peytavi U, Tan J, Tennstedt D, Boralevi F, Fabbrocini G, Torrelo A, et al. Fragility of epidermis in newborns, children and adolescents. J Eur Acad Dermatol Venereol 2016;30(Suppl. 4):3–56.

[44] Al-Nuaimi Y, Sherratt MJ, Griffiths CE. Skin health in older age. Maturitas 2014;79(3):256–64.

[45] Farage MA, Miller KW, Elsner P, Maibach HI. Functional and physiological characteristics of the aging skin. Aging Clin Exp Res 2008;20(3):195–200.

[46] Ledoux WR, Pai S, Shofer JB, Wang YN. The association between mechanical and biochemical/histological characteristics in diabetic and non-diabetic plantar soft tissue. J Biomech 2016;49(14):3328–33.

[47] Argyropoulos AJ, Robichaud P, Balimunkwe RM, Fisher GJ, Hammerberg C, Yan Y, et al. Alterations of dermal connective tissue collagen in diabetes: molecular basis of aged-appearing skin. PLoS One 2016;11(4):e0153806.

[48] Oe M, Sanada H, Nagase T, Minematsu T, Ohashi Y, Kadono T, et al. Factors associated with deep foot fissures in diabetic patients: a cross-sectional observational study. Int J Nurs Stud 2012;49(6):739–46.

[49] Hamatani M, Mori T, Oe M, Noguchi H, Takehara K, Amemiya A, et al. Factors associated with callus in patients with diabetes, focused on plantar shear stress during gait. J Diabetes Sci Technol 2016;10(6):1353–9.

[50] Pavicic T, Korting HC. Xerosis and callus formation as a key to the diabetic foot syndrome: dermatologic view of the problem and its management. J Dtsch Dermatol Ges 2006;4(11):935–41.

[51] Mori S, Shiraishi A, Epplen K, Butcher D, Murase D, Yasuda Y, et al. Characterization of skin function associated with obesity and specific correlation to local/systemic parameters in American women. Lipids Health Dis 2017;16(1):214.

[52] Loffler H, Aramaki JU, Effendy I. The influence of body mass index on skin susceptibility to sodium lauryl sulphate. Skin Res Technol 2002;8(1):19–22.
[53] Yosipovitch G, DeVore A, Dawn A. Obesity and the skin: skin physiology and skin manifestations of obesity. J Am Acad Dermatol 2007;56(6):901–16. quiz 17–20.
[54] Hahnel E, Blume-Peytavi U, Trojahn C, Kottner J. Associations between skin barrier characteristics, skin conditions and health of aged nursing home residents: a multi-center prevalence and correlational study. BMC Geriatr 2017;17(1):263.
[55] Martin SF. Contact dermatitis: from pathomechanisms to immunotoxicology. Exp Dermatol 2012;21(5):382–9.
[56] Vierkotter A, Krutmann J. Environmental influences on skin aging and ethnic-specific manifestations. Dermatoendocrinol 2012;4(3):227–31.
[57] Kaya G, Saurat JH. Dermatoporosis: a chronic cutaneous insufficiency/fragility syndrome. clinicopathological features, mechanisms, prevention and potential treatments. Dermatology 2007;215(4):284–94.
[58] Soltanipoor M, Stilla T, Riethmuller C, Thyssen JP, Sluiter JK, Rustemeyer T, et al. Specific barrier response profiles after experimentally induced skin irritation in vivo. Contact Dermatitis 2018 Apr 2. https://doi.org/10.1111/cod.12981. is still in press.
[59] Brasch J, Becker D, Aberer W, Bircher A, Kranke B, Jung K, et al. Guideline contact dermatitis: S1-guidelines of the German contact allergy group (DKG) of the German dermatology society (DDG), the information network of dermatological clinics (IVDK), the German society for allergology and clinical immunology (DGAKI), the working group for occupational and environmental dermatology (ABD) of the DDG, the medical association of German allergologists (AeDA), the professional association of German dermatologists (BVDD) and the DDG. Allergo J Int 2014;23(4):126–38.
[60] Kottner J, Beeckman D. Incontinence-associated dermatitis and pressure ulcers in geriatric patients. G Ital Dermatol Venereol 2015;150(6):717–29.
[61] Hahnel E, Blume-Peytavi U, Trojahn C, Dobos G, Jahnke I, Kanti V, et al. Prevalence and associated factors of skin diseases in aged nursing home residents: a multicentre prevalence study. BMJ Open 2017;7(9):e018283.
[62] Bader DL, Worsley PR. Technologies to monitor the health of loaded skin tissues. Biomed Eng Online 2018;17(1):40.
[63] Farage MA, Miller KW, Elsner P, Maibach HI. Intrinsic and extrinsic factors in skin ageing: a review. Int J Cosmet Sci 2008;30(2):87–95.
[64] Fukuda S, Baba S, Akasaka T. Psychological stress has the potential to cause a decline in the epidermal permeability barrier function of the horny layer. Int J Cosmet Sci 2015;37(1):63–9.
[65] Sonenblum SE, Sprigle SH. Buttock tissue response to loading in men with spinal cord injury. PLoS One 2018;13(2):e0191868.
[66] White-Chu EF, Reddy M. Dry skin in the elderly: complexities of a common problem. Clin Dermatol 2011;29(1):37–42.
[67] Elsner P, Holzle E, Diepgen T, Grether-Beck S, Honigsmann H, Krutmann J, et al. Recommendation: daily sun protection in the prevention of chronic UV-induced skin damage. J Dtsch Dermatol Ges 2007;5(2):166–73.
[68] Trakatelli M, Morton C, Nagore E, Ulrich C, Del Marmol V, Peris K, et al. Update of the European guidelines for basal cell carcinoma management. Eur J Dermatol 2014;24(3):312–29.
[69] Krutmann J, Liu W, Li L, Pan X, Crawford M, Sore G, et al. Pollution and skin: from epidemiological and mechanistic studies to clinical implications. J Dermatol Sci 2014;76(3):163–8.
[70] Danby SG, Chalmers J, Brown K, Williams HC, Cork MJ. A functional mechanistic study of the effect of emollients on the structure and function of the skin barrier. Br J Dermatol 2016;175(5):1011–9.
[71] Shim JH, Park JH, Lee JH, Lee DY, Lee JH, Yang JM. Moisturizers are effective in the treatment of xerosis irrespectively from their particular formulation: results from a prospective, randomized, double-blind controlled trial. J Eur Acad Dermatol Venereol 2016;30(2):276–81.

[72] Hahnel E, Blume-Peytavi U, Trojahn C, Dobos G, Stroux A, Garcia Bartels N, et al. The effectiveness of standardized skin care regimens on skin dryness in nursing home residents: a randomized controlled parallel-group pragmatic trial. Int J Nurs Stud 2017;70:1–10.
[73] Tomova-Simitchieva T, Lichterfeld-Kottner A, Blume-Peytavi U, Kottner J. Comparing the effects of 3 different pressure ulcer prevention support surfaces on the structure and function of heel and sacral skin: an exploratory cross-over trial. Int Wound J 2018 Jun;15(3):429–37.
[74] Zouboulis CC, Krutmann J. [Geriatric dermatology]. Hautarzt 2012;63(12):929–30.
[75] Proksch E, Schunck M, Zague V, Segger D, Degwert J, Oesser S. Oral intake of specific bioactive collagen peptides reduces skin wrinkles and increases dermal matrix synthesis. Skin Pharmacol Physiol 2014;27(3):113–9.
[76] Blaak J, Kaup O, Hoppe W, Baron-Ruppert G, Langheim H, Staib P, et al. A long-term study to evaluate acidic skin care treatment in nursing home residents: impact on epidermal barrier function and microflora in aged skin. Skin Pharmacol Physiol 2015;28(5):269–79.
[77] Angelova-Fischer I, Fischer TW, Abels C, Zillikens D. Accelerated barrier recovery and enhancement of the barrier integrity and properties by topical application of a pH 4 compared to a pH 5.8 w/o emulsion in aged skin. Br J Dermatol 2018 Aug;179(2):471–7.
[78] 8 tips to protect and preserve aging skin. Moisturizer isn't the only solution for more supple skin. Harv Womens Health Watch 2014;21(8):5.

CHAPTER 12

Animal models in chronic wound healing research: for innovations and emerging technologies in wound care

Noelle Remoué, Christelle Bonod, Bérengère Fromy, Dominique Sigaudo-Roussel

Laboratory of Tissue Biologie and Therapeutic Engineering, UMR 5305, CNRS, University of Lyon 1, Lyon, France

Chapter Outline

Innovations and Emerging Technologies in Wound Care. https://doi.org/10.1016/B978-0-12-815028-3.00012-2

1. Introduction

Wound healing is an important physiological process for the integrity of the skin after different trauma. The normal wound healing process involves four overlapping events: hemostasis, inflammation, cellular proliferation, and extracellular matrix (ECM) remodeling [1] to restore tissues and maintain skin integrity [2].

When wound healing is compromised resulting in wound healing delay or no healing at all, it is called a chronic wound, and it puts at risk the body to infection and further health complications. Aging and pathological conditions such as diabetes are the main cause of chronic wounds. Chronic wounds are a major concern, not only for elderly but also for patients with diabetes mellitus. A diabetic foot infection remains one of the major causes of lower extremity amputation and thus constitutes a major burden for patients and clinicians [2].

In addition, as a result of the global increase in chronic diseases due to population aging, it is foreseeable that there will be a concomitant increase in the number of people with wounds secondary to chronic disease [3,4]. Therefore, with a rising prevalence of chronic wounds bearing notable effects on patient morbidity including amputations, appropriate and effective intervention to treat these debilitating wounds can make a significant clinical impact.

Until now, efficient and targeted treatments for chronic wounds are still lacking, which, in part, is due to our limited understanding of the biological mechanisms underlying these diseases [5]. Therefore, the understanding of the possible mechanisms that would promote skin regeneration and function restoration is crucial to develop effective therapeutics for the treatment of impaired wound healings [6].

Consequently, the development of drugs to improve wound-healing process is of great therapeutic interest, and additional knowledge about the molecular mechanism underlying the healing process is needed [2]. Several biological models have been used to comprehend skin healing process from in vitro, ex vivo, and in silico to in vivo studies [7,8]. Each of these models provides valuable information about one or more components/ factors involved in the pathological healing process. However, none of them reproduce all the characteristics of the pathological human skin regarding the mechanical environment, biophysics, and structure.

The objective of this chapter is to focus on experimental chronic wound healing in animal models and the resulting progress in potential innovative therapeutics.

2. From acute to chronic wound healing process

There are numerous etiologies for wounds from the most common such as traumatic, surgical, infectious, skin tears to pressure ulcers (PUs), thermal injuries, vascular ulcers, and wounds also systemic and chronic disease. Wounds are either acute or chronic.

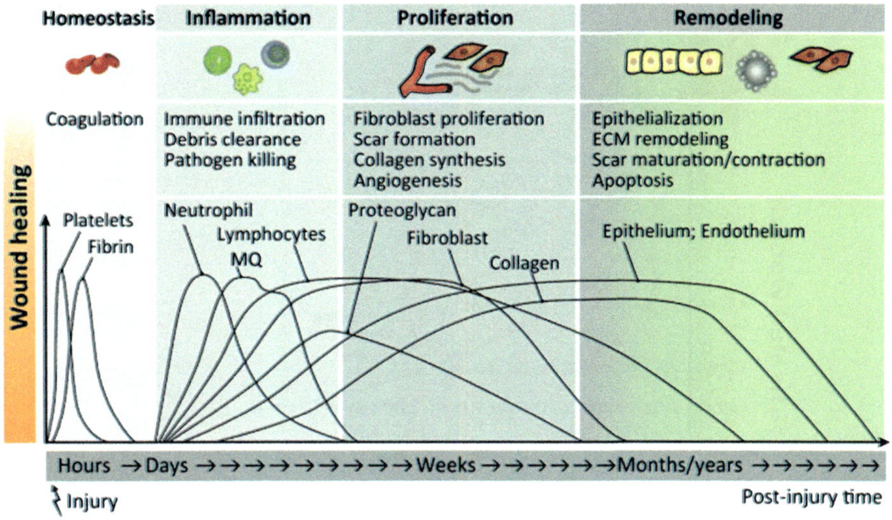

Figure 12.1

Timeline of normal wound healing process. *From Mele E. Electrospinning of natural polymers for advanced wound care: towards responsive and adaptive dressings. J Mater Chem B 2016;4:4801–4812.*

Acute wounds are typically due to some form of trauma that could be blunt or penetrating such as incision, excision, and burn. The tissue damage then stimulates a coordinated physiological response described into overlapping stages: (1) the hemostatic phase to quickly cover the site of injury and prevent infection; (2) inflammation phase with immune cells infiltration to clean cell debris; (3) proliferation phase with tissue regeneration to replace the lost cells; and finally (4) remodeling phase with wound resolution to restore the diverse functions of the skin and remodel the new matrix [9].(Fig. 12.1).

After skin injury occurs, the hemostatic phase takes place with platelets that are activated and recruited at the site of injury to promote clot formation to stop blood loss. The beginning of the inflammatory phase starts also with the activation of the platelets that attract immune cells from the blood flow into the wound due to vascular permeability. Neutrophils are first to arrive to clean the wound by producing high levels of reactive oxygen species (ROS), proteases, and proinflammatory cytokines that kill bacteria. Then monocytes differentiate into macrophages [11] to help in further cleaning debris by releasing matrix metalloproteinase and bacterial phagocytosis [12,13]. During the proliferative phase fibroblasts, endothelial cells and keratinocytes proliferate and migrate to regenerate the lost tissue [14,15]. The ECM, called the granulation tissue, is laid down, and keratinocytes migrate on it to close the wound. Reepithelialization is a critical phase and begins within hours after injury and lasts until complete healing [16,17]. Then occurs the final phase of wound healing, the remodeling of the granulation tissue, where fibroblasts produce ECM and allows ECM network to mature [17] and increase its

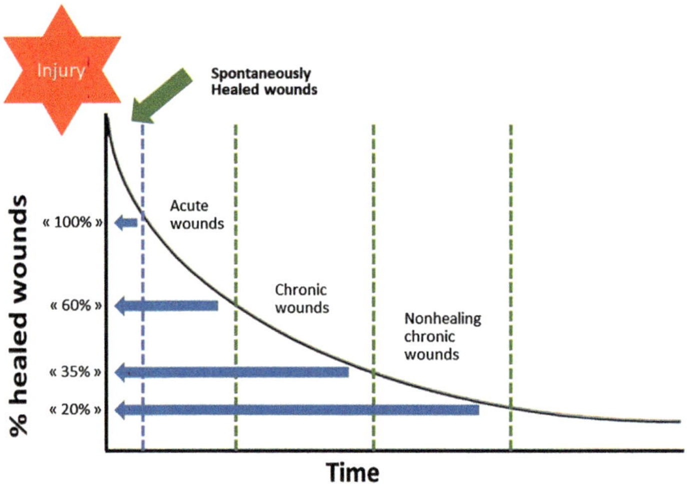

Figure 12.2

Regression curve showing the % of healed wounds over time. *From Bosanquet DC, Harding KG. Wound duration and healing rates: cause or effect? Wound Repair Regen 2014;22(2):143–150.*

mechanical strength [18]. Wound healing is complete following apoptosis of myofibroblasts, vascular cells, and macrophages, leaving an acellular matrix rich in collagen and elastin [12,15,19].

Therefore, acute wound is considered to go through each wound healing stages within the expected period. However, comorbid conditions, chronicity, and confounding factors can impair the velocity and the mechanisms with which repair should occur.

Chronic wounds are typically wounds that demonstrate delayed healing 12 weeks after the initial insult often caused by pathological conditions such as diabetes or because of aging processes (Fig. 12.2).

There are numerous data supporting the hypothesis that chronic wounds display excessive inflammation with a greater number of inflammatory cells, such as neutrophils, lymphocytes, and macrophages [20]. In chronic wounds, the early impairment of the macrophage switch from proinflammatory to antiinflammatory phenotypes is believed to contribute to the impairment in tissue repair [19,21]. The resultant inflammatory environment contains high levels of inflammatory cytokines, proteases, and ROS [22,23] and low levels of protease inhibitors, such as metalloproteinases inhibitors (TIMPs) [12,24], and growth factors such as Vascular endothelial growth factor (VEGF) and basic fibroblast growth factor [25]. The consequence is an excessive degradation of the ECM and absence of both proliferative and remodeling phase [12,19], resulting in failure of progression of wound healing [20]. (Fig. 12.3).

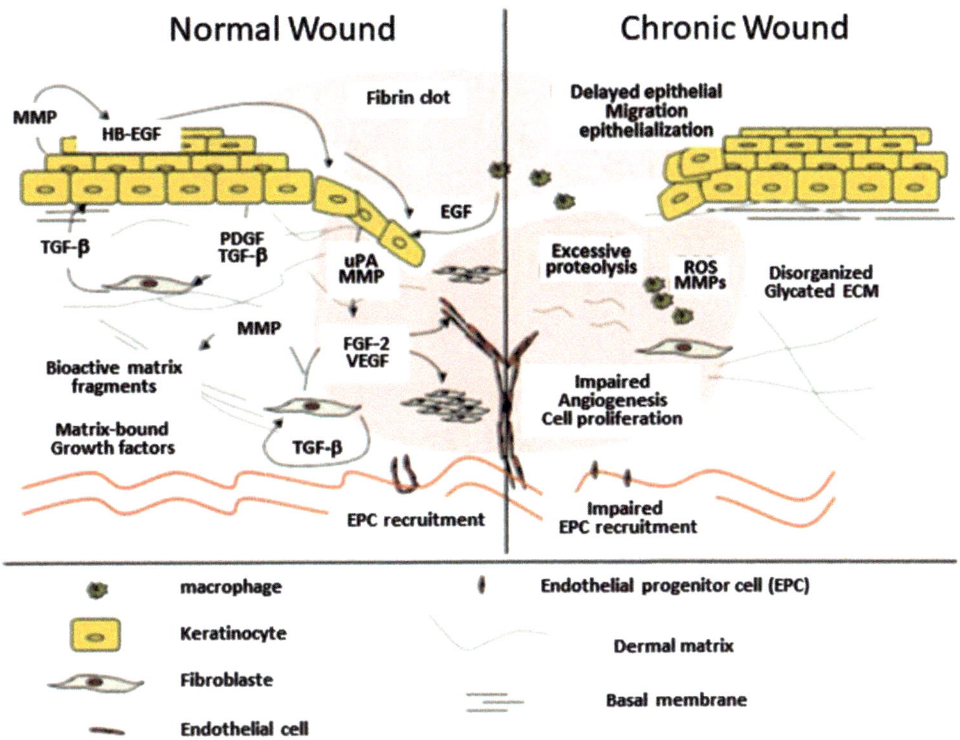

Figure 12.3

Comparison of the microenvironment in the epidermis and the dermis between a normal wound and a chronic wound. *From Demidova-Rice TN, Hamblin MR, Herman IM. Acute and impaired wound healing: pathophysiology and current methods for drug delivery, part 2: role of growth factors in normal and pathological wound healing: therapeutic potential and methods of delivery. Adv Skin Wound Care 2012;25(8):349–370.*

Therefore, in pathological conditions, any wound can become chronic resulting in the disorder of the original tissue structure and homeostasis.

Regardless the etiology of the wound, the main repair processes are common, but there could be some difference in duration of wound healing and in cellular mechanism depending on the following:

- The size and the depth of the lesion as well as the extend of any necrotic tissue to be cleaned. In addition, the wound depth has to be taken into consideration regarding epithelialization time [13,27].
- The type of injury. Indeed, for incisional repair the majeor healing process is a reintegration of the separated tissue sections. In contrast, excisional wound repair is a regeneration of tissue mass to replace the lost tissue to heal the wound. Thus, the granulation

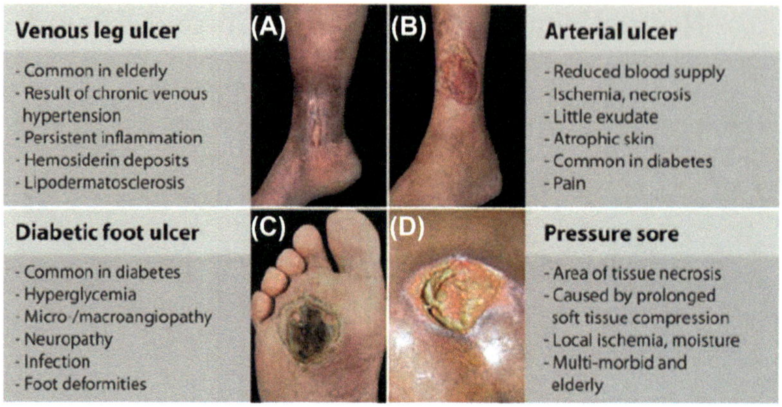

Figure 12.4

Clinical presentation of various chronic wound pathologies. *From Eming SA, Martin P, Tomic-Canic M. Wound repair and regeneration: mechanisms, signaling, and translation. Sci Transl Med 2014;6(265):265sr6.*

tissue response that marks excisional repair is largely absent during incisional repair [8,9].

Among the chronic wounds etiology, there are specific chronic wounds, which include venous leg ulcers (VLUs), diabetic foot ulcers (DFUs), arterial insufficiency, and PUs. The main causes of human chronic wounds are vascular compromise and repeated trauma [28]. However, several other factors are also involved in human chronic wounds such as the following:

- Diabetes and aging through the pathological changes occurring such as damages to the peripheral nervous system decreasing, vascular function, and skin homeostasis;
- Sedentarity by increasing the duration of tissue compression and friction;
- Polymicrobial wound infection with an ensuing biofilm further contributes to the development of chronic wounds and impaired wound healing.

In addition, the other big issues of chronic wounds, after having first struggled to manage to heal the wound, are the reulceration of healed wounds. In excisional healing, rapid reepithelialization could cause premature transition to the resolution phase [29,30] resulting in a thin and weaker dermis that would predispose to reulceration [9]. Aberration of wound healing such as hypertrophic scar and keloid will not be discussed in this chapter (Fig. 12.4).

3. Animal models

In vivo animal studies require good practice with ethical agreements and by promoting the 3R guidelines to replace the use of animal with alternative techniques, reduce the number of animals used to a minimum, and refine the way experiments are carried out to improve animal welfare. Therefore, procedures for the maintenance and use of the experimental animals are carried out in accordance with the principles of each country's directive and have to be approved by an ethic committee for animal experiments and validated by the administration in charge.

Animal model has to be chosen appropriately depending on the studies as none of the animal skin exhibits all the human skin properties. As very well discussed by Wells and colleagues [9], human skin presents an infrequently used architecture that is shared only with other primates and pigs.

There are also differences in life span, size, and genetics in all the animal models, differences in chronic disease pathophysiological manifestations, and aging process compared to human [9]. However, various animal wound healing models have been successfully developed to better understand the basic mechanisms of chronic wound healing process and its pathophysiology. Mouse, rat, porcine, and rabbit are the main animal models assessed for wound healing studies, but porcine skin has major similarities with human skin.

Furthermore, in efforts to induce a wound healing delay, a pathological chronic wound environment is reproduced by submitting animal models to several pathological conditions such as diabetes, ischemia, ischemia–reperfusion (IR), wound biofilm, or mechanical pressure [31].

3.1 Porcine

Porcine models have emerged as promising models to study wound healing as pigs have anatomically and physiologically similar skin to humans and have been used to study many other diseases as reviewed by Seaton [13]. Similar to humans, pigs have a thick epidermis, dense elastic fibers in the dermis [32], similar immune cells, and biochemical structure of collagen to human skin [33]. Nevertheless, there are some differences with human skin, and among them, we can notice a reduction in microvascular density in the dermis as well as apocrine sweat glands as compared to eccrine sweat glands in human skin, which are major reservoirs of keratinocytes during wound epithelialization [34].

Nonetheless, pig wound models appear to be an appropriate model to comprehend cutaneous wound healing in humans as reported by Sullivan et al. [35], showing that pig models were 78% concordant with human studies regarding wound therapies.

Although pig models have been developed to study several chronic wound healing pathologies, they are expensive to purchase and feed, difficult to manage, difficult to render transgenic, and difficult to supply antibodies and PCR probes compared with other animal species [13].

3.2 Rodents

Mice are commonly used animal models for wound healing studies; nevertheless, there are some differences with human skin [36].

Contraction is one of the main mechanisms involved in murine skin healing in contrast to reepithelialization that occurs in human to close the wound. However, recently, it has been reported that both contraction and reepithelialization were involved in skin healing [37]. In addition, by applying a silicone splint around the cutaneous wound in rodents, it facilitates "humanized" less contracted wound healing through the processes of granulation and reepithelialization [38].

Mice have more densely packed hair follicles and thinner epidermal and dermal layers compared to human skin [139] with substantial differences in immune systems compared to humans (Mestas and Hughes 2004). Mice lacking either hair follicles or hair follicle stem cells show delayed wound healing [39,40]. Mice have been used to study most of the chronic wound healing pathologies. Rats have also loose skin because of their skin elasticity and lack of adherence to the underlying structures as mice but not human skin [41]. Particularly, in wound healing studies, it is important to note that rodents have the ability to synthetize vitamin C that plays a key role in collagen synthesis [42].

In a recent review, Isakson et al. [38] reported that rodent models were used in 78% of the papers assessed, with incisional, excisional, and burn wounds used to assess healing (see figure). Rodents are mainly used as animal models for preclinical wound healing studies because of high availability, low cost, and ease of handling [38]. Nevertheless, they clearly cannot be considered a gold standard for preclinical studies, which is why pig test is a preferred preclinical test before translation to human studies [38] (Fig. 12.5).

3.3 Rabbits

Rabbits are large animals with easily accessible ear lobes, which makes logistics of skin as well as the vascular system easier than in smaller animals. They have a thicker skin and a lack of sebaceous glands compared to human skin. The wound does not heal by contraction but heals from the wound edge [43]. Finally, animals with minimal genome variations as well as with specific genetic differences are commercially available [44].

3.4 Diseased animals

To be able to study chronic wound healing or at least delayed wound healing, we need to perform experiments on models that exhibit those conditions. Aging or diabetes is the well-known factors promoting wound healing delay and conditions suffering from chronic nonhealing wound.

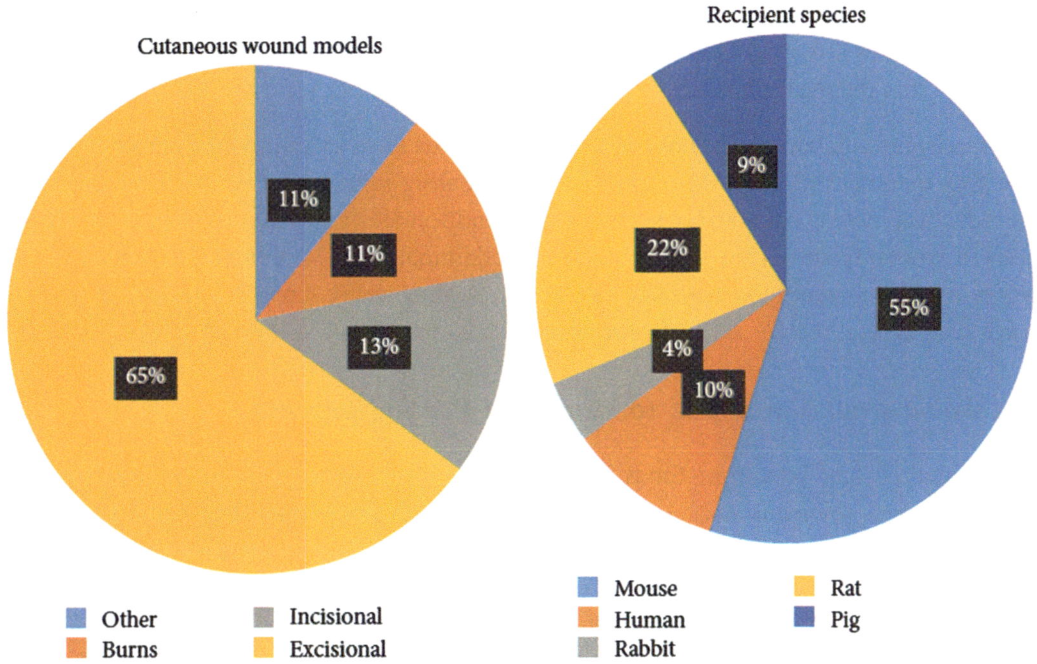

Figure 12.5

Cutaneous wound models studied and animal models used for cutaneous wound healing studies. *From Isakson M, de Blacam C, Whelan D, McArdle A, Clover AJ. Mesenchymal stem cells and cutaneous wound healing: current evidence and future potential. Stem Cell Int 2015;2015:831095.*

Diabetic mice are either generated by genetic mutation that results in a diabetes-like phenotype (db/db, NOD mice), induced by chemical injection with streptozocin [45] and/or by using diverse hypercaloric diet [46]. Rats, rabbits, and pigs can also be rendered diabetic using either streptozotocin or alloxan [43,47,48]. Not all the diabetic human complications are present or similar in diabetic model animals. Indeed, arteriolosclerosis observed in the human diabetic pathology is not present in diabetic rodent models, and although hyperglycemia and advanced glycosylated end products are present, they are more expressed than in the human pathology [9,12].

Nonetheless, diabetic mice wounds share several characteristics in common with chronic wounds in diabetic patients such as nerve density, angiogenesis, granulation tissue formation, and collagen content [49]. However, although diabetic mice exhibit a wound healing delay than wild-type mice, they do finally heal without any additional treatment in contrast to human diabetic chronic nonhealing wound. Therefore, diabetic mice are not an ideal model for chronic wound studies, but since they do share some similarities with human diabetic pathology that makes them as widely accepted model in current wound healing research [12,50].Indeed, delayed healing process is observed in the presence of peripheral neuropathy using Streptozotocin (STZ) in the context of diabetes as well as selective damages affecting C and Aδ fibers due to resiniferatoxin-mediated small fiber neuropathy that delayed granulation tissue remodeling after burn in rats [51-53].

In comparison to diabetic mice only few murine studies have examined the effect of age on healing outcomes [54,55]. Age-related wound healing impairment seems to be caused by a reduction in oestrogen levels in both males and females that have an impact on all phases of wound healing [56] and specially on inflammation [57]. In addition, social isolation related to ageing is associated to wound healing delay through a reduction of growth factors required for skin cell function such as proliferation [58,59].

4. Chronic wound methods to study the healing process

Experimental chronic wound methods have been developed through the years from the most simple ones (incision, excision) to the most complex ones (burn, PU, and VLU) to better understand the pathophysiology of delayed wound healing to improve the treatment. In this field, the skin is the main organ assessed, but some studies also focused on skeletal muscle that can be impaired in a severe form of PU (deep tissue injury) without any visible damage on the skin [60]. In addition, we will also review numerous models that have been employed in animals using friction, temperature, biofilm, and muscle damage, which are associated to chronic wound (Fig. 12.6).

4.1 Excisional wounds

The most commonly employed method model used to study wound healing is the excisional wound model that can be superficial, partial thickness, or full thickness [13]. Many studies have been performed in healthy and diseased rats, mice, rabbits, and pigs using this method to better understand chronic wound healing [58]. The wound is mainly created on the dorsum of the animal and can extend through the panniculus carnosus for the rodents, except in rabbits, where the wounds are mainly performed on ears. In mice, rats, and rabbits, the size of excision can vary from 6 to 12 mm of diameter, whereas in pigs from 2 to 5 cm of diameter.

Wound closure is then photographed regularly and calculated based on wound size relative to the original wound dimensions [58]. To limit the wound contraction observed in rodents, the addition of a silicone ring onto the cutaneous wound in rodents has allowed for an accurate, reproducible model of wound healing with less contraction [38].

4.2 Pressure wounds

PUs can start superficially or deep within the tissues. Superficial ulcers form within the skin and may progress downward, whereas deep ulcers arise in muscle layers covering bony prominences and are mainly caused by sustained compression of the tissues [61,62].

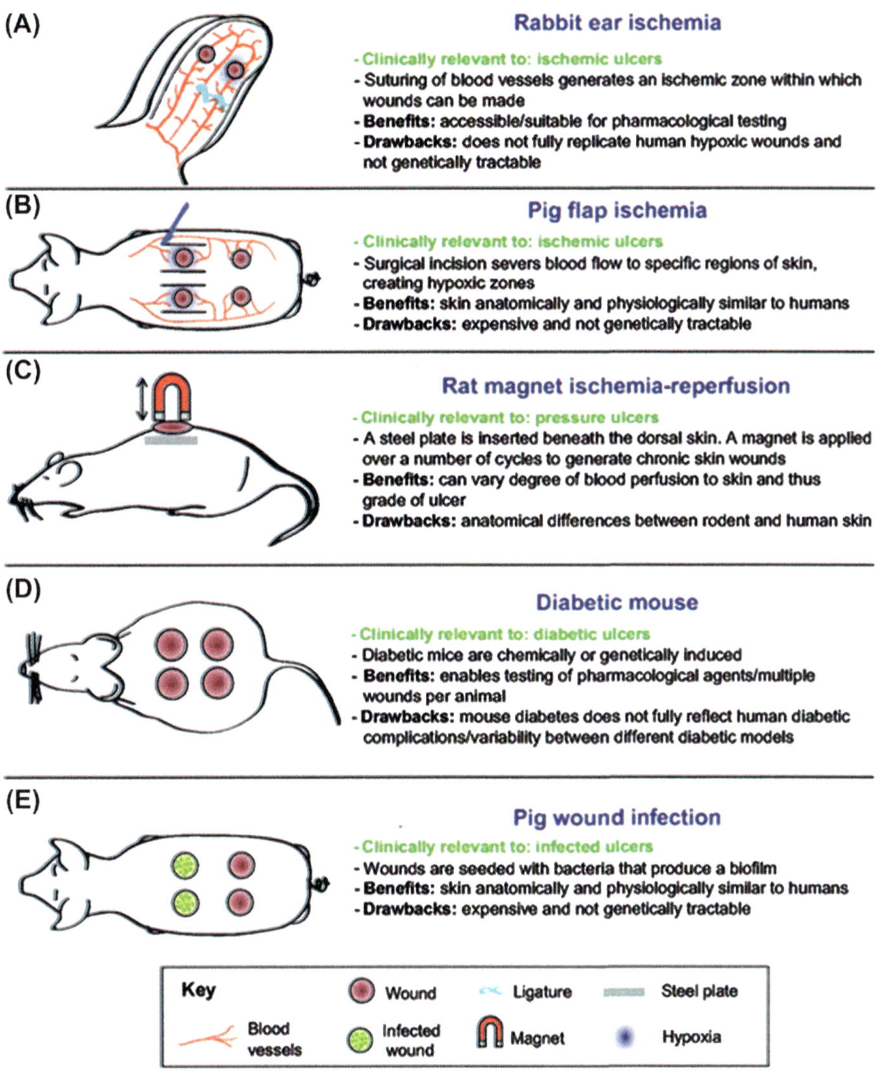

Dis Model Mech. 2014 Nov; 7(11): 1205–1213.

Figure 12.6

Different animal models and methods to study chronic-skin-wound. *From Nunan R, Harding KG, Martin P. Clinical challenges of chronic wounds: searching for an optimal animal model to recapitulate their complexity. Dis Model Mech 2014;7(11):1205–1213*

Several methods have been developed to induce and explore PU in animals to identify specific aspects of wound development, quantify the extent of lesions, assess wound healing mechanism, and assess outcomes for extrapolation to clinical relevance [63].

Prolonged mechanical loading will lead to capillary occlusion, resulting in tissue ischemia with associated localized hypoxia, and associated-lymphatic occlusion will lead to accumulation of toxic metabolites leading to local cell death and local breakdown of soft tissues [64–66].

Although the fundamental mechanisms responsible for healing in PUs are not fully understood, skin injury models provide valuable clues. Several methods have been developed to apply skin pressure and study either the incidence of PU or the healing process [58,63,67,68]. Models based on single or cyclical magnetic compression [58] allowed researchers to examine wound inflammation, apoptosis, and reactive oxygen-mediated signaling following injury. When the skin is at risk, a single pressure application is able to induce a skin lesion due to ischemia and hypoxia [69].

The use of a metal plate underneath the skin requires a surgery from 2 days to 1 week before applying a magnet on the skin. However, this method does not allow following up the healing process as the injured skin tends to stick to the metal plate inducing some confounding effects. The method used to both induce a pressure injury and follow up the healing process requires two magnets on both sides of a fold of skin on the back of the animals under short isoflurane anesthesia. This leads to the induction of two PUs.

Protocols used either a single or cyclic magnet application. Cyclic magnet application provides more severe skin injury than single application, but it all depends on the pressure applied, which can vary from 50 mmHg to 500 mmH. Some extreme pressures such as 800–2000 mmHg have been used in some studies [70,71]. The duration can also vary from 2 to 15h. The interest of the magnet method is to calibrate the pressure/duration in function of the question to answer. Those protocols are mostly performed in rodents but, in rabbit magnet with attached string loop, can be placed on the left ear ventral surface and a second identical magnet attached to the dorsal side to complete a magnetic seal. The magnets (49 mmHg/cm^2) can be kept on the right ear for 90 min before being removed [44].

The pathogenesis of PUs and diabetic foot is thought to be mediated through injurious cycles of IR. IR injury is a major factor in the pathogenesis of chronic skin wounds. IR is due to positional changes and/or mechanical loading and unloading of the affected areas, which leads to repetitive IR with a major impact on cutaneous blood flow, and can be related to venous stasis ulcers, diabetic ulcers, and PUs. Studies suggest that repeated cycles of IR, impairing blood and lymph flow locally, lead to cell necrosis in the epidermis, dermis, subcutaneous adipose, and skeletal muscle tissues from Refs. [50,72] together with the molecular derangements induced by local tissue hypoxia, cellular senescence, and bacterial colonization [72].

Proinflammatory cytokine expression seems to be cycle responsive and correlates in some reports with the influx of inflammatory cells that, in turn, mediate tissue destruction

[72,73]. The loss of necrotic tissue is essential to the closure of PUs. In PUs, the delay in the keratinocyte migration compromises reepithelialization impairing necrotic skin loss and subsequent phases of wound healing [74,75]. However, when dermal adipose tissue is increased, repeated IR cycle induced major skin compared to single pressure application [76,77].

Reactive hyperemia is an important physiological mechanism in response to the release of mechanical loading of the skin resulting in increased cutaneous blood flow that can be measured noninvasively by a laser Doppler flowmeter, laser Doppler perfusion imaging, and dynamic laser speckle imaging techniques [64,69,78,79].

To improve the pressure model and to avoid the problem of mice skin, recently, Bujan's group was able to check the stability of dermal scaffold in a mouse long-term transplanted human skin and its retained capacity to react after a PU. The fibrillar microfibrillar proteins of the human dermis act as a scaffold, and their redistribution close to the dermoepidermal junction guarantees the stability of the epidermis. This novel model will facilitate the development of new clinically relevant therapeutic strategies that can be tested directly on human skin [80,81].

Beside diabetes induction or aging studies, other animal models can be assessed under additional contributing factors to PU development. For example, cigarette smoke is also reported to be a risk factor for PUs through endothelial dysfunction and atherosclerosis [82]. Five IR cycles, with 2 h of ischemia and 30 min of reperfusion, were performed in rats that received intraperitoneal nicotine injections. The results suggest that the increased risk of PUs due to cigarette smoking is mediated, in part, by nicotine. They also indicated that the effect of nicotine is not mediated by a change in blood pressure, but it is more likely elicited via an increase of inflammatory mediators [82].

For many years, it has been suggested that floor type is the key factor for the occurrence of footpad injuries in rabbits, and alternatives to wire mesh flooring were developed several years. However, footpad injuries are commonly observed [83,84] and could represent an alternative model to study pressure injury incidence that occurs at the the foot level like in diabetic patients for example. Indeed, wire grates rapidly induce hindpaw tactile hyperesthesia and plantar neuropathy in rats [85].

4.3 Friction

In contrast to pressure that represents the normal force per unit area acting on the skin, shear forces are forces acting parallel to the skin at the interface with a support surface [64]. Emerging evidence has demonstrated the clinical significance of shear to foot ulceration [86,87].

For a long time, friction/shear stress research work suffered from a lack of robust measurement techniques [64,88], but new apparatus have resolved this major point [64]. The presence of shear forces in association with pressure can lead to an increasing rate of cell death caused by enhanced tissue distortion and enhanced occlusion of blood vessels [89]. Indeed, shear forces when added to pressure application cause ulcers for low level pressures in a swine model [64,90]. This enhanced internal deformations in the presence of shear at the skin surface were also observed in a computational model [64,89].

New apparatus were specifically designed to apply a combination of pressure and shear on the skin surface of not only animal models but also in humans. Ohura developed an experimental model consisting of porcine skin and a Predia sensor capable of measuring pressure and shear force simultaneously. External force was applied to the skin model using a 1 kg weight consisting of metal beads. The purpose of this study was to measure the impact of external shear force and pressure on a superficial layer of skin and subcutaneous layer with an underlying bony prominence and also to verify how the influence of these external forces can be reduced after dressings are applied [88].

When shear force is added to pressure, the cutaneous blood flow decreases and showed a significant decrease in cutoff pressure of the microvascular blood supply of the skin in humans [91]. Concentrating pressure and shear stress in deep tissues, while minimizing pressure and shear stress in the overlying skin, produced an effective rat model of deep tissue injury deterioration [92].

4.4 Deep tissue injury

Deep tissue injury (DTI) is a severe form of PU where tissue damage starts in deep tissues, such as skeletal muscle, underneath intact skin rendering it often not detected until it has reached the skin surface. DTI is a recently categorized type of PU and is considered to originate from deep tissue damage that expands outwards to the overlying skin (bottom-up theory), whereas conventional PUs originate in superficial tissues and extend to deeper structures [93].

The etiology of DTI involves several factors, such as ischemia due to tissue compression [71,94] and local tissue deformations [66,95,96], especially during short-duration loading periods when ischemia is not expected to cause any significant skeletal muscle damage [97,98].

Previous studies also showed that muscle tissue is more susceptible to mechanical loading than skin [61]. To date, no study has shown the exact mechanisms of DTI deterioration. In animal experiments of Refs. [99,100], the effect of mechanical loading on the development of muscle damage was investigated by compressing the tibialis anterior muscle of rats with an indenter for 2 h. This revealed large differences in the amount and locations of muscle damage between the animals, partly due to the inability to reproduce the loading conditions between experiments. Another method was described with a full-thickness skin/muscle PU model using a metal disk implanted beneath the mouse gluteal muscle and cyclical pressure applied with an external magnet [97,101].

In another study, tangent elastic moduli and strain energy densities of rat gracilis muscles were exposed to apply pressure in vivo (11.5, 35, or 70 kPa for 2, 4, or 6 h) [102]. Histological evaluation showed that muscle stiffening under strain accompanied extensive necrotic damage and thus information on internal muscle stresses was required to establish new criteria for pressure sore prevention. A rat model was also used to distinguish between the different factors that contribute to muscle tissue damage related to deep PUs that develop after compressive loading [98]. MRI and histology were performed before and after tissue compression and ischemic loading. Uniaxial loading was applied to the tibialis anterior by means of an indenter, and ischemic loading was accomplished with an inflatable tourniquet. The separate and combined effects of ischemia and deformation were studied and revealed that large deformation with ischemia was responsible for irreversible muscle damage [98]. However, it has been reported in a muscle mechanical loading study that animal and finite element model results suggest that acute ischemia does not develop in skeletal muscles under physiological load levels within a time frame of 40 min [89]. As there is evidence that DTI develops within a shorter time, ischemia is unlikely to be the only factor causing DTI.

Severe pressure loading on flank skin induced early increases in muscle damage associated to elevated expression and activation of Hypoxia-Inducible Factor-1 (HIF1) in skeletal muscles [103]. Apoptotic and/or proteolytic roles of HIF1 might be a part of the mechanism of DTI deterioration [92,104].

In addition, MRI readouts and histological analyses of the damaged muscle tissue revealed that damage occurred at some distance from the deformation and lasted up to 14 days after a single indentation period [105].

However, all those studies have been performed in healthy animals, and further experiments have to be done to understand the behavior of diseased skeletal muscle under compression (Fig. 12.7).

4.5 Temperature

Temperature and humidity can affect the structure of the skin, and under pressure, they can exacerbate the risk of wound occurrence [106]. However, because of several differences in sebaceous glands and thermoregulation in animal skin compared to the human skin, there are only few studies interested in the temperature effect. One study developed an apparatus and procedure to simultaneously apply 12 metal discs (each with a

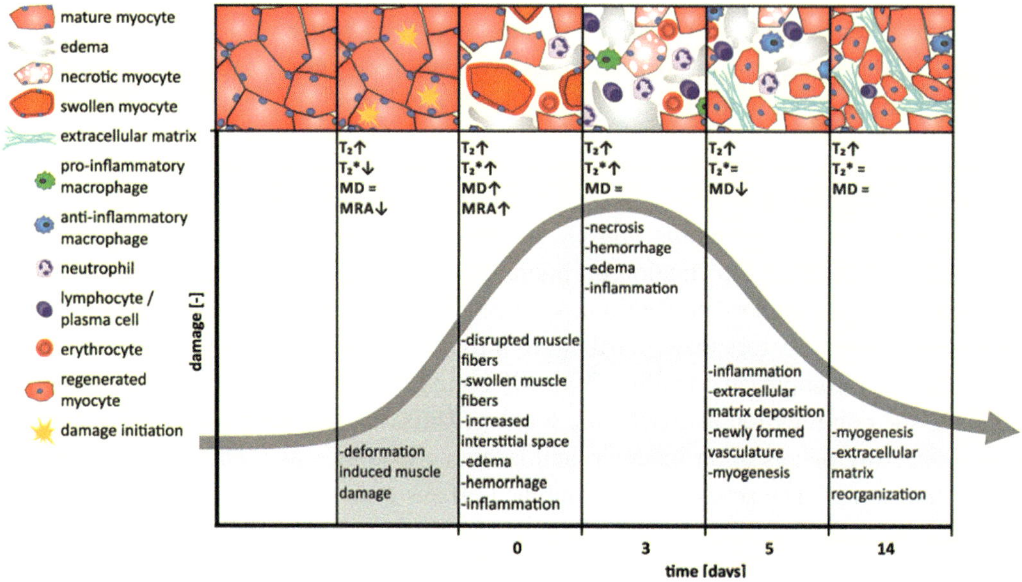

Figure 12.7

Time course of tissue damage and remodeling in the rat model of deep tissue injury. *From Nelissen JL, Traa WA, de Boer HH, de Graaf L, Mazzoli V, Savci-Heijink CD, et al. An advanced magnetic resonance imaging perspective on the etiology of deep tissue injury. J Appl Physiol (1985). 2018;124(6):1580–1596.*

diameter of 51 mm) on the dorsal aspect of the swine, all at an equal pressure of 100 mmHg, for a 5-h period, while servocontrolling disc temperatures at 25, 35, 40, or 45°C. This reliable porcine model was developed to facilitate investigations of PU formation, healing, and prevention. The severity of the resultant tissue injuries correlated with an increase in applied temperature [107].

4.6 Ischemic wounds

The rabbit ear ischemic model was created using a minimally invasive surgical technique [108]. On one ear, three small skin incisions were made on the vascular pedicles about 1 cm from the base of the ear. The central and cranial arteries were ligated and divided along with their accompanying nerves. A circumferential subcutaneous tunnel is made through the incisions to cut subcutaneous tissues, muscles, nerves, and small vessels [109]. Thus, it creates a full-thickness wound that lacks a vascular base and has very little lateral vascular supply [50,108].

Previous models used an H- or U-shaped cutaneous flap on the back of mice or rats, which is usually used as a simple ischemic skin wound model. However, the degree of ischemia to each flap varies with the size and dimension of the flap, new vascular channels develop quickly within few days, and blood perfusion returns to normal within about 2 weeks. There also exist natural craniocaudal differences in granular tissue formation in small animals such as mice or rats, which adds to the complexity in comparison [109].

The ischemic ulcer model is reproducible and quantifiable and is potentially a useful model for examining agents to improve ulcer healing where blood flow is decreased [108]. Previously, it has been established in the rabbit ear ischemia model that the impact of age and ischemia is additive and together produce a profound impairment in wound healing sufficient to produce a chronic wound and a lack of response to transforming growth factor (TGF)-1. The rabbit ear ulcer does not heal by skin contraction. This is advantageous over cutaneous wounds in rodents, which heal by contraction [43]. Several adaptations of the ear ischemic model have been done as the development of a reversible IR clamp and the effects of cyclic IR injury [72]).

Ischemic models are also performed on the back of rat skin more than mice due to the size [110]. A novel addition to this ischemic skin flap model is the placement of a silicone sheet beneath the flap that functions as a barrier and a splint to prevent revascularization and reduce contraction as the wounds heal [50].

4.7 Burn

Burns are a global public health problem with an estimation of 180,000 deaths every year in the world. Although burn wound management has been largely improved by using epithelial cells and sheets, there is still a need to better understand the healing process. A number of animal models have been described to assess the impact of burns on healing after injury [111]. The animal model has to be reproducible and as close as possible to burn lesions occurring in humans, taking into account the different structural properties of the skin between different animal species and humans. Most of the animal models described so far were set up many years ago and are still currently used [111].

Seaton et al. published a very thorough review on the different techniques to use to study burn wounds [13]. A contact burn can be made by applying a heated metal bar or container filled with hot water directly to the skin. Alternatively, scalding and flame burn mechanisms have been reported. Reproducibility of burn depth is lower than excisional or pressure models. However, the specificity of burn model is to evaluate processes that convert ischemic zones to hyperemic or coagulated zones [13]. One of the limitations reported in using mice in burn wound models is that they cannot tolerate more than 30% total body surface area burns, and those burn injuries do not provoke the hypermetabolic response seen after large human burns [41,112].

4.8 Infected wounds: biofilm

The importance of biofilms on chronic wounds has only recently gained attention. About 60%–90% of chronic wounds have a biofilm versus only 6% of acute wounds [113]. Microbial biofilms are complex communities of microorganisms that attach to surfaces and secrete extracellular polymeric substances (EPS). The secretion of EPS acts as a biochemical barrier, protecting the biofilm from both mechanical and shear forces, and more importantly, the penetration of antimicrobial substances [114].

Bacterial species found to reside in chronic wounds biofilm are mainly *Staphylococcus aureus* followed by *Staphylococcus epidermidis*, *Pseudomonas aeruginosa*, *Proteus mirabilis*, and *Escherichia coli* [115–118]. Therefore, novel cutaneous infected wound models were developed by inoculating one or mixed bacterial species such as *S. aureus*, *S. epidermidis*, or *P. aeruginosa* in mice or pigs [119,120].

Older mice excision wound model inoculated with *S. aureus* failed to clear the infection and showed less immune cells infiltration and increased bacterial colonization [121,122] with delayed reepithelialization than the young mice [119]. Similar results have been observed in diabetic mice with *P. aeruginosa*–infected chronic wounds displaying impaired bacterial clearing and wound closure in comparison with their nondiabetic littermates [123]. It is suggested that biofilm bacteria secretion could also inhibit cell proliferation and migration delaying even further the wound healing process.

In contrast to mice observations, in porcine burn wounds infected with biofilms, closure rates were normal, but they did exhibit failure in barrier function with epidermal water loss resulting from malfunctioning tight junction through biofilm-inducible microRNAs (miRNAs) that specifically silence tight junction proteins [120]. Therefore, biofilms could lead to postclosure complications, including infections and the recurrence of skin breakdown contributing to the chronicity of the wound [50,120].

4.9 Innovation

From all those animal models and chronic wound studies, many results arose, providing better comprehension of the pathological process and several new paths to improve the healing process. Therapeutic strategies that emerged are either based on physical or biological stimulations. Physical therapy approaches have been designed using different protocols of either electrical, ultrasound, or light stimulations [124]. Biological therapies are based on molecular stimulation and target cell replacement, and both need delivery technologies such as dressings or nanocarriers and alternatively onboard cutaneous preventive system.

4.10 Molecular target

Several molecules have been administrated to the wound area to improve the healing process, and for many years, growth factor—based therapies have been assessed. Several growth factors including platelet-derived growth factor, fibroblast growth factor, TGF, and epidermal growth factor (EGF) have been introduced as potential adjunctive therapies to promote ulcer healing [125]. Recently, a first multicenter, double-blind, randomized, placebo-controlled trial showed the efficacy and safety of a recombinant human EGF spray by decreasing the healing duration of DFUs [52].

4.11 Living cell advanced therapy

Interest in the use of stem cells has grown in the last decade by using them to regenerate the skin during chronic wound. Stem cells can be distinguished in epithelial and mesenchymal cells. Epithelial stem cells are adult tissue—resident stem cells such as interfollicular epidermal stem cells and hair follicle stem cells, while mesenchymal stem cells come from bone marrow's adipose tissue and umbilical blood cord [126]. Several experimental studies and some clinical studies have proved the efficacy of transplanted cultured epithelial cells or epithelial allografts to treat chronic wounds. Further clinical trials will be necessary to determine the safety from isolation to administration and efficacy of such cell treatments [126].

Very recently, it has been showed that it was also possible to reprogram wound-resident mesenchymal cells that enable all regions of the wound to reepithelialize and lead to a regenerative functional response in the endogenous skin. Before clinical applications of this potentially transformative, nonsurgical technology can be realized, further improvements must be made [127].

Major improvement enhances cell survival following transplantation through improving the delivery system, e.g., 3D printing or environment, e.g., biomimetic scaffold delivery system [38].

4.12 MicroRNAs

miRNAs are noncoding RNAs that act as posttranscriptional regulators of multiple proteins and associated pathways. Thus, identification of the appropriate miRNA involved in the different phases of wound healing could reveal an effective genetic therapy in chronic wound care. Several miRNAs have been shown to be upregulated or downregulated during the wound healing process, and many studies have suggested that some miRNAs may be dysregulated in different phase of diabetic wound healing contributing to diabetic foot wound [6,128].

In addition, it has been highlighted that wound biofilm could induce some miRNAs (miR-146a and miR-106b) [120]. The observation that wound biofilm may induce host skin miRNAs is novel and particularly significant because these miRNAs target the same tight junction proteins of the ZO family, which are known to be critically required for skin barrier function [120].

4.13 Exosomes

Microparticles and nanocarriers have also been developed, allowing spatial control and enhanced biodistribution of the pharmaceutical agents to the affected areas [2,129,130].

Exosomes are nanovesicles secreted by a variety of cells and have emerged as a new intercellular communication system for transporting proteins and RNAs. Exosomes secreted by human adipose stem cells are easily obtained and can be effectively used in research and clinical treatment. Exosomes can optimize the characteristics of fibroblasts, such as promoting the migration, proliferation, and collagen synthesis of fibroblasts, thereby accelerating wound healing of soft tissue. It has been shown that exosomes secreted by adipose stem cells may represent a novel therapeutic tool in soft tissue wound healing [131]. The cell action of exosomes is going through its molecular content that can vary depending on its cell origin. Important components of the exosomal cargo secreted by fibroblasts can be heat shock protein 90a, total and activated signal transducer and activator of transcription 3, proangiogenic (miR-126, miR-130a, miR-132) and antiinflammatory (miR124a, miR- 125b) miRNAs, and miRNA regulating collagen deposition (miR-21), which have shown to have beneficial effect on wound healing [132].

Further studies are under way to perform a complete analysis of the exosome cargo and to develop a manufacturing process for the production of exosomes in the quantity and quality required for testability in clinical trials [132].

4.14 Dressings

Therapeutic strategies for chronic wound healing, such as polymeric matrices loaded with drugs, have been designed, aiming to target the whole cascade more specifically [2,129,130,133,134]. These wound dressings encompass a wide range of products containing synthetic matrix scaffolds, animal-derived matrices, and human tissue [9].

Smart dressings can also monitor physical and chemical parameters important in wound healing such as some biomarkers such as PH, infection using specific sensors, and technology to avoid removal of the dressing several times and reduce pain and discomfort [135,136]. New ideas and technologies have emerged such as the graphene electronic tattoo sensors to measure several in vivo parameters from skin temperature and hydration to electrocardiogram, electromyogram, and electroencephalogram [137]. Further development could inform on more specific wound healing parameters.

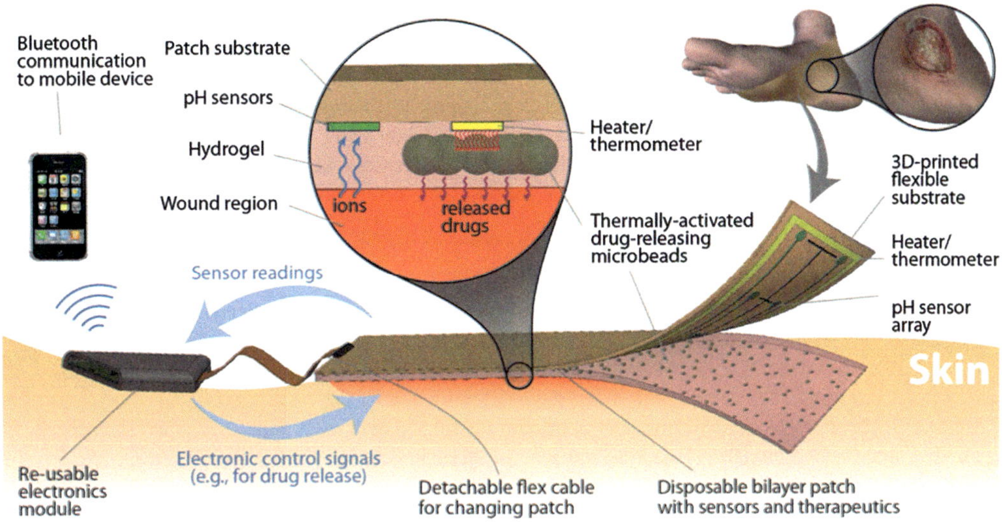

Figure 12.8
A smart bandage with wound-covering component (right), containing sensors and a drug carrier, and a microprocessor (left) that interprets sensor input and triggers drug delivery. *Credits: Tufts University.*

Effective smart bandages should be mechanically flexible, breathable, easy to apply, and capable of reporting quantitative information about the wound status in real time to guide.

3D cell imprinted substrates and nanoscale-architectured surfaces have emerged with beneficial properties and features [138]. The addition of exosomes, miRNA, and the biotechnology-related advances such as powerful gene-editing techniques including CRISPR-Cas–based systems will create a milestone in the therapy of various diseases, particularly with major effects on treatment of chronic wound healing [138] (Fig. 12.8).

5. Conclusions

The challenge is now to improve the delivery of effective factors at a specific time/phase of the wound healing that could be performed in cascade with multiple time release. There is also a requirement to consider different needs related to specific chronic wounds/individuals.

Many research groups are working very hard on nanotechnologies and applications to nanomedicine, which should lead to new and innovative approaches to the treatment of chronic wounds. Prevention should not be forgotten, and indeed smart sensors could be very useful not only for monitoring the wound healing process but also to inform on cutaneous pressure wound risk.

References

[1] Shaw TJ, Martin P. Wound repair at a glance. J Cell Sci 2009;122(Pt 18):3209–13.

[2] Kaplani K, Koutsi S, Armenis V, Skondra FG, Karantzelis N, Champeris Tsaniras S, et al. Wound healing related agents: ongoing research and perspectives. Adv Drug Deliv Rev 2018;129:242–53.

[3] Choi EP, Chin WY, Wan EY, Lam CL. Evaluation of the internal and external responsiveness of the Pressure Ulcer Scale for Healing (PUSH) tool for assessing acute and chronic wounds. J Adv Nurs 2016;72(5):1134–43.

[4] Sgonc R, Gruber J. Age-related aspects of cutaneous wound healing: a mini-review. Gerontology 2013;59(2):159–64.

[5] Eming SA, Martin P, Tomic-Canic M. Wound repair and regeneration: mechanisms, signaling, and translation. Sci Transl Med 2014;6(265):265sr6.

[6] Meng Z, Zhou D, Gao Y, Zeng M, Wang W. miRNA delivery for skin wound healing. Adv Drug Deliv Rev 2018;129:308–18.

[7] Ud-Din S, Bayat A. Non-animal models of wound healing in cutaneous repair: in silico, in vitro, ex vivo, and in vivo models of wounds and scars in human skin. Wound Repair Regen 2017;25(2):164–76.

[8] Davidson JM. Animal models for wound repair. Arch Dermatol Res 1998;290(Suppl):S1–11.

[9] Wells A, Nuschke A, Yates CC. Skin tissue repair: matrix microenvironmental influences. Matrix Biol 2016;49:25–36.

[10] Mele E. Electrospinning of natural polymers for advanced wound care: towards responsive and adaptive dressings. J Mater Chem B 2016;4:4801–12.

[11] Sindrilaru A, Scharffetter-Kochanek K. Disclosure of the culprits: macrophages-versatile regulators of wound healing. Adv Wound Care (New Rochelle) 2013;2(7):357–68.

[12] Krzyszczyk P, Schloss R, Palmer A, Berthiaume F. The role of macrophages in acute and chronic wound healing and interventions to promote pro-wound healing phenotypes. Front Physiol 2018;9:419.

[13] Seaton M, Hocking A, Gibran NS. Porcine models of cutaneous wound healing. ILAR J 2015;56(1):127–38.

[14] Baum CL, Arpey CJ. Normal cutaneous wound healing: clinical correlation with cellular and molecular events. Dermatol Surg 2005;31(6):674–86. Discussion 86.

[15] Gurtner GC, Werner S, Barrandon Y, Longaker MT. Wound repair and regeneration. Nature 2008;453(7193):314–21.

[16] Rousselle P, Braye F, Dayan G. Re-epithelialization of adult skin wounds: cellular mechanisms and therapeutic strategies. Adv Drug Deliv Rev 2018 Jul 5. pii: S0169-409X(18)30158-3.

[17] Rousselle P, Montmasson M, Garnier C. Extracellular matrix contribution to skin wound re-epithelialization. Matrix Biol 2018.

[18] Falanga V. Wound healing and its impairment in the diabetic foot. Lancet 2005;366(9498):1736–43.

[19] Zhao R, Liang H, Clarke E, Jackson C, Xue M. Inflammation in chronic wounds. Int J Mol Sci 2016;17(12).

[20] Bosanquet DC, Harding KG. Wound duration and healing rates: cause or effect? Wound Repair Regen 2014;22(2):143–50.

[21] Hesketh M, Sahin KB, West ZE, Murray RZ. Macrophage phenotypes regulate scar formation and chronic wound healing. Int J Mol Sci 2017;18(7).

[22] Roy S, Khanna S, Nallu K, Hunt TK, Sen CK. Dermal wound healing is subject to redox control. Mol Ther 2006;13(1):211–20.

[23] Schafer M, Werner S. Oxidative stress in normal and impaired wound repair. Pharmacol Res 2008;58(2):165–71.

[24] Frykberg RG, Banks J. Challenges in the treatment of chronic wounds. Adv Wound Care (New Rochelle) 2015;4(9):560–82.

[25] Jiang LP, Tu Q, Wang Y, Zhang E. Ischemia-reperfusion injury-induced histological changes affecting early stage pressure ulcer development in a rat model. Ostomy Wound Manage 2011;57(2):55–60.
[26] Demidova-Rice TN, Hamblin MR, Herman IM. Acute and impaired wound healing: pathophysiology and current methods for drug delivery, part 2: role of growth factors in normal and pathological wound healing: therapeutic potential and methods of delivery. Adv Skin Wound Care 2012;25(8):349–70.
[27] Singer AJ, McClain SA. Development of a porcine excisional wound model. Acad Emerg Med 2003;10(10):1029–33.
[28] Mustoe TA, O'Shaughnessy K, Kloeters O. Chronic wound pathogenesis and current treatment strategies: a unifying hypothesis. Plast Reconstr Surg 2006;117(7 Suppl):35S–41S.
[29] Clark RA. Epithelial-mesenchymal networks in wounds: a hierarchical view. J Investig Dermatol 2003;120(6):ix–xi.
[30] Gallant-Behm CL, Du P, Lin SM, Marucha PT, DiPietro LA, Mustoe TA. Epithelial regulation of mesenchymal tissue behavior. J Investig Dermatol 2011;131(4):892–9.
[31] Pastar I, Wong LL, Egger AN, Tomic-Canic M. Descriptive vs mechanistic scientific approach to study wound healing and its inhibition: is there a value of translational research involving human subjects? Exp Dermatol 2018;27(5):551–62.
[32] Lindblad WJ. Considerations for selecting the correct animal model for dermal wound-healing studies. J Biomater Sci Polym Ed 2008;19(8):1087–96.
[33] Summerfield A, Meurens F, Ricklin ME. The immunology of the porcine skin and its value as a model for human skin. Mol Immunol 2015;66(1):14–21.
[34] Rittie L, Sachs DL, Orringer JS, Voorhees JJ, Fisher GJ. Eccrine sweat glands are major contributors to reepithelialization of human wounds. Am J Pathol 2013;182(1):163–71.
[35] Sullivan TP, Eaglstein WH, Davis SC, Mertz P. The pig as a model for human wound healing. Wound Repair Regen 2001;9(2):66–76.
[36] Murray PJ. Macrophage polarization. Annu Rev Physiol 2017;79:541–66.
[37] Chen L, Mirza R, Kwon Y, DiPietro LA, Koh TJ. The murine excisional wound model: contraction revisited. Wound Repair Regen 2015;23(6):874–7.
[38] Isakson M, de Blacam C, Whelan D, McArdle A, Clover AJ. Mesenchymal stem cells and cutaneous wound healing: current evidence and future potential. Stem Cell Int 2015;2015:831095.
[39] Hsu YC, Li L, Fuchs E. Emerging interactions between skin stem cells and their niches. Nat Med 2014;20(8):847–56.
[40] Langton AK, Herrick SE, Headon DJ. An extended epidermal response heals cutaneous wounds in the absence of a hair follicle stem cell contribution. J Investig Dermatol 2008;128(5):1311–8.
[41] Abdullahi A, Amini-Nik S, Jeschke MG. Animal models in burn research. Cell Mol Life Sci 2014;71(17). 3241–3255.
[42] Peterkofsky B. Ascorbate requirement for hydroxylation and secretion of procollagen: relationship to inhibition of collagen synthesis in scurvy. Am J Clin Nutr 1991;54(6 Suppl):1135S–40S.
[43] O'Loughlin A, Kulkarni M, Vaughan EE, Creane M, Liew A, Dockery P, et al. Autologous circulating angiogenic cells treated with osteopontin and delivered via a collagen scaffold enhance wound healing in the alloxan-induced diabetic rabbit ear ulcer model. Stem Cell Res Ther 2013;4(6):158.
[44] Schivo M, Aksenov AA, Pasamontes A, Cumeras R, Weisker S, Oberbauer AM, et al. A rabbit model for assessment of volatile metabolite changes observed from skin: a pressure ulcer case study. J Breath Res 2017;11(1):016007.
[45] Sigaudo-Roussel D, Demiot C, Fromy B, Koitka A, Leftheriotis G, Abraham P, et al. Early endothelial dysfunction severely impairs skin blood flow response to local pressure application in streptozotocin-induced diabetic mice. Diabetes 2004;53(6):1564–9.
[46] Heydemann A. An overview of murine high fat diet as a model for type 2 diabetes mellitus. J Diabetes Res 2016;2016:2902351.

[47] Abadir P, Hosseini S, Faghih M, Ansari A, Lay F, Smith B, et al. Topical reformulation of valsartan for treatment of chronic diabetic wounds. J Investig Dermatol 2018;138(2):434–43.
[48] Hsieh J, Longuet C, Maida A, Bahrami J, Xu E, Baker CL, et al. Glucagon-like peptide-2 increases intestinal lipid absorption and chylomicron production via CD36. Gastroenterology 2009;137(3):997–1005. e1-4.
[49] Blakytny R, Jude E. The molecular biology of chronic wounds and delayed healing in diabetes. Diabet Med 2006;23(6):594–608.
[50] Nunan R, Harding KG, Martin P. Clinical challenges of chronic wounds: searching for an optimal animal model to recapitulate their complexity. Dis Model Mech 2014;7(11):1205–13.
[51] Strong AL, Bowles AC, MacCrimmon CP, Frazier TP, Lee SJ, Wu X, et al. Adipose stromal cells repair pressure ulcers in both young and elderly mice: potential role of adipogenesis in skin repair. Stem Cells Transl Med 2015;4(6):632–42.
[52] Park KH, Han SH, Hong JP, Han SK, Lee DH, Kim BS, et al. Topical epidermal growth factor spray for the treatment of chronic diabetic foot ulcers: a phase III multicenter, double-blind, randomized, placebo-controlled trial. Diabetes Res Clin Pract 2018;142:335–44.
[53] Thornton MJ. Estrogens and aging skin. Dermatoendocrinol 2013;5(2):264–70.
[54] Wilkinson HN, Hardman MJ. The role of estrogen in cutaneous ageing and repair. Maturitas 2017;103:60–4.
[55] Pyter LM, Yang L, da Rocha JM, Engeland CG. The effects of social isolation on wound healing mechanisms in female mice. Physiol Behav 2014;127:64–70.
[56] Keyes BE, Liu S, Asare A, Naik S, Levorse J, Polak L, et al. Impaired epidermal to dendritic T cell signaling slows wound repair in aged skin. Cell 2016;167(5):1323–1338 e14.
[57] Fang RC, Mustoe TA. Animal models of wound healing: utility in transgenic mice. J Biomater Sci Polym Ed 2008;19(8):989–1005.
[58] Wong VW, Sorkin M, Glotzbach JP, Longaker MT, Gurtner GC. Surgical approaches to create murine models of human wound healing. J Biomed Biotechnol 2011;2011:969618.
[59] Laverdet B, Girard D, Bayout A, Bordeau N, Demiot C, Desmouliere A. Effects of small-fiber neuropathy induced by resiniferatoxin on skin healing and axonal regrowth after burn. Burns 2017;43(3):562–72.
[60] Kim DJ, Mustoe T, Clark RA. Cutaneous wound healing in aging small mammals: a systematic review. Wound Repair Regen 2015;23(3):318–39.
[61] Bouten CV, Oomens CW, Baaijens FP, Bader DL. The etiology of pressure ulcers: skin deep or muscle bound? Arch Phys Med Rehabil 2003;84(4):616–9.
[62] Stekelenburg A, Oomens CW, Strijkers GJ, Nicolay K, Bader DL. Compression-induced deep tissue injury examined with magnetic resonance imaging and histology. J Appl Physiol (1985) 2006;100(6):1946–54.
[63] Salcido R, Popescu A, Ahn C. Animal models in pressure ulcer research. J Spinal Cord Med 2007;30(2):107–16.
[64] de Wert LA, Bader DL, Oomens CW, Schoonhoven L, Poeze M, Bouvy ND. A new method to evaluate the effects of shear on the skin. Wound Repair Regen 2015;23(6):885–90.
[65] Oomens CW, Bader DL, Loerakker S, Baaijens F. Pressure induced deep tissue injury explained. Ann Biomed Eng 2015;43(2):297–305.
[66] Gawlitta D, Oomens CW, Bader DL, Baaijens FP, Bouten CV. Temporal differences in the influence of ischemic factors and deformation on the metabolism of engineered skeletal muscle. J Appl Physiol (1985) 2007;103(2):464–73.
[67] Peirce SM, Skalak TC, Rodeheaver GT. Ischemia-reperfusion injury in chronic pressure ulcer formation: a skin model in the rat. Wound Repair Regen 2000;8(1):68–76.
[68] Reid RR, Sull AC, Mogford JE, Roy N, Mustoe TA. A novel murine model of cyclical cutaneous ischemia-reperfusion injury. J Surg Res 2004;116(1):172–80.

[69] Fromy B, Lingueglia E, Sigaudo-Roussel D, Saumet JL, Lazdunski M. Asic3 is a neuronal mechanosensor for pressure-induced vasodilation that protects against pressure ulcers. Nat Med 2012;18:1205–7.

[70] Danigo A, Magy L, Richard L, Desmouliere A, Bourthoumieu S, Funalot B, et al. Neuroprotective effect of erythropoietin against pressure ulcer in a mouse model of small fiber neuropathy. PLoS One 2014;9(11):e113454.

[71] Daniel RK, Priest DL, Wheatley DC. Etiologic factors in pressure sores: an experimental model. Arch Phys Med Rehabil 1981;62(10):492–8.

[72] Steinberg JP, Gurjala AN, Jia S, Hong SJ, Galiano RD, Mustoe TA. Evaluating the effects of subclinical, cyclic ischemia-reperfusion injury on wound healing using a novel device in the rabbit ear. Ann Plast Surg 2014;72(6):698–705.

[73] Saito Y, Hasegawa M, Fujimoto M, Matsushita T, Horikawa M, Takenaka M, et al. The loss of MCP-1 attenuates cutaneous ischemia-reperfusion injury in a mouse model of pressure ulcer. J Investig Dermatol 2008;128(7):1838–51.

[74] Donato-Trancoso A, Monte-Alto-Costa A, Romana-Souza B. Olive oil-induced reduction of oxidative damage and inflammation promotes wound healing of pressure ulcers in mice. J Dermatol Sci 2016;83(1):60–9.

[75] Romana-Souza B, Santos JS, Bandeira LG, Monte-Alto-Costa A. Selective inhibition of COX-2 improves cutaneous wound healing of pressure ulcers in mice through reduction of iNOS expression. Life Sci 2016;153:82–92.

[76] Gust MJ, Hong SJ, Fang RC, Lanier ST, Buck 2nd DW, Nunez JM, et al. Adipose tissue drives response to ischemia-reperfusion injury in a murine pressure sore model. Plast Reconstr Surg 2017;139(5):1128e–38e.

[77] Nguyen-Tu MS, Begey AL, Decorps J, Boizot J, Sommer P, Fromy B, et al. Skin microvascular response to pressure load in obese mice. Microvasc Res 2013;90:138–43.

[78] Nixon J, Cranny G, Bond S. Pathology, diagnosis, and classification of pressure ulcers: comparing clinical and imaging techniques. Wound Repair Regen 2005;13(4):365–72.

[79] Bergstrand S, Kallman U, Ek AC, Lindberg LG, Engstrom M, Sjoberg F, et al. Pressure-induced vasodilation and reactive hyperemia at different depths in sacral tissue under clinically relevant conditions. Microcirculation 2014;21(8):761–71.

[80] Cristobal L, Ortega MA, Asunsolo A, Romero B, Alvarez-Mon M, Bujan J, et al. Human skin model for mimic dermal studies in pathology with a clinical implication in pressure ulcers. Histol Histopathol 2018:11990.

[81] Maldonado AA, Cristobal L, Martin-Lopez J, Mallen M, Garcia-Honduvilla N, Bujan J. A novel model of human skin pressure ulcers in mice. PLoS One 2014;9(10):e109003.

[82] Tsutakawa S, Kobayashi D, Kusama M, Moriya T, Nakahata N. Nicotine enhances skin necrosis and expression of inflammatory mediators in a rat pressure ulcer model. Br J Dermatol 2009;161(5):1020–7.

[83] Buijs S, Hermans K, Maertens L, Van Caelenberg A, Tuyttens FA. Effects of semi-group housing and floor type on pododermatitis, spinal deformation and bone quality in rabbit does. Animal 2014;8(10):1728–34.

[84] Rommers JM, de Jong IC. Technical note: plastic mats prevent footpad injuries in rabbit does. World Rabbit Science 2011;19:233–7.

[85] Mizisin AP, Kalichman MW, Garrett RS, Dines KC. Tactile hyperesthesia, altered epidermal innervation and plantar nerve injury in the hindfeet of rats housed on wire grates. Brain Res 1998;788(1–2):13–9.

[86] Yavuz M, Ersen A, Hartos J, Schwarz B, Garrett AG, Lavery LA, et al. Plantar shear stress in individuals with a history of diabetic foot ulcer: an emerging predictive marker for foot ulceration. Diabetes Care 2017;40(2):e14–5.

[87] Delbridge L, Ctercteko G, Fowler C, Reeve TS, Le Quesne LP. The aetiology of diabetic neuropathic ulceration of the foot. Br J Surg 1985;72(1):1–6.

[88] Ohura T, Takahashi M, Ohura Jr N. Influence of external forces (pressure and shear force) on superficial layer and subcutis of porcine skin and effects of dressing materials: are dressing materials beneficial for reducing pressure and shear force in tissues? Wound Repair Regen 2008;16(1):102–7.
[89] Linder-Ganz E, Gefen A. The effects of pressure and shear on capillary closure in the microstructure of skeletal muscles. Ann Biomed Eng 2007;35(12):2095–107.
[90] Goldstein B, Sanders J. Skin response to repetitive mechanical stress: a new experimental model in pig. Arch Phys Med Rehabil 1998;79(3):265–72.
[91] Goossens RH, Zegers R, Hoek van Dijke GA, Snijders CJ. Influence of shear on skin oxygen tension. Clin Physiol 1994;14(1):111–8.
[92] Sari Y, Minematsu T, Huang L, Noguchi H, Mori T, Nakagami G, et al. Establishment of a novel rat model for deep tissue injury deterioration. Int Wound J 2015;12(2):202–9.
[93] Gefen A. Deep tissue injury from a bioengineering point of view. Ostomy Wound Manage 2009;55(4):26–36.
[94] Kosiak M. Etiology and pathology of ischemic ulcers. Arch Phys Med Rehabil 1959;40(2):62–9.
[95] Ceelen KK, Stekelenburg A, Loerakker S, Strijkers GJ, Bader DL, Nicolay K, et al. Compression-induced damage and internal tissue strains are related. J Biomech 2008;41(16):3399–404.
[96] Linder-Ganz E, Shabshin N, Itzchak Y, Yizhar Z, Siev-Ner I, Gefen A. Strains and stresses in sub-dermal tissues of the buttocks are greater in paraplegics than in healthy during sitting. J Biomech 2008;41(3):567–80.
[97] Loerakker S, Bader DL, Baaijens FP, Oomens CW. Which factors influence the ability of a computational model to predict the in vivo deformation behaviour of skeletal muscle? Comput Methods Biomech Biomed Eng 2013;16(3):338–45.
[98] Stekelenburg A, Strijkers GJ, Parusel H, Bader DL, Nicolay K, Oomens CW. Role of ischemia and deformation in the onset of compression-induced deep tissue injury: MRI-based studies in a rat model. J Appl Physiol (1985) 2007;102(5):2002–11.
[99] Bosboom EM, Bouten CV, Oomens CW, Baaijens FP, Nicolay K. Quantifying pressure sore-related muscle damage using high-resolution MRI. J Appl Physiol (1985) 2003;95(6):2235–40.
[100] Oomens CW, Bressers OF, Bosboom EM, Bouten CV, Blader DL. Can loaded interface characteristics influence strain distributions in muscle adjacent to bony prominences? Comput Methods Biomech Biomed Eng 2003;6(3):171–80.
[101] Wassermann E, van Griensven M, Gstaltner K, Oehlinger W, Schrei K, Redl H. A chronic pressure ulcer model in the nude mouse. Wound Repair Regen 2009;17(4):480–4.
[102] Linder-Ganz E, Gefen A. Mechanical compression-induced pressure sores in rat hindlimb: muscle stiffness, histology, and computational models. J Appl Physiol (1985) 2004;96(6):2034–49.
[103] Minematsu T, Nakagami G, Sari Y, Akase T, Sugama J, Nagase T, et al. Candidate biomarkers for deep tissue damage from molecular biological and biochemical aspects. J Tissue Viability 2010;19(2): 77–83.
[104] Sari Y, Nagase T, Minematsu T, Akase T, Nakagami G, Sanada H, et al. Hypoxia is involved in deep tissue injury formation in a rat model. Wounds 2010;22(2):44–51.
[105] Nelissen JL, Traa WA, de Boer HH, de Graaf L, Mazzoli V, Savci-Heijink CD, et al. An advanced magnetic resonance imaging perspective on the etiology of deep tissue injury. J Appl Physiol (1985) 2018;124(6):1580–96.
[106] Kottner J, Black J, Call E, Gefen A, Santamaria N. Microclimate: a critical review in the context of pressure ulcer prevention. Clin Biomech (Bristol, Avon) 2018;59:62–70.
[107] Kokate JY, Leland KJ, Held AM, Hansen GL, Kveen GL, Johnson BA, et al. Temperature-modulated pressure ulcers: a porcine model. Arch Phys Med Rehabil 1995;76(7):666–73.
[108] Ahn ST, Mustoe TA. Effects of ischemia on ulcer wound healing: a new model in the rabbit ear. Ann Plast Surg 1990;24(1):17–23.
[109] Chien S. Ischemic rabbit ear model created by minimally invasive surgery. Wound Repair Regen 2007;15(6):928–35.

[110] Trujillo AN, Kesl SL, Sherwood J, Wu M, Gould LJ. Demonstration of the rat ischemic skin wound model. J Vis Exp 2015;98:e52637.
[111] Domergue S, Jorgensen C, Noel D. Advances in research in animal models of burn-related hypertrophic scarring. J Burn Care Res 2015;36(5):e259–66.
[112] Pereira CT, Herndon DN. The pharmacologic modulation of the hypermetabolic response to burns. Adv Surg 2005;39:245–61.
[113] Wolcott RD, Hanson JD, Rees EJ, Koenig LD, Phillips CD, Wolcott RA, et al. Analysis of the chronic wound microbiota of 2,963 patients by 16S rDNA pyrosequencing. Wound Repair Regen 2016;24(1):163–74.
[114] Suleman L, Percival SL. Biofilm-infected pressure ulcers: current knowledge and emerging treatment strategies. Adv Exp Med Biol 2015;831:29–43.
[115] Krishnaswamy VR, Mintz D, Sagi I. Matrix metalloproteinases: the sculptors of chronic cutaneous wounds. Biochim Biophys Acta 2017;1864(11 Pt B):2220–7.
[116] James GA, Swogger E, Wolcott R, Pulcini E, Secor P, Sestrich J, et al. Biofilms in chronic wounds. Wound Repair Regen 2008;16(1):37–44.
[117] Kirketerp-Moller K, Jensen PO, Fazli M, Madsen KG, Pedersen J, Moser C, et al. Distribution, organization, and ecology of bacteria in chronic wounds. J Clin Microbiol 2008;46(8):2717–22.
[118] Bessa LJ, Fazii P, Di Giulio M, Cellini L. Bacterial isolates from infected wounds and their antibiotic susceptibility pattern: some remarks about wound infection. Int Wound J 2015;12(1):47–52.
[119] Schierle CF, De la Garza M, Mustoe TA, Galiano RD. Staphylococcal biofilms impair wound healing by delaying reepithelialization in a murine cutaneous wound model. Wound Repair Regen 2009;17(3):354–9.
[120] Roy S, Elgharably H, Sinha M, Ganesh K, Chaney S, Mann E, et al. Mixed-species biofilm compromises wound healing by disrupting epidermal barrier function. J Pathol 2014;233(4):331–43.
[121] Brubaker AL, Rendon JL, Ramirez L, Choudhry MA, Kovacs EJ. Reduced neutrophil chemotaxis and infiltration contributes to delayed resolution of cutaneous wound infection with advanced age. J Immunol 2013;190(4):1746–57.
[122] Gould L, Abadir P, Brem H, Carter M, Conner-Kerr T, Davidson J, et al. Chronic wound repair and healing in older adults: current status and future research. J Am Geriatr Soc 2015;63(3):427–38.
[123] Watters C, DeLeon K, Trivedi U, Griswold JA, Lyte M, Hampel KJ, et al. *Pseudomonas aeruginosa* biofilms perturb wound resolution and antibiotic tolerance in diabetic mice. Med Microbiol Immunol 2013;202(2):131–41.
[124] Zarrintaj P, Moghaddam AS, Manouchehri S, Atoufi Z, Amiri A, Amirkhani MA, et al. Can regenerative medicine and nanotechnology combine to heal wounds? The search for the ideal wound dressing. Nanomedicine (Lond) 2017;12(19):2403–22.
[125] Barrientos S, Brem H, Stojadinovic O, Tomic-Canic M. Clinical application of growth factors and cytokines in wound healing. Wound Repair Regen 2014;22(5):569–78.
[126] Strong AL, Neumeister MW, Levi B. Stem cells and tissue engineering: regeneration of the skin and its contents. Clin Plast Surg 2017;44(3):635–50.
[127] Kurita M, Araoka T, Hishida T, O'Keefe DD, Takahashi Y, Sakamoto A, et al. In vivo reprogramming of wound-resident cells generates skin epithelial tissue. Nature 2018;561(7722):243–7.
[128] Mulholland EJ, Dunne N, McCarthy HO. MicroRNA as therapeutic targets for chronic wound healing. Mol Ther Nucleic Acids 2017;8:46–55.
[129] Ranjan S, Fontana F, Ullah H, Hirvonen J, Santos HA. Microparticles to enhance delivery of drugs and growth factors into wound sites. Ther Deliv 2016;7(10):711–32.
[130] Gagliardi M. Novel biodegradable nanocarriers for enhanced drug delivery. Ther Deliv 2016;7(12):809–26.
[131] Hu L, Wang J, Zhou X, Xiong Z, Zhao J, Yu R, et al. Exosomes derived from human adipose mensenchymal stem cells accelerates cutaneous wound healing via optimizing the characteristics of fibroblasts. Sci Rep 2016;6:32993.

[132] Geiger A, Walker A, Nissen E. Human fibrocyte-derived exosomes accelerate wound healing in genetically diabetic mice. Biochem Biophys Res Commun 2015;467(2):303–9.

[133] Gagliardi M, Bertero A, Bardi G, Bifone A. A poly(ether-ester) copolymer for the preparation of nanocarriers with improved degradation and drug delivery kinetics. Mater Sci Eng C Mater Biol Appl 2016;59:488–99.

[134] Anish S. Skin substitutes in dermatology. Indian J Dermatol Venereol Leprol 2015;81(2):175–8.

[135] Jankowska DA, Bannwarth MB, Schulenburg C, Faccio G, Maniura-Weber K, Rossi RM, et al. Simultaneous detection of pH value and glucose concentrations for wound monitoring applications. Biosens Bioelectron 2017;87:312–9.

[136] Mostafalu P, Tamayol A, Rahimi R, Ochoa M, Khalilpour A, Kiaee G, et al. Smart bandage for monitoring and treatment of chronic wounds. Small 2018:e1703509.

[137] Kabiri Ameri S, Ho R, Jang H, Tao L, Wang Y, Wang L, et al. Graphene electronic tattoo sensors. ACS Nano 2017;11(8):7634–41.

[138] Mofazzal Jahromi MA, Sahandi Zangabad P, Moosavi Basri SM, Sahandi Zangabad K, Ghamarypour A, Aref AR, et al. Nanomedicine and advanced technologies for burns: preventing infection and facilitating wound healing. Adv Drug Deliv Rev 2018;123:33–64.

[139] Pasparakis M, Haase I, Nestle FO. Mechanisms regulating skin immunity and inflammation. Nat Rev Immunol 2014;14:5 289–301.

CHAPTER 13

Smart technologies in wound prevention and care

Ana Scalamandré BS [1], Kath M. Bogie D. Phil [1,2]

[1]Department of Biomedical Engineering, Case Western Reserve University, Cleveland, Ohio, United States; [2]Louis Stokes Cleveland Veterans Affairs Medical Center(LSCVAMC), Cleveland, Ohio, United States

Chapter Outline

1. Background

A soft tissuewound is considered clinically healed when there is restoration of normal anatomic continuity to a previously disrupted area of tissue. Normal healing occurs when there is systematic progress along the healing pathway, from early hemostasis to inflammation, proliferation/closure, and remodeling. However, many wounds fail to heal normally, exhibiting impaired or delayed healing. When a wound fails to progress toward healing for 3 weeks, it becomes chronic. These chronic wounds occur in all patient populations. There are many types of chronic ischemic wounds, including diabetic ulcers,

Innovations and Emerging Technologies in Wound Care. https://doi.org/10.1016/B978-0-12-815028-3.00013-4

pressure injuries (PrI), venous ulcers, arterial ulcers, and postsurgical complications. In the United States, PrI alone were responsible for 60,000 deaths in 2018, which accounts for more casualties than car accidents [1]. While all patient populations are at some level of risk, those with increased risk include persons with reduced mobility and/or sensation. The elderly, persons with spinal cord injury (SCI) or cerebrovascular accident/hemiparesis and those with uncontrolled diabetes, are chronically at increased risk, i.e., throughout their lifetimes [2,3]. Those with infectious diseases in the ICU or having surgery lasting more than 2 h without changing positions are acutely at risk [4,5]. Development of a chronic, nonhealing wound has a profoundly detrimental impact on the care and rehabilitation of affected individuals [6].

The effective management of chronic wounds may be considered under the World Health Organization (WHO) primary, secondary, and tertiary prevention models [7]. Primary chronic wound prevention addresses reducing incidence by identifying high-risk patients before wounds develop. This remains challenging, due in no small part to the multiple factors that play into both the specific condition and the complex clinical profiles of individuals at risk [8]. Secondary prevention aims to reduce the impact of chronic wounds by providing early and effective treatments and reduce recurrence. Tertiary prevention seeks to minimize the incidence of further related complications.

Not only do wounds cause significant pain and disability for affected individuals but they also incur a major economic burden. The total costs of wound management encompass primary, secondary, and tertiary prevention. While the true societal costs remain unknown, in the United States, clinical treatment costs in excess of $25 billion per annum in general wound care were reported in 2009, with another $12 billion for treatment of skin scarring [9]. In the United Kingdom, PrI alone were reported in 2018 to cost the National Health Service an estimate of £1.4–£2.1billion, which represents 78% of their wound treatment budget [1]. Out-of-pocket cost for self-funded patients should also be taken into consideration. An Australian study in 2017 found that individuals who suffer from chronic wound can spend 10% on their income on wound care products [10]. This causes as financial burden as further negatively impacts quality of life (QoL). As the global population ages and the number of people with chronic conditions such as obesity and diabetes increase, the amount spent on wound care will continue to escalate.

Unfortunately, wound management is one of the few areas where little reduction in incidence has been seen in recent years, demonstrating that new approaches to prevention remain critically important. The development of smart technologies to monitor risk, treat incident wounds, and report wound status to clinicians is critically needed to impact wound management in all domains of the WHO model. It has been proposed that resources should be allocated into proactive measures [11], as well as for newer technologies that can provide efficient and cost-effective treatment.

The multifactorial nature of chronic wound care strongly implies that there is highly unlikely to be a universal panacea. Many disciplines not traditionally associated with clinical care are developing devices, materials, and methodologies that could impact wound management. There is a wealth of development in smart technologies and health-care wearables, which have the potential to reverse the current trends for increasing costs and incidence. The ultimate goal to improve the care of individuals with or at risk for chronic wounds should always be borne in mind.

2. Smart technologies in wound prevention (primary prevention)

Effective primary prevention requires accurately identifying at-risk individuals, implementing appropriate preventative measures, and identifying early stage tissue damage. Proactive measures are less detrimental to the individual's QoL and reduce societal costs. However, even for hospital-acquired PrI, the current literature on economic evaluations of primary prevention strategies is limited. There remains a need for high-quality studies to compare current and emerging approaches to primary prevention [12].

Healthy able-bodied individuals rarely develop chronic wounds. However, anyone can become at acute risk following trauma or during prolonged elective surgery [13]. Chronically at-risk individuals may have intrinsic risk factors and/or be exposed to extrinsic risk factors [14]. Intrinsic factors refer to conditions and habits that are within the individual such as mobility impairment, chronic illness, nutritional status, and aging [14]. Extrinsic factors derive from the individual's surroundings and include pressure, friction, humidity, and the shear force between the skin and the individual's environment, such as their bed [14].

Education is a key component of primary prevention of chronic wounds. However, to be effective, the information provided must be valid and tailored to the individual's level of understanding. To evaluate efficacy, quantifiable objective input and output measures must be used, which have predictive accuracy over an appropriate time frame. Current risk assessment scales can provide some short-term predictive accuracy, i.e., within 2–3 days, but long-term predictive value is limited [15,16]. To provide more tailored education, real-time personalized tools are needed.

2.1 Primary prevention and extrinsic risk factors

Primary PrI prevention traditionally relies on regular turning and repositioning, based on the premise of maintaining a safe level of applied pressure for a safe duration. Although regular turning schedules in bed and regular pressure relief while sitting are essential, there is limited evidence for selecting a specific pressure relief schedule [17]. In addition, there is no universal safe interface pressure threshold [18]. It is well established that the widely used metric of interface pressure measurement cannot provide an adequate predictor of PrI risk when used alone [19]. Internal tissue forces cannot be measured [20]. Surface interface pressures can be similar, whereas internal mechanics change because of changes in muscle composition that compromise tissue resilience [21].

Power wheelchair users can also be at risk of PrI development due to foot mispositioning on the footplate. This can occur when the wheelchair is at rest or during driving but may be occurring most frequently during or after tilt for pressure relief when the wheels are not moving. Majerus et al. reported on the development of a smart footplate sensor system that an array of force-sensing resistors and infrared distance sensors to detect the pressure and location of the foot within the immediate confines of the footplate in real time, as the wheelchair is driven [22]. The system shown in Fig. 13.1 can also provide hitherto unavailable alerts to increase user awareness of excessive speeds and ramp inclines or declines, thus enhancing safety during advanced wheelchair operation.

Advanced technologies present new assessment tools for the at-risk population that have the potential to enable early assessment of tissue damage before it becomes visible to the human eye. Surface impedance has been evaluated in a preclinical model as an alternative

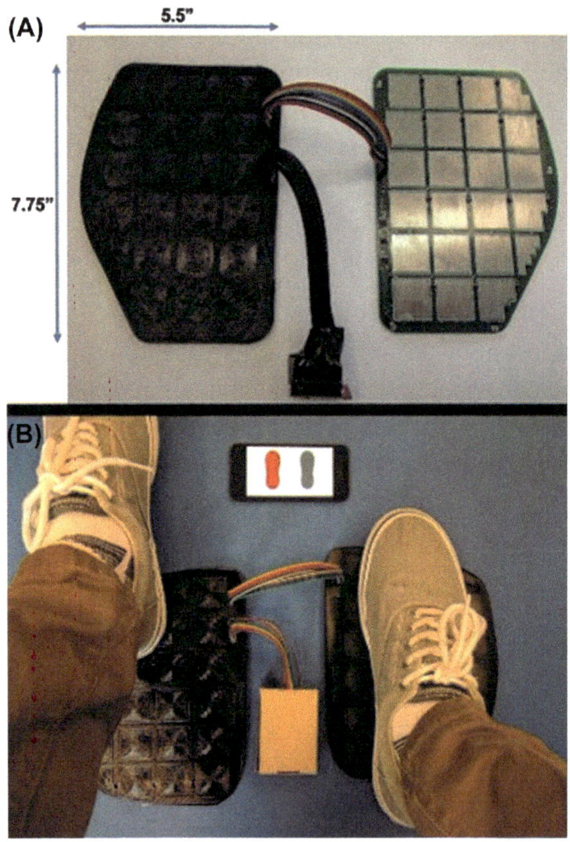

Figure 13.1: Smart footplate sensor system (Henzel et al.).
(Top) Right side printed circuit board showing force-sensing resistors and infrared distance sensors. Left side showing encapsulated footplate. (Bottom) Smartphone application in graphical alarm mode showing foot displacement.

to interface pressure measurement; Swisher et al. developed an inkjet-printed system including a flexible polyethylene naphthalate substrate [23]. The system included an array of gold contact electrode points and outputs a surface impedance map. Preclinical testing found that impedance mapping detected tissue damage before it was visible to the naked eye. However, the impermeable nature of the substrate would limit clinical use. Subepidermal moisture levels may provide an early indication of early stage PrI [24,25]. This approach has been reported as a potential technique for detecting early PrI in individuals with dark skin [26], thus improving care for this population. Multispectral imaging has also been reported as a potential method for detecting erythema in individuals with dark skin pigment [27]. Initial evaluations have also been performed using diagnostic ultrasound [28–31], elastography [32], and assessment of tissue viscoelasticity [33]. A 2017 review by Oliveira et al. found that while subepidermal moisture measurement and ultrasound are promising modalities for early PrI detection, further pragmatic study is needed to clarify the potential of these methodologies for widespread clinical use [34].

A smart bed equipped with a mesh of optic sensors has been developed to measure patient movement, among other parameters [35]. Duvall et al. recently reported on a detection and classification methodology using simple scales placed under each leg of a hospital bed [36]. It was reported that the methodology enabled movement classification with >94% accuracy. These systems could be utilized to alert health-care professionals that the patient has not moved based on standardized turning protocols. To date, the systems have only been tested under controlled conditions with able-bodied participants. The impact on primary PrI prevention is yet to be evaluated.

Bates-Jensen et al. utilized a dermal phase meter to evaluate the subepidermal moisture levels of a wound located in the torso of elderly patients [37]. All participants of this trial were first evaluated using the Braden scale and were classified as being at risk (score 15.6). By using a threshold value, clinicians were able to predict 41% of actual formed wounds, while visual-only assessments predicted 27%. The dermal phase meter is an emerging option for assessing wound risk [38].

With respect to prevention of diabetic foot ulcers (DFU), it is well established that primary prevention is based on regular foot and nail care and adherence to wearing well-fitted shoes, always worn with socks [39]. Gait analysis is a valuable tool to assess walking patterns. However, clinical gait analysis can currently only be carried out in a specialized Gait Lab. It has been found that outputs are prone to incorrect results, as patients are often uncomfortable in such environment [40]. Smart insoles embedded with gait analysis sensors have the potential to facilitate more reliable data collection under real-life use situations. Bae et al. developed a force-sensing resistive sensor embedded in an insole [41], while Liu et al. has used a three-axis accelerometer and gyroscope [42]. Xu et al. developed a smart insole capable of such measurements in an affordable and practical

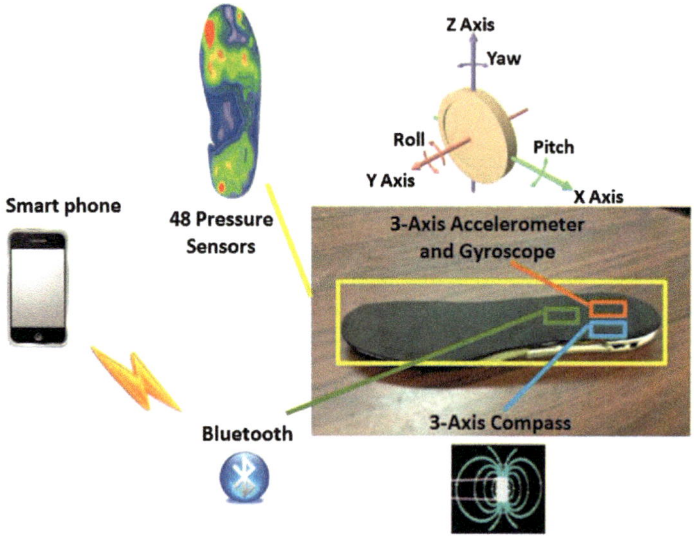

Figure 13.2

Smart insole embedded with sensors and communication module. *Publication: Xu W, Huang M-C, Amini N, Liu JJ, He L, Sarrafzadeh M. Smart insole: A wearable system for gait analysis. Proc 5th International Conference on PErvasive Technologies Related to Assistive Environments, Heraklion, Crete, Greece, June 2012. https://doi.org/10.1145/2413097.2413120.*

manner [43]. The insole included a 48-element pressure sensor array, 3-axis gyroscope, 3-axis accelerometer, and 3-axis compass and can be seen in Fig. 13.2. It was also equipped with a Bluetooth module to send the data to external devices. These approaches provide valuable preliminary insight for proof-of-concept testing. Further development is required for clinical applications.

These smart technologies and algorithms may provide earlier indications of Prl development in individuals already considered to be at risk. However true primary prevention seeks to accurately identify risk before any tissue damage occurs.

2.2 Primary prevention and intrinsic risk factors

There are limited technological approaches to primary PrI prevention by addressing intrinsic risk factors. Clinical practice guidelines (CPG) recommend weight shifting every 20 min. Gluteal muscle can rapidly be replaced by up to 80% intramuscular adipose tissue (IMAT). If the gluteal muscle tissue is replaced by high levels of IMAT, individuals are at increased risk of pressure ulcer/deep tissue injury development [19]. These intrinsic changes in muscle composition cannot be altered by passive exercise. Interventions for at-risk individuals typically focus on extrinsic factors, such as regular weight shifting. However, persons with limited mobility often cannot or do not carry out regular pressure relief as recommended by CPG. Primary prevention thus remains a major challenge, which gluteal electrical stimulation (ES) can address. Unfortunately, surface gluteal ES has limited efficacy [44]: reliable daily placement of surface electrodes is difficult, the gluteal motor point is deep, and increased subcutaneous fat raises the stimulus current needed for a contractile response [45]. In contrast, Intermittent gluteal neuromuscular electric stimulation (iGSTIM) has been reported as an intrinsic prophylactic treatment to combat PrI formation [3,46]. Bogie et al. utilized a semiimplanted system, consisting of four implanted percutaneous electrodes and an external stimulator [3]. Short- and long-term results were evaluated in a patient with SCI. Tissue health assessments were performed every 3 months during the first year of treatment and then annually for the remainder of the study. After 6 weeks of treatment, seated interface pressure distribution reduced significantly. The 40-week assessment showed sustained improvements in tissue health. The study participant's sitting tolerance increased from 6 to 12 hours a day with regular use of iGSTIM. Using bilaterally implanted electrodes provides effective, regular, and automatic weight shifting, which improves muscle quality and maintains improved tissue health over time. Further development is required to develop a fully implanted iGSTIM system for reliable and repeatable weight shifting. A fully implanted iGSTIM system will improve muscle quality and reduce the risk of PrI development due to sustained seating loads. This smart technology will impact QoL for persons with SCI, allowing them to maintain active and engaged lives.

2.3 Primary prevention summary

There is increasing evidence that assessment of extrinsic risk factors alone may not be adequate to achieve primary prevention. Further research is needed to establish reliability and efficacy and provide a sound basis for clinical implementation of smart technologies for effective primary prevention for all individuals at risk for PrI development.

Ideally, assessment of intrinsic and extrinsic factors should be combined: effective primary prevention could include smart support surfaces for at-risk patients to monitor tissue health in contact with the support system. A threshold algorithm for compromised tissue heath could trigger the delivery of treatment, such as iGSTIM, thus preventing PrI formation at that specific location. A smart approach would be less labor-intensive. Tissue health sensor information could be sent to the primary care physician, as well as the nursing staff for remote monitoring. This approach could also be used during lengthy surgeries to reduce the acute risk of PrI formation.

The enhancement of patient/clinician interactions, with personalized near real-time feedback, could facilitate both primary and secondary PrI prevention. Research is needed to evaluate whether technologies such as home telehealth, social messaging platforms, text messages, or "blogs" can promote adoption of protective behaviors and/or early notification of tissue damage. Assistive technology and telehealth can also provide support for primary and secondary prevention by providing access to information as well as equipment and interventions for people with SCI who live distant from or who lack transportation to centers of expertise [8].

3. Smart technologies in wound care (secondary prevention)

Individuals with chronic wounds often have multiple comorbidities. Clinical management includes frequent outpatient clinic visits and often leads to hospital admission for inpatient care, limiting QoL for prolonged periods of time. Treatment options generally require health-care providers to assess the wound healing subjectively. Telehealth wound care clinics can enable remote wound monitoring [47,48]. This approach reduces the burden on patients who live in remote areas and can enable chronic wound, such as a PrI, at home or by attending local medical facilities. It has been reported that time to healing and travel costs can be reduced using wound care telehealth clinics. However, wounds require evaluation by a clinical expert, most often a wound care nurse. Assessment may be in real-time or by store-and-forward secure data transmission. Current telehealth approaches lack the ability to objectively monitor the wound status continuously or remotely. Thus, while travel costs may be reduced, a pragmatic study has shown that the introduction of telehealth does not appear to produce reliable improvement in PrI treatment [49]. A therapeutic approach combined with smart wound assessment and monitoring of treatment efficacy could lead to improved treatment as well as reduced costs.

3.1 Wound staging and measurement

Current approaches to assess wound severity and healing status require clinical expertise and are predominantly subjective. The most widely utilized tool in PrI assessment is the National Pressure Ulcer Advisory Panel grading system, which categorizes wounds into four stages together with “unstageable” and “deep tissue pressure injury” [50]. More complex scales take into account multiple aspects of wound healing. For example, the Pressure Sore Status Tool incorporated 15 variables and generated a wound score from 1 (healthy tissue) to 65 (severe wound) [51]. The TIME framework is used to provide a standardized description of the chronic wound, specifically, tissue (nonviable or deficient), infection (or inflammation), moisture balance, and edge of the wound (nonadvancing or undermined) [52].

Similarly, there are multiple approaches to classify DFU; however, there is no single universally accepted method [53]. The Wagner-Meggitt classification categorizes the wound into six depths [54]. Grade 0 is intact skin, while 5 is the most severe, which categorizes whole foot gangrene. Other methods include the University of Texas system classification, which takes into account wound depth and the presence of infection and ischemia [55].

Manual wound measurement using paper rulers is both rudimentary and inaccurate. This somewhat subjective approach leads to inconsistent outcomes measures that can impact treatment planning. Haghpanah et al. found that the only way to achieve reliability in multiple manual measurements is to use the same observer [56]. The healing of a chronic wound can be a slow and intermittent process. The healing time frame is hard to predict. Several tests may be used to monitor healing; however, metaanalysis indicates that models based on clinical measurement data currently have limited precision due to a low quality of evidence [57,58]. This can be detrimental not only to the patient but also to caretakers who might change the course of the treatment if results are not seen within an expected time frame.

The use of smart technologies may provide detailed point-of-care evaluation using noncontact approaches to obtain more detailed and objective wound evaluation over time. Xu et al. developed a model capable of predicting personalized healing time utilizing detailed stereophotogrammetric imaging [59]. This model includes both surface area and volume changes over time. It illustrates that accurate monitoring of wound geometry can provide personalized predictions of wound healing time and adaptively predict healing progression. However, measurement using imaging techniques can only assess the wound bed area that can be directly visualized, i.e., these approaches cannot measure areas of tunneling or epibole. Other smart approaches to wound assessment combine wound measurement with other critical factors such as periwound temperature [60]. However, these methods are limited by the need to remove all wound dressings to image the wound. Thus, real-time, at-home monitoring using noncontact approaches is limited to use by the clinical care provider.

3.2 Smart wound dressings: critical design factors and clinical requirements

Chronic wounds vary in exudate level, depth, cleanliness, and infection and inflammation levels. A critical design factor for any chronic wound dressing is that it protects the wound from the external environment, serving as a barrier against external bacteria as well as protection against further mechanical trauma. Bandage materials should be hypoallergenic and nontoxic to ensure that the wound dressing is suitable for people with sensitive and/or delicate skin. In addition, clinical studies have established that a moist wound bed is essential for effective wound healing [61]. The wound should be kept moist to optimize the healing process and minimize pain. However, a wound that is soaked with exudate is detrimental to wound healing [62]. Bandage changes should be minimized to avoid disruption of the delicate granulating wound bed. To promote healthy granulation tissue growth, proper wound oxygenation is vital [63]. Although the role of oxygen in wound healing is not fully understood, allowing the wound to gas exchange is essential for effective wound healing [64]. Thus, clinical requirements for an effective wound dressing include that it provides an appropriate moisture balance on the wound bed and be vapor permeable.

3.2.1 Smart wound dressings: measurable parameters

There are many aspects of wound healing that are amenable to monitoring. Measurable parameters, such as temperature, pH, and impedance, provide valuable insight on the progression of wound healing. Open-loop– or closed-loop–sensorized wound dressings can enhance current treatment options enable personalized therapy. An open-loop smart wound dressing would record measurable parameters, characterize the wound status, and inform clinical providers who would decide the treatment plan. A closed-loop smart dressing would be capable of sensing wound status and optimizing treatment based on measured inputs. A closed-loop smart bandage would also monitor the healing process and provide real-time feedback to clinical providers. Real-time continuous monitoring would enable early detection of incipient infections, which could prevent biofilm formation, a major inhibitor of healing in chronic wounds [65]. Table 18.1 shows a list of measurable parameters.

To sense these parameters, there are application-specific sensor requirements that make manufacturing challenging. It is an essential requirement that any wound status will operate reliably within a physiologically appropriate range over time so that clinicians receive valid status information. Wound sensors must be flexible and conform to body shape. They must also be able to withstand the harsh wound environment, which has pH fluctuations and is rich in protein and enzymes [64]. For ease of use, wound status sensors should be fully integrated with the wound dressing so that real-time monitoring can be implemented. Current technologies are outlined below.

Table 13.1: Summary of measurable wound healing parameters

Parameter	Importance
Temperature	Provides insight regarding inflammation/infection and also ischemia
pH	Chronic wounds usually have lower pH than healthy skin. Moreover, pH can provide information regarding the rate of healing and the presence of an infection.
Impedance	Impedance of open wound is an order of magnitude lower than intact skin.
Moisture	Moisture can be an enhancing or degrading factor: wound bed should be moist to optimize healing; however, increased moisture levels can be detrimental.
Oxygen	Hypoxia may impede the healing process by causing tissue loss. The rate of angiogenesis can also be detected by oxygen levels.

3.2.2 pH sensors for wound care

The pH of healthy skin is acidic, ranging from 4.8 to 6.0. This acidity can be attributed to the skin's natural barrier to the external environment, called the "acid mantle" [66]. The chronic wound pH is, however, more basic, ranging from 7.15 to 8.9 [67]. Sharpe et al. reported that more alkaline pH values promoted the proliferation and migration of keratinocyte and fibroblasts in an ex vivo skin model [68]. A direct correlation has been shown between a wound's pH and its healing time; a higher, more alkaline pH is associated with a slower rate of healing for both acute and chronic wounds [69]. Wound pH is also related to the oxygenation levels in the wound bed. It has been found that lowering the pH of the wound by 0.6 units can increase the amount of oxygen by 50% [67]. Lowering the pH can also reduce the toxicity of by-product's bacterial activity and promote angiogenesis [70], which also contributes to faster wound healing.

Wound bed pH has the potential to be a powerful tool to assess the healing process. Colorimetric measurements using surface applied dyes or films can be achieved with relatively straightforward image processing [71]. Using these indicators, pH may be even assessed using the naked eye if the coloration change is significant and the wound bed is exposed [72]. However, there are also associated risks of damage to the delicate wound bed and the separation of the dye to wound tissue. Indirect pH measurements either sense the pH at the wound surface of the wound or evaluate wound exudate. Tamayol et al. developed pH-responsive hydrogel fibers that can provide continuous real-time pH monitoring of intact skin [73]. The approach loaded alginate-based microfibers with pH-responsive beads. Short-term monitoring showed reliable stable measurements at 30 min. The long-term stability and reliably of a hydrogel-based pH sensor in the moist wound environment has yet to be determined. Another approach created potentiometric pH sensors using cotton threads coated with carbon/polyaniline and silver/silver chloride to provide working and reference electrodes [74]. The prototype device for benchtop testing can be seen in Fig. 13.3. Prototype sensors had a sensitivity of 54 mV/pH different buffer solutions with pH values ranging from 4 to 8. In vivo functionality of this sensor in a relevant chronic model has yet to be reported. pH measurement shows promise for evaluation of wound healing, further clinical studies are required to validate these techniques [75].

3.2.3 Temperature sensors for smart wound care

Elevated temperature is a recognized parameter in assessing inflammation and infection [76]. Reduced temperatures are indicative of local ischemia [77] and are also detrimental to wound healing. Periwound temperature can provide insight to localized blood flow and infection status [78]. Continuous temperature monitoring shows potential as an effective approach for assessing wound status and healing path [79].

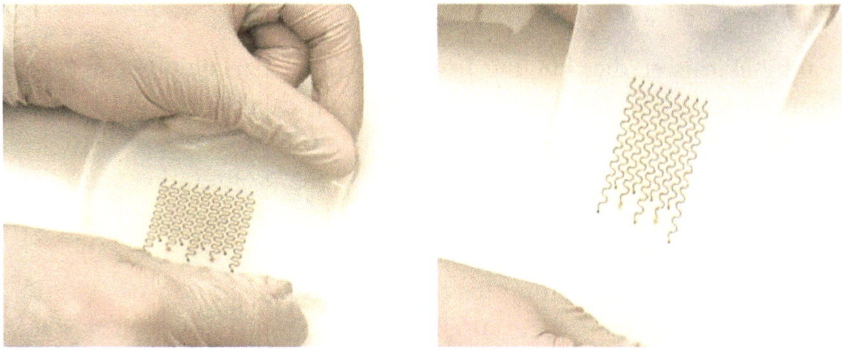

Figure 13.3
Highly stretchable pH sensor. *Published in Rahimi R, Ochoa M, Tamayol A, Khalili S, Khademhosseini A, Ziaie B. Highly stretchable potentiometric pH sensor fabricated via laser carbonization and machining of carbon-polyaniline composite. ACS Appl Mater Interfaces. 2017 Mar 15; 9(10):9015–9023 Image Copyright © 2017 American Chemical Society. Do not include: Permissions info: https://pubs.acs.org/page/copyright/permissions.html. Permission/License is Granted For Your Order at no Charge (03/11/19).*

There are multiple potential approaches to temperature measurement; however, the majority of sensors use a resistive change over a relevant temperature range. The material used needs to a high temperature coefficient of resistance (TCR) with sensitivity within a physiologically appropriate range.

Inkjet printing, a low-cost additive manufacturing process with minimal waste, has shown great promise for use in sensor technology [80,81]. Because of the low cost, it is also feasible for disposable applications, such as single-use bandages [82]. Courbat et al. [80] created temperature sensors using nanoparticle silver traces on a paper substrate, passivated by a layer of parylene, to ensure stability. A TCR of $0.0011°C^{-1}$ was observed. The sensor's flexibility makes it a good candidate to be implemented into a smart bandage; however, it has not yet been evaluated for any clinical application at this time. Vuorinen et al. created temperature sensors using inkjet-printed graphene/PEDOT:PSS on a flexible polyurethane bandage [81]. Sensor sensitivity was greater than 0.06%, with a negative TCR, meaning that an increase in temperature decreased in the sensor's resistance. The system was designed for evaluation of intact temperature over intact skin. More recently, Sui et al. reported on a temperature sensor that uses a novel silver-based which effectively applies traces to flexible substrates using low-temperature inkjet printing [82]. They also found that the sensor sensitivity was independent of the substrate material used. This approach has great potential for the design of customized sensors because the pitch and width of the conductive traces can readily be adjusted to meet functional requirements. The use of biocompatible silver also enables sensor printing for smart biomedical applications (Fig. 13.4). In vivo testing of inkjet-printed temperature sensors in the wound environment is still needed.

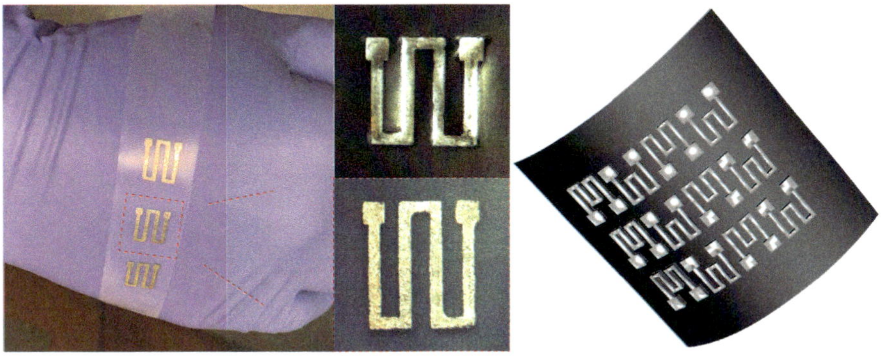

Figure 13.4
Temperature sensor inkjet printed on paper (Sui et al.).

3.3 Smart bandages with drug release

Drug delivery may be defined as passive or active, depending on how the drug is released over time. A bandage with passive drug delivery provides a predetermined dose independent of external factors [83,84]. This open-loop approach is more straightforward to integrate into wound care devices. However, the wound environment is dynamic throughout the healing process; thus, a treatment that can become unsuitable over time. A smart bandage with active drug delivery would have the capability to modulate the dosage throughout a period of time or even at a specific location of the wound [85]. Some common active mechanisms for drug delivery include piezoelectric pumping, thermo-responsive systems, and iontophoretic transport [84].

Nabavinia et al. created a bandage capable of delivering controlled dosage of drugs at specific target locations [86]. The dressing, which is wearable and flexible, is equipped with threads with conductive core involved in PEGDA-ALG hydrogel and thermally triggered drug carriers. The drug carriers embedded on the hydrogel can be selected depending on the applications, and each fiber thread can be independently triggered, thereby giving this bandage flexibility of drug delivery location. A microcontroller is responsible for transmitting and receiving data wirelessly from external devices. The assessment of this system effectiveness was done in vitro, where it was possible to exterminate bacteria using antibiotics and VEGF, and also in vivo utilizing a mice model with a diabetic wound model. The device showed promising results.

3.4 Smart technologies in wound care—future developments

Ideally, a bandage would be able to both sense the status of the wound healing and also treat it accordingly. Fig. 13.5 shows a concept for a smart bandage that can both actively sense wound healing status and treat concurrently.

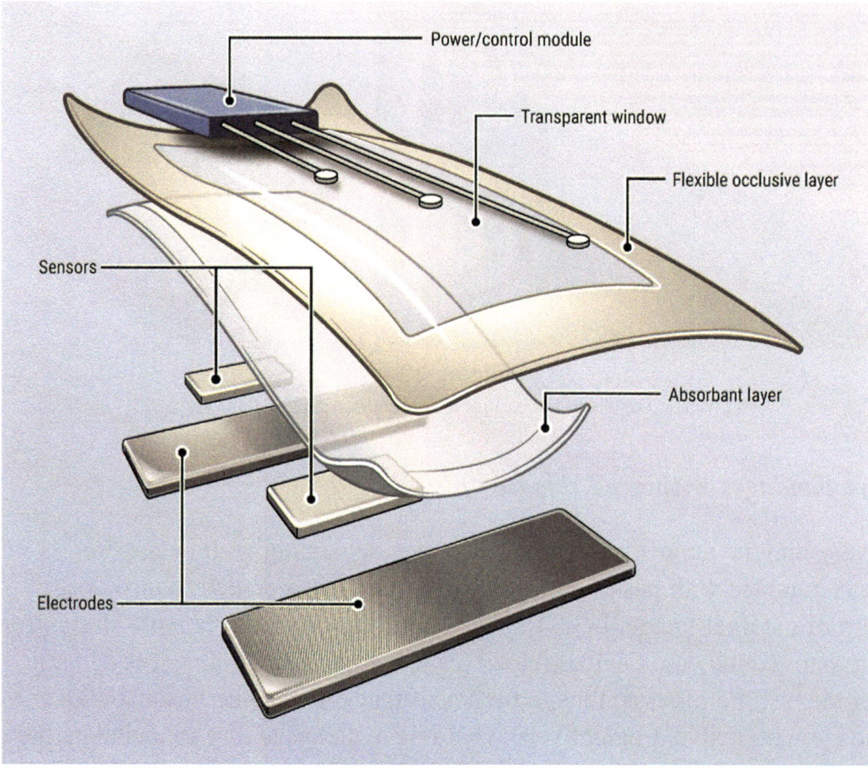

Figure 13.5
Smart bandage prototype concept. *Image created by Grace Gongaware.*

The bandage material would allow gas exchange and be hypoallergenic, while still protecting the wound from external environment. Embedded in it, an absorbent material would make sure the wound environment is kept moist, and at the same time, it is responsible for removing excess exudate, which would decrease the rate of wound healing.

The bandage would be equipped with two categories of sensors: the first would be responsible for monitoring the wound status, called wound sensors, whereas the second would sense the outside environment, named environment sensors.

Temperature, pH, oxygen, and impedance can serve as measurable parameters to assess wound healing. Although wounds present different geometries and sizes, it is difficult to sense a single point in a wound; therefore, using a sensor array would accommodate any size fluctuation.

The impedance of intact skin is of around 10 kΩ [4]. Once tissue is damaged, this value decreases by an order of magnitude, to a value of 1–5Ω [87], partially due to wound exudate, which keeps the wound moist [88]. During the healing process, impedance across the wound bed increases, until full reepithelization takes place. The two electrodes pictured in Fig. 13.5 would serve as impedance sensors.

Oxygen is tightly related to the wound healing process. While hypoxia is needed to start wound healing, higher oxygen concentration is thought to be related to the rate of angiogenesis [63] Thus, real-time oxygen sensing also provides with valuable insight into progressive healing status. Temperature, pH, and oxygen sensors could be alternated between the testing points to get a map of the wound bed. All sensors would need to be reliable, flexible, miniaturized, and able to withstand the wound environment for up to 7 days.

As temperature and oxygen level readings can be affected by the external environment, it is necessary to equip the bandage with environment sensors, so that the parameters can be accurately read. Therefore, sensors should be placed on the outside of the wound bed, but still within the bandage. Adjacent to them could be compartments filled with drugs, which could be accurately and easily controlled by an outside device.

The smart bandage would be equipped with the necessary electronics to perform accurate data collection, signal processing, and communication with an external module. All electronics would be flexible to ensure that the patient is comfortable and also that all sensors are making contact to the skin. The accompanying Wi-Fi module could send statistics and alerts to caretakers, monitoring the status of the treatment and also ensuring the correct dosage of drugs is being released into the body.

In addition, to reduce overall costs and increase accessibility, the electronics on the smart bandage would be reusable, while the sensors and substrate material would be disposed after a week of use. As the bandage could stay in place for up to 7 days before being discarded, it would not perturb the delicate wound environment. The device would be powered by a rechargeable battery.

4. Summary

Chronic wounds remain a threatening and growing epidemic. They are a burden for people who suffer from them and for the health-care system. The number of patients affected by chronic wound is likely to rise with increases in the elderly population and the number of diabetic patients.

Smart wound management technology as a primary prevention tool would make it possible to sense wound formation before their appearance and implement proactive prevention strategies.

Once chronic wounds have formed, smart bandages can serve as a bridge between patients and caretakers, thus decreasing the cost of treatment and optimizing healing time. The use of quantified wound status would make it is possible to standardize the reporting of wound healing and improve the efficiency of wound treatment. The application of closed-loop smart wound care technology, fully integrating sensing, and delivery of treatment will enhance point-of-care therapy and improve personalized care.

The future of wound care will be revolutionized by the application of smart technology to prevent, treat, and monitor chronic wounds.

References

[1] Padula WV, Pronovost PJ. Addressing the multisectoral impact of pressure injuries in the USA, UK and abroad. BMJ Qual Saf 2018;27(3):171–3. PMID: 28928166.

[2] Järbrink K, Ni G, Sönnergren H, Schmidtchen A, Pang C, Bajpai R, et al. The humanistic and economic burden of chronic wounds: a protocol for a systematic review. Syst Rev 2017;6(1):1–8. Available from: https://doi.org/10.1186/s13643-016-0400-8.

[3] Bogie KM, Wang X, Triolo RJ. Long-term prevention of pressure ulcers in high-risk patients: a single case study of the use of gluteal neuromuscular electric stimulation. Arch Phys Med Rehabil 2006;87(4):585–91. PMID: 16571402.

[4] National Pressure Ulcer Advisory Panel. European pressure ulcer advisory panel and Pan Pacific pressure injury alliance. In: Haesler E, editor. Prevention and treatment of pressure ulcers: clinical practice guideline. Osborne Park (Western Australia): Cambridge Media; 2014.

[5] Peerless J, Allyn D, Klein D, Yu D. Skin complications in the intensive care unit. Clin Chest Med 1999;20(2):453–67.

[6] González-Consuegra RV, Verdú J. Quality of life in people with venous leg ulcers: an integrative review. J Adv Nurs 2011;67(5):926–44. PMID: 21241355.

[7] Primary, Secondary and Tertiary Prevention. Institute for work and health 2015. p. 1–3 [Internet]. Available from: https://www.iwh.on.ca/wrmb/primary-secondary-and-tertiary-prevention.

[8] Henzel MK, Bogie KM, Guihan M, Ho CH. Pressure ulcer management and research priorities for patients with spinal cord injury: consensus opinion from SCI QUERI expert panel on pressure ulcer research implementation. J Rehabil Res Dev 2011;48(3):xi–xxxii. PMID: 21480093.

[9] Sen CK, Gordillo GM, Roy S, Kirsner R, Lambert L, Hunt TK, et al. Human skin wounds: a major and snowballing threat to public health and the economy. Wound Repair Regen 2009;17(6):763–71. PMID: 19903300; PMCID: PMC2810192.

[10] Kapp S, Santamaria N. The financial and quality-of-life cost to patients living with a chronic wound in the community. Int Wound J 2017;14(6):1108–19. PMID: 28635188.

[11] Cullum N, Deeks J, Sheldon TA, Song F, Fletcher AW. Beds, mattresses and cushions for pressure sore prevention and treatment. J Tissue Viability 1999;9(4):138. PMID: 10808844.

[12] Ocampo W, Cheung A, Baylis B, Clayden N, Conly JM, Ghali WA, et al. Economic evaluations of strategies to prevent hospital-acquired pressure injuries. Adv Skin Wound Care 2017;30(7). 319–33 PMID: 28617751; PMCID: PMC5482558.

[13] Spruce L. Back to basics: preventing perioperative pressure injuries. AORN J 2017 Jan;105(1):92–9. PMID: 28034406.

[14] Magalhães MG, Gragnani A, Veiga DF, Blanes L, Galhardo VA, Kállas H, et al. Risk factors for pressure ulcers in hospitalized elderly without significant cognitive impairment. Wounds January 2007;19(1):20–4. PMID: 26111408.

[15] Krishnan S, Brick RS, Karg PE, Tzen YT, Garber SL, Sowa GA, et al. Predictive validity of the Spinal Cord Injury Pressure Ulcer Scale (SCIPUS) in acute care and inpatient rehabilitation in individuals with traumatic spinal cord injury. NeuroRehabilitation April 6, 2016;38(4):401–9. PMID: 27061168.

[16] Delparte JJ, Scovil CY, Flett HM, Higgins J, Laramée MT, Burns AS. Psychometric properties of the Spinal Cord Injury Pressure Ulcer Scale (SCIPUS) for pressure ulcer risk assessment during inpatient rehabilitation. Arch Phys Med Rehabil 2015; Nov;96(11). 1980–5. PMID: 26205694.

[17] Groah SL, Schladen M, Pineda CG, Hsieh CH. Prevention of pressure ulcers among people with spinal cord injury: a systematic review. PM R June 2015;7(6):613–36. PMID: 25529614.

[18] Oomens CW, Loerakker S, Bader DL. The importance of internal strain as opposed to interface pressure in the prevention of pressure related deep tissue injury. J Tissue Viability May 2010;19(2):35–42. Epub 2009Dec 16. PMID: 20005716.

[19] Lemmer D, Alvarado N, Henzel K, Richmond MA, McDaniel J, Graebert J, et al. What lies beneath: why some pressure injuries may be unpreventable for individuals with spinal cord injury. Arch Phys Med Rehabil 2019 Jun;100(6):1042–9. Epub 2018 Dec 6. PMID:30529322.

[20] Gefen A, Levine J. The false premise in measuring body-support interface pressures for preventing serious pressure ulcers. J Med Eng Technol 2007 ;31(5):375–80. PMID: 17701783.

[21] Bogie K, Henzel K, Richmond M, Alvarado N, McDaniel J, Graebert J, et al. Shifting the paradigm: why some pressure injuries may be unpreventable for individuals with spinal cord injury. Topics in SCI Rehab 2018;24(1):61.

[22] Majerus SJA, Lerchbacker J, Barbaro D, Mitchell SJ, Bogie KM, Henzel MK. Power wheelchair footplate pressure and positioning sensor. In: 40th annual international conference of the IEEE Engineering in Medicine and Biology Society (EMBC), Honolulu, HI, 2018; 2018. p. 4367–70. https://doi.org/10.1109/EMBC.2018.8513373.

[23] Swisher SL, Lin MC, Liao A, Leeflang EJ, Khan Y, Pavinatto FJ, et al. Impedance sensing device enables early detection of pressure ulcers in vivo. Nat Commun March 17, 2015;6:6575. PMID: 25779688.

[24] Kim CG, Park S, Ko JW, Jo S. The relationship of subepidermal moisture and early stage pressure injury by visual skin assessment. J Tissue Viability August 2018;27(3):130–4. Epub 2018 May 8. PMID: 29773438.

[25] Moore Z, Patton D, Rhodes SL, O'Connor T. Subepidermal moisture (SEM) and bioimpedance: a literature review of a novel method for early detection of pressure-induced tissue damage (pressure ulcers). Int Wound J April 2017;14(2):331–7. Epub 2016 Apr 29. PMID: 27126794.

[26] Bates-Jensen BM, McCreath HE, Pongquan V. Subepidermal moisture is associated with early pressure ulcer damage in nursing home residents with dark skin tones: pilot findings. J Wound Ostomy Cont Nurs 2009 ;36(3):277–84. PMID: 19448508; PMCID: PMC5656450.

[27] Sprigle S, Zhang L, Duckworth M. Detection of skin erythema in darkly pigmented skin using multispectral images. Adv Skin Wound Care April 2009;22(4):172–9. PMID: 19325277.

[28] Quintavalle PR, Lyder CH, Mertz PJ, Phillips-Jones C, Dyson M. Use of high-resolution, high-frequency diagnostic ultrasound to investigate the pathogenesis of pressure ulcer development. Adv Skin Wound Care 2006 ;19(9):498–505. PMID: 17132955.

[29] Helvig EI, Nichols LW. Use of high-frequency ultrasound to detect heel pressure injury in elders. J Wound Ostomy Cont Nurs 2012 ;39(5):500–8. PMID: 22864192.

[30] Scheiner J, Farid K, Raden M, Demisse S. Ultrasound to detect pressure-related deep tissue injuries in adults admitted via the emergency department: a prospective, descriptive, pilot study. Ostomy Wound Manage March 2017;63(3):36–46. PMID: 28355138.

[31] Swaine JM, Breidahl W, Bader DL, Oomens CWJ, O'Loughlin E, Santamaria N, et al. Ultrasonography detects deep tissue injuries in the subcutaneous layers of the buttocks following spinal cord injury. Top Spinal Cord Inj Rehabil 2018 Fall;24(4):371–8. Epub 2018 Feb 12. PMID:30459500; PMCID: PMC6241223.

[32] Schäfer G, Dobos G, Lünnemann L, Blume-Peytavi U, Fischer T, Kottner J. Using ultrasound elastography to monitor human soft tissue behaviour during prolonged loading: a clinical explorative study. J Tissue Viability November 2015;24(4):165–72. Epub 2015 Jun 26. PMID: 26165202.

[33] Fiedler M, Gerhardt LC, Derler S, Bischofberger G, Hürny C, Münzer T. Assessment of biophysical skin properties at different body sites in hospitalized old patients: results of a pilot study. Gerontology 2012;58(6):513–7. Epub 2012 Apr 5. PMID: 22487874.

[34] Oliveira AL, Moore Z, O Connor T, Patton D. Accuracy of ultrasound, thermography and subepidermal moisture in predicting pressure ulcers: a systematic review. J Wound Care May 2, 2017;26(5):199–215. PMID: 28475447.
[35] Spillman WB, Mayer M, Bennett J, Gong J, Meissner KE, Davis B, et al. A "smart" bed for non-intrusive monitoring of patient physiological factors. Meas Sci Technol 2004;15(8):1614–20.
[36] Duvall J, Karg P, Brienza D, Pearlman J. Detection and classification methodology for movements in the bed that supports continuous pressure injury risk assessment and repositioning compliance. Tissue Viability February 2019;28(1):7–13. Epub 2018 Dec 24. PMID: 30598376; PMCID: PMC6382541.
[37] Bates-Jensen BM, McCreath HE, Patlan A. Subepidermal moisture detection of pressure induced tissue damage on the trunk: the pressure ulcer detection study outcomes. Wound Repair Regen May 2017;25(3):502–11. Epub 2017 May 31. PMID: 28494507; PMCID: PMC5568916.
[38] Okonkwo H, Milne J, Bryant R. Evaluation of a novel device using capacitance of the detection of early pressure ulcers (PU), a multi-site longitudinal study. Proc Wounds UK Ann Conf 2017;R40. Accessed 02/22/19.
[39] Iraj B, Khorvash F, Ebneshahidi A, Askari G. Prevention of diabetic foot ulcer. Int J Prev Med March 2013;4(3):373–6. PMID: 23626896; PMCID: PMC3634178.
[40] Lin F, Wang A, Zhuang Y, Tomita MR, Xu W. Smart insole: a wearable sensor device for unobtrusive gait monitoring in daily life. IEEE Trans Ind Inform December 2016;12(6):2281–91. https://doi.org/10.1109/TII.2016.2585643.
[41] Bae J, Kong K, Byl N, Tomizuka M. A mobile gait monitoring system for gait analysis. IEEE Intern Conf Rehabil Robot June 2006:50–7.
[42] Liu T, Inoue Y, Shibata K. A wearable ground reaction force sensor system and its application to the measurement of extrinsic gait variability. Sensors (Basel). 2010;10(11):10240–55. Epub 2010 Nov 16. PMID: 22163468; PMCID: PMC3231031.
[43] Xu W, Huang M-C, Amini N, Liu JJ, He L, Sarrafzadeh M. Smart insole: a wearable system for gait analysis. In: Proc 5th international conference on PErvasive technologies related to assistive environments. Crete, Greece: Heraklion; June 2012. https://doi.org/10.1145/2413097.2413120.
[44] Bogie KM, Triolo RJ. Effects of regular use of neuromuscular electrical stimulation on tissue health. J Rehabil Res Dev 2003 ;40(6):469–75. PMID: 15077659.
[45] Ahmetović A, Mushahwar VK, Sommer R, Schnepf D, Kawasaki L, Warwaruk-Rogers R, et al. Safety and feasibility of intermittent electrical stimulation for the prevention of deep tissue injury. Adv Wound Care (New Rochelle) March 1, 2015;4(3):192–201. PMID: 25785240.
[46] Bogie KM, Reger SI, Levine SP. Electrical stimulation for pressure sore prevention and wound healing. Assist Technol 2000;12(1):50–66. PMID: 11067577.
[47] Sood A, Granick MS, Trial C, Lano J, Palmier S, Ribal E, et al. The role of telemedicine in wound care: a review and analysis of a database of 5,795 patients from a mobile wound-healing center in Languedoc-Roussillon, France. Plast Reconstr Surg September 2016;138(3 Suppl). 248S-56S. PMID: 27556769.
[48] Le Goff-Pronost M, Mourgeon B, Blanchère JP, Teot L, Benateau H, Dompmartin A. Real-world clinical evaluation and costs of telemedicine for chronic wound management. Int J Technol Assess Health Care January 2018;34(6):567–75. Epub 2018 Oct 29. PMID: 30369340.
[49] Stern A, Mitsakakis N, Paulden M, Alibhai S, Wong J, Tomlinson G, et al. Pressure ulcer multidisciplinary teams via telemedicine: a pragmatic cluster randomized stepped wedge trial in long term care. BMC Health Serv Res February 24, 2014;14:83. PMID: 24559218; PMC4104322.
[50] National Pressure Ulcer Advisory Panel. NPUAP Press Injury Stages 2016;29(6):269–76. http://www.npuap.org/resources/educational-and-clinical-resources/npuap-pressure-injury-stages.
[51] Bates-Jensen BM. The pressure sore status tool a few thousand assessments later. Adv Wound Care September 1997;10(5):65–73. PubMed PMID: 9362584.
[52] Harries RL, Bosanquet DC, Harding KG. Wound bed preparation: TIME for an update. Int Wound J September 2016;13(Suppl. 3):8–14. PMID: 27547958.
[53] Doupis J, Veves A. Classification, diagnosis, and treatment of diabetic foot ulcers May 2008;20(5):117–26. PMID: 25942412.

[54] James WB. Classification of foot lesions in diabetic patients. Levin and O'Neals the diabetic foot. 9; 2008. p. 221–6.
[55] Oyibo SO, Jude EB, Tarawneh I, Nguyen HC, Harkless LB, Boulton AJ. A comparison of two diabetic foot ulcer classification systems: the Wagner and the University of Texas wound classification systems. Diabetes Care January 2001;24(1):84–8. PMID: 11194247.
[56] Haghpanah S, Bogie K, Wang X, Banks PG, Ho CH. Reliability of electronic versus manual wound measurement techniques. Arch Phys Med Rehabil October 2006;87(10):1396–402. PMID: 17023252.
[57] Wang Z, Hasan R, Firwana B, Elraiyah T, Tsapas A, Prokop L, et al. A systematic review and meta-analysis of tests to predict wound healing in diabetic foot. J Vasc Surg February 2016;63(2 Suppl. l):29S–36S. e1-2. PMID: 26804365.
[58] Pan X, Chen G, Wu P, Han C, Ho JK. Skin perfusion pressure as a predictor of ischemic wound healing potential. Biomed Rep April 2018;8(4):330–4. Epub 2018 Feb 13. PMID: 29541454; PMCID: PMC5838289.
[59] Xu Y, Sun J, Carter RR, Bogie KM. Personalized prediction of chronic wound healing: an exponential mixed effects model using stereophotogrammetric measurement. J Tissue Viability May 2014;23(2):48–59. Epub 2014 Apr 18. PMID: 24810677.
[60] Wang SC, Anderson JAE, Evans R, Woo K, Beland B, Sasseville D, et al. Point-of-care wound visioning technology: reproducibility and accuracy of a wound measurement app. PLoS One August 17, 2017;12(8):e0183139. eCollection 2017. PMID: 28817649; PMCID: PMC5560698.
[61] Rippon M, Davies P, White R. Taking the trauma out of wound care: the importance of undisturbed healing. J Wound Care August 2012;21(8):359–60. 362, 364-8. PMID: 22885308.
[62] Spear M. Wound exudate—the good, the bad, and the ugly. Plast Surg Nurs 2012 ;32(2):77–9. PMID: 22647956.
[63] Kimmel HM, Grant A, Ditata J. The presence of oxygen in wound healing. Wounds August 2016;28(8):264–70. PMID: 27560469.
[64] Kruse CR, Nuutila K, Lee CC, Kiwanuka E, Singh M, Caterson EJ, et al. The external microenvironment of healing skin wounds. Wound Repair Regen 2015 ;23(4):456–64. Epub 2015 Jul 14. PMID: 25857996.
[65] Zhao G, Usui ML, Lippman SI, James GA, Stewart PS, Fleckman P, et al. Biofilms and inflammation in chronic wounds. Adv Wound Care (New Rochelle) September 2013;2(7):389–99. PMID: 24527355; PMCID: PMC3763221.
[66] Gethin G. The significance of surface pH in chronic wounds. Wounds UK 2007;3(3):52–5. Available from: https://www.wounds-uk.com/journals/issue/11/article-details/the-significance-of-surface-ph-in-chronic-wounds-1.
[67] Tsukada K, Tokunga K, Iwama T, Mishima Y. The pH changes of pressure ulcers related to the healing process of wounds. Wounds 1992;4(1):16–20.
[68] Sharpe JR, Harris KL, Jubin K, Bainbridge NJ, Jordan NR. The effect of pH in modulating skin cell behaviour. Br J Dermatol September 2009;161(3):671–3. Epub 2009 Apr 29. PMID: 19438462.
[69] Nagoba BS, Suryawanshi NM, Wadher B, Selkar S. Acidic environment and wound healing: a review. Wounds 2015;27(1):5–11.
[70] Percival SL, McCarty S, Hunt JA, Woods EJ. The effects of pH on wound healing, biofilms, and antimicrobial efficacy. Wound Repair Regen March–April 2014;22(2):174–86. Epub 2014 Feb 24. PMID: 24611.
[71] Punjiya M, Rezaei H, Zeeshan MA, Sonkusale S. A flexible pH sensing smart bandage with wireless CMOS readout for chronic wound monitoring. TRANSDUCERS 2017 – 19th Int Conf Solid-State Sensors, Actuators Microsystems. 2017. 1700–2.
[72] Schreml S, Meier RJ, Weiß KT, Cattani J, Flittner D, Gehmert S, et al. A sprayable luminescent pH sensor and its use for wound imaging in vivo. Exp Dermatol December 2012;21(12):951–3. PMID: 23171458.
[73] Tamayol A, Akbari M, Zilberman Y, Comotto M, Lesha E, Serex L, et al. Flexible pH-Sensing hydrogel fibers for epidermal applications. Adv Healthc Mater March 2016;5(6):711–9. Epub 2016 Jan 22. PMID: 26799457; PMCID: PMC4805432.

[74] Rahimi R, Ochoa M, Tamayol A, Khalili S, Khademhosseini A, Ziaie B. Highly stretchable potentiometric pH sensor fabricated via laser carbonization and machining of carbon-polyaniline composite. ACS Appl Mater Interfaces March 15, 2017;9(10):9015–23. Epub 2017 Mar 6. PMID: 28224783.
[75] Power G, Moore Z, O'Connor T. Measurement of pH, exudate composition and temperature in wound healing: a systematic review. J Wound Care July 2, 2017;26(7):381–97. PMID: 28704150.
[76] Symptoms of infection after a skin injury. WebMD [Online].
[77] Gaiziunas AG, Hast MH. Temperature gradients and prediction of flap viability. J Otolaryngol October 1976;5(5):399–402. PMID: 792468.
[78] Fierheller M, Sibbald RG. A clinical investigation into the relationship between increased periwound skin temperature and local wound infection in patients with chronic leg ulcers. Adv Skin Wound Care August 2010;23(8):369–79. quiz 380-1. PMID: 20631603.
[79] Salvo P, Dini V, Kirchhain A, Janowska A, Oranges T, Chiricozzi A, et al. Sensors and Biosensors for C-reactive protein, temperature and pH, and their applications for monitoring wound healing: a review. Sensors (Basel) December 19, 2017;(12):17. pii: E2952. PMID: 29257113; PMCID: PMC5750823.
[80] Courbat J, Kim YB, Briand D, De Rooij NF. In: Inkjet printing on paper for the realization of humidity and temperature sensors. 2011 16th int solid-state sensors, actuators microsystems conf TRANSDUCERS'11; 2011. p. 1356–9.
[81] Vuorinen T, Niittynen J, Kankkunen T, Kraft TM, Mäntysalo M. Inkjet-printed graphene/PEDOT:PSS temperature sensors on a skin-conformable polyurethane substrate. Sci Rep October 18, 2016;6:35289. PMID: 27752050; PMCID: PMC5082757.
[82] Sui Y, Kreider LP, Bogie KM, Zorman CA. Fabrication of a silver-based thermistor on flexible, temperature-sensitive substrates using a low-temperature inkjet printing technique. IEEE Sensors Lett Feb. 2019;3(2):1–4. https://doi.org/10.1109/LSENS.2019.2893741. Art no. 2500704.
[83] Gainza G, Villullas S, Pedraz JL, Hernandez RM, Igartua M. Advances in drug delivery systems (DDSs) to release growth factors for wound healing and skin regeneration. Nanomedicine August 2015;11(6):1551–73. Epub 2015 Mar 21. PMID: 25804415.
[84] Saghazadeh S, Rinoldi C, Schot M, Kashaf SS, Sharifi F, Jalilian E, et al. Drug delivery systems and materials for wound healing applications. Adv Drug Deliv Rev March 1, 2018;127:138–66. Epub 2018 Apr 5. PMID: 29626550; PMCID: PMC6003879.
[85] Bagherifard S, Tamayol A, Mostafalu P, Akbari M, Comotto M, Annabi N, et al. Dermal patch with integrated flexible heater for on demand drug delivery. Adv Healthc Mater January 7, 2016;5(1):175–84. Epub 2015 Oct 26. PMID: 26501166.
[86] Nabavinia M, Dokmeci MR, Khademhosseini A, Sonkusale S, et al. A textile dressing for temporal and dosage controlled drug delivery. Adv Funct Mater 2017;27(41):1702399.
[87] Yamamoto Y. Measurement and analysis of skin electrical impedance. Acta Derm Venereol Suppl (Stockh) 1994;185:34–8. PMID: 8091925.
[88] McColl D, Cartlidge B, Connolly P. Real-time monitoring of moisture levels in wound dressings in vitro: an experimental study. Int J Surg 2007;5(5):316–22.

CHAPTER 14

Method for improving skin color accuracy of three-dimensional printed training models for early pressure ulcer recognition

Louis W.M.M. Krieger[1], Tom Ridge[1], Alex Demers[1], Karthik Vishwanath[2], Jessica L. Sparks[1]

[1]Department of Chemical, Paper and Biomedical Engineering, Miami University, Oxford, OH, United States; [2]Department of Physics, Miami University, Oxford, OH, United States

Chapter Outline

Innovations and Emerging Technologies in Wound Care. https://doi.org/10.1016/B978-0-12-815028-3.00014-6

1. Introduction

1.1 Pressure injury in skin

Pressure ulcers, also called decubitus ulcers or bed sores, are localized injuries to the epidermis and underlying tissue that develop because of a combination of pressure and shear stress on tissues for prolonged periods of time [1]. Application of pressure to an area of skin for extended periods of time results in a lesion due to the lack of oxygenated blood flow to the area [2,3]. This is most prevalent in tissue regions surrounding a bony prominence such as the tailbone, elbow, or hip [4]. Immobile populations, such as those limited to bed rest or wheel chairs for extended periods of time, are especially susceptible to pressure ulcer formation [5]. At the time of diagnosis, pressure ulcers are classified as stages 1−4, with stage 4 representing the most severe cases. Pressure ulcers classified as stage 1 are commonly characterized by an inflammatory or bruising resemblance at the injury site, with the skin largely intact and notable hyperemia [4,6]. Stage 4 pressure ulcers have extensive skin damage, tissue loss, and necrosis in extreme cases [7]. Pressure ulcers are a significant cause of morbidity and mortality in long-term care facilities, impairing quality of life for patients and draining resources from the health-care system [8]. Around 60,000 patients die from pressure ulcers each year [9,10]. In the United States alone, 2.5 million people annually are afflicted with pressure ulcer diagnoses with treatment costs of $20,900−$151,700 per pressure ulcer constituting an overall cost of $9.1−$11.6 billion per year in the United States ([9,11]).

The current paradigm for pressure ulcer detection is visual inspection by nursing staff at hospitals and long-term care facilities during routine skin assessments for immobile patients [4]. There can be uncertainty associated with the diagnosis of injuries, which requires consultation with wound care specialists and physicians for further evaluation [9].

The resulting delays before a diagnosis is confirmed increase the risk of the pressure ulcer becoming more severe. To circumvent such delays, extensive efforts have been directed toward developing thorough continuing education modules for nurses and caregivers to expedite early pressure ulcer detection.

Current training methods involve the presentation of pressure ulcer injuries in image format to illustrate different stages and severities of patient conditions. These images are typically presented in two-dimensional modalities such as PowerPoint presentations, printed handouts, or digital images on training websites. Two types of images are commonly used to discuss and teach the anatomy of pressure ulcer injuries: cross-sectional anatomical models produced from drawing software, and medically acquired images of late-stage pressure ulcer sites on specific patients. Because of the need for clear delineation of pressure ulcer injuries in an academic setting, pressure ulcer images are most often obtained from populations with lighter skin complexions. The contrast between light skin tones and pigmentation of the injury site allows for the clear definition of the pressure ulcer with respect to the surrounding tissue [7].

Despite the clear illustration of pressure ulcer progression in current training models, limitations associated with these methods surface in the assessment of patients with darker complexions and skin tones. This is primarily attributed to the lack of visual cues and color gradients inherent with darker skin pigmentation [12]. Additionally, the angle of image acquisition for pressure ulcer training models often skews the perspective of geometries at the injury site as the image attempts to capture the full area of pressure ulcer development. While didactic presentation of pressure ulcers provides the necessary understanding of the injury's progression [13], diagnostic proficiency primarily stems from residency and experience associated with training at long-term care facilities and hospitals. The ability to successfully diagnose pressure ulcers is highly dependent on recognition and exposure to high quantities of patient cases during hands-on training. If health-care students are not presented with sufficient pressure ulcer cases during their residency, gaps in their understanding of the injury may present later in their career. With the global population rapidly increasing in age and the diversity of patient populations continually expanding, it is necessary to incorporate new training methods for pressure ulcer detection. Limitations inherent to current training models can be mitigated through the use of newly available technologies, thus allowing for increased diagnostics and improved health for those susceptible to pressure ulcers.

The goal of this study was to develop a method for producing color-accurate skin models representative of human skin tones through the application of three-dimensional (3D) printing, 3D scanning, and color science. In future, this methodology could be applied to create more intricate and life-like training models for educating health-care workers on recognition of early stage pressure ulcers in patients with diverse skin tones. To illustrate applicability of the method to printing human skin tones, color 3D scans were collected from healthy volunteers in accordance with an Institutional Review Board-approved protocol. These were used to create color-corrected digital and physical 3D models of the skin. The methodology developed in this study can be applied in the future to create digital and physical 3D models of early stage pressure ulcers, pending future acquisition of appropriate scans of pressure injury in patients. It is postulated that incorporation of these 3D models will increase the understanding of pressure ulcers by health-care professional and enhance their ability to recognize these injuries at an early stage to minimize injury progression.

1.2 Three-dimensional printing of skin tones

As noted in previous studies [14,15], color error and geometrical error associated with scanning visual data and 3D printing a color 3D model can lead to the production of imperfect physical products. In the research of color-accurate facial prosthetics by Xiao et al., color management techniques were developed to account for the variation of color profiles for a 3D camera system and a powder-based 3D printer manufactured by ZCorp [15]. The protocols associated with the study by Xiao et al. sought to develop individual device color profiles using correlation of spectrophotometrically acquired CIELAB values for a standardized color set with respect to the corresponding red, green, and blue (RGB) values for the captured digital image [15]. CIELAB refers to a system of color measurement developed by the Commission International d'Eclairage (CIE), where LAB indicates the L^*, a^*, and b^* coordinates that define a given color within an approximately uniform 3D space [19]. With the definition of device color profiles as reported by Xiao et al., the capacity to execute a transfer function exists to correct input RGB values for the 3D printed image to produce colors representative of original object colors [15]. In the present study, we are pursuing a similar objective using different transfer function derivation methods based on different image acquisition equipment (color 3D scanner vs. photography/photogrammetry) and different 3D printing technology (paper-based 3D printer vs. powder-based 3D printer).

The 3D printer used in this research was the Mcor Iris (Fig. 14.1) (Mcor Technologies; Dublin, Ireland). This printer uses paper as the building material. To create a print, a 3D digital model file is loaded into a build setup interface, where the model can be reoriented to a convenient reference frame. The model is sliced into virtual layers using the built-in SliceIt software. Then the layers are printed via a regular Epson color printer, which prints the color outline for each slice. Afterward, the stack of printed paper is loaded into the Mcor Iris, where one piece of paper at a time is pulled onto a base plate, cut with a knife that extends to only the thickness of a single layer of paper, and glued. This pattern is repeated until the model is finished. Once complete, the model must be excavated from the paper stack by carefully removing the cut paper surrounding the model.

Figure 14.1
Mcor Iris color 3D printer.

1.3 Color measurement background

Several factors influence the color appearance of an object and thus affect the methods used to measure color. The first is the nature of the light source used. Although an object may appear as one color when viewed in sunlight, it may take on a different appearance when viewed under an LED lamp. Thus, the light source must be taken into account when measuring the color of an object [16]. A second factor is modification of incident light by interaction with the object itself. For example, an object's surface and orientation can alter the color that is perceived by the viewer [17]. The third important factor is that the human vision system is trichromatic (three channeled), meaning that a single perceived color may be regarded as the net effect of three separate stimuli on the visual cortex [18,19]. This allows humans to perceive more subtle differences in color, and it informs many of the color spaces currently used in color quantification.

The CIE color system has defined standard conditions to address each factor noted above, so that describing the appearance of a color in terms of CIELAB coordinates provides an objective quantification of the visual perception of that color. In addition, color specification in terms of CIELAB values enables the measurement of perceived color differences or changes. An excellent overview of the CIE color system and its application to skin color measurement is available in Ref. [19] and is summarized briefly here. The relative spectral energy distributions of CIE standard illuminants have been specified and are available as published tables that are incorporated into the software provided with various color measurement reflectance spectrophotometers. To measure how illuminant light is modified by interaction with the object, a reflectance spectrophotometer is used with an optical configuration corresponding to CIE recommendations. A reflectance spectrum is thus generated, which is expressed as the percentage of the incident light intensity reflected by an object at each wavelength, where visible light wavelengths correspond to the range from approximately 400 to 700 nm. Lastly, the trichromatic nature of color vision is accounted for via the "CIE standard observer," which is a table of published values that represents a numerical description of human color vision [19]. For convenience, these tabulated values have also been incorporated into typical software provided with color measurement reflectance spectrophotometers.

For the purposes of this study, two color systems were used to quantify color of 3D printed models. These systems included CIELAB and RGB color coordinates. As described above, the CIELAB color system is based on human perception of colors and is broken down into three coordinate axes: L^*, a^*, and b^* [20]. In this system, the L^* value corresponds to the lightness of a given color and is defined on a dark to bright scale quantified by the range 0–100 [21]. The terms a^* and b^* quantify the hue by characterizing the degree of red/green and blue/yellow in the given color through the use of vector quantities [21]. The degree of each of these color compositions is evaluated by the direction and magnitude of the terms a^* and b^*, with $-a^*$ values equivalent to green, $+a^*$ values equal to red, $-b^*$ values representative of blue, and $+b^*$ values indicative of yellow [21]. In contrast, the RGB color system is linked to digital color values with respect to the degree of red, green, and blue hues making up a given color on a scale of 0–255 [21].

Assessing the difference between the colors of the 3D printed object and the original item requires the acquisition of CIELAB values via spectrophotometric measurement [15,20]. These CIELAB values are then compared using a color difference metric known as ΔE^*_{ab} [14,22]. The difference in color values, ΔE^*_{ab}, represents a Euclidean distance between two colors found in the color space. The original ΔE^*_{ab} formula was reported in 1976 [19,23] and is defined in Eq. (14.1):

$$\Delta E^*_{ab} = \sqrt{\Delta L^{*^2} + \Delta a^{*^2} + \Delta b^{*^2}} \tag{14.1}$$

where L^*, a^*, and b^* are the three coordinates in the CIELAB space.

As a rough approximation, when ΔE^*_{ab} has a magnitude of 2 or lower, the color difference between them is indistinguishable by the average human eye [24]. Fig. 14.2 represents an attempt to illustrate the degree of color difference for ΔE^*_{ab} values of 3 and 10, respectively, for a grayscale color (with equal L^*, a^*, and b^* values) and for a randomly selected shade of purple. However, please bear in mind that the perceived color for these

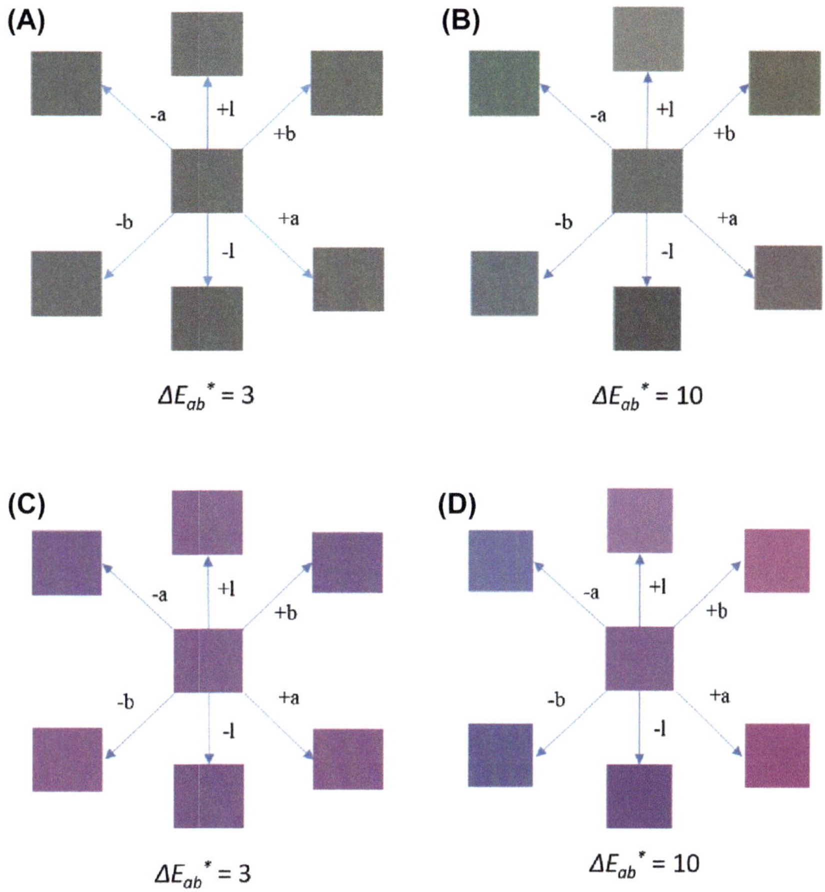

Figure 14.2

Illustration of the magnitude of color differences, for variations along the L^*, a^*, and b^* axes of the CIELAB color space. (A) Color difference of $\Delta E^*_{ab} = 3$ for a grayscale color; (B) color difference of $\Delta E^*_{ab} = 10$ for a grayscale color; (C) color difference of $\Delta E^*_{ab} = 3$ for a shade of purple; (D) color difference of $\Delta E^*_{ab} = 10$ for a shade of purple.

images will be impacted by electronic monitor display settings, color printer settings, lighting conditions, and other factors.

Although many studies have drawn on the CIE 1976 ΔE^*_{ab} equation, given by Eq. (14.1), an improved formula (ΔE^*_{94}) was developed by the CIE in 1994 to provide a more accurate measure of perceived color differences. The ΔE^*_{94} equation takes into account the fact that the CIELAB color space is not perfectly uniform and that the human eye is better able to detect some color differences than others. For example, for colors of high chroma, the CIE 1976 ΔE^*_{ab} color difference formula values color differences too strongly compared to experimental results from color perception experiments [22,25].

The ΔE_{94}^* equation is defined in Eq. (14.2) [22], which is expressed in polar coordinates of lightness (L^*), chroma (C_{ab}^*), and hue (H_{ab}^*) rather than rectangular L^*, a^*, and b^* coordinates:

$$\Delta E_{94}^* = \sqrt{\left(\frac{\Delta L^*}{k_L S_L}\right)^2 + \left(\frac{\Delta C_{ab}^*}{k_C S_C}\right)^2 + \left(\frac{\Delta H_{ab}^*}{k_H S_H}\right)^2} \tag{14.2}$$

Chroma differences between color 1 and color 2 are calculated as

$$\Delta C_{ab}^* = \sqrt{a_1^{*2} + b_1^{*2}} - \sqrt{a_2^{*2} + b_2^{*2}} \tag{14.3}$$

Hue differences are then found as

$$\Delta H_{ab}^* = \sqrt{\Delta a^{*2} + \Delta b^{*2} - \Delta C_{ab}^{*2}} \tag{14.4}$$

The k_L, k_C, and k_H terms are factors to match the perception of background conditions, and S_L, S_C, and S_H are linear functions of C_{ab}^* [22]. Standard reference values were assumed for the k and S terms as listed below [22]:

$$S_c = 1 + .045C_{ab}^* \tag{14.5}$$

$$S_H = 1 + .015C_{ab}^* \tag{14.6}$$

$$S_L = 1 \tag{14.7}$$

$$k_L = k_C = k_H = 1 \tag{14.8}$$

2. Methods

2.1 Overview

Using an Artec Spider 3D color scanner (Fig. 14.3) (Artec 3D; Luxembourg, Luxembourg), scans were collected from inanimate test objects and from a limited number of healthy human volunteers in accordance with a protocol reviewed by the Institutional Review Board of Miami University. Scans were rendered into digital 3D models in the Artec software. From there, if desired, it is possible to manipulate and visualize the digital 3D models from different angles while recording the monitor view with various screen capture recording software. In future, such digital 3D models of human skin could be readily incorporated into online training modules, for training and education purposes. Alternatively, the digital 3D models can be printed using, for example, the Mcor Iris color 3D printer to create a physical 3D model, with potential applications as hands-on training aids. The present study focused on evaluating the color accuracy of the digital 3D models and of the 3D printed objects

Figure 14.3
Artec Spider 3D color scanner (Artec 3D; Luxembourg, Luxembourg).

created from the digital 3D models. Color differences can occur between original object and final model. Texture modifications are required to compensate for these errors.

A texture file is one of the files generated from the color 3D scan of an object. A texture file is "wrapped around" the geometry of the 3D digital model and contains the color information used by the color 3D printer. However, there is no perfect correspondence between the color of the original object, the color information encoded in the texture file, and the color of the 3D printed object generated by sending the 3D digital model (including its associated texture file) through the 3D printer.

In this study, a color calibration method was developed and tested to compensate for color errors introduced by the 3D color scanning process, for the specific scanner model used in the present work. Second, a transfer function was derived to perform the required texture modifications to compensate for color errors introduced via the 3D printing process, for the specific model of color 3D printer used in the present work. The effectiveness of the transfer function was then tested by measuring color errors for a set of 3D printed color blocks fabricated with and without the transfer function color correction method. Lastly, a feasibility study was conducted to demonstrate the application of the color correction algorithm to 3D printed skin models generated from 3D color scans of healthy human subjects of varying skin tones.

2.2 Scanner color accuracy assessment

To assess the degree of variation between the color of a physical object and that of the 3D texture file produced by the scanner, validation procedures were executed as follows. Through the use of an Ocean Optics STS spectrophotometer (Fig. 14.4) (Oxford, UK), CIELAB values were obtained for a set of 100 matte Pantone (Pantone; Carlstadt, NJ, USA) color cards according to the protocol described in Section 2.2.1. Matte color cards were used to minimize any reflective interference with spectrophotometric measurements associated with gloss finishes. The color cards were then scanned with the scanner, and the digital texture file was interrogated in Adobe Photoshop 2017 (Adobe; San Jose, CA, USA), using the protocol described in Section 2.2.2, to determine mean RGB values associated with each color. These RGB values were converted to CIELAB values via a MATLAB script (Mathworks, Natick, MA, USA), and then the CIELAB values from the digital model were compared against the CIELAB values from the spectrophotometer to calculate the color error using the ΔE_{94}^{*} method.

2.2.1 Spectrometer color measurements

For all color evaluation procedures, CIELAB values were measured with an Ocean Optics STS spectrophotometer (Oxford, UK) equipped with a tungsten halogen light source (Fig. 14.4). To obtain consistent measurements, the light source of the spectrophotometer was turned on and allowed to warm up for 5 minutes. To further promote accurate

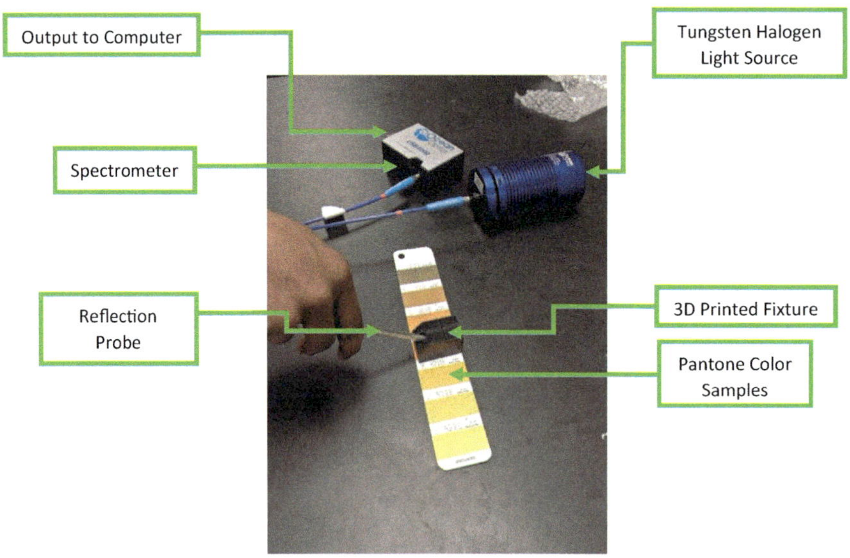

Figure 14.4
Color measurement setup for Pantone color cards.

measurement, calibration procedures for the spectrophotometer were completed for the hardware and corresponding Spectrasuite software (Ocean Optics; Largo, FL, USA) according to the manufacturer's recommendations. Calibration was a onetime procedure executed at the beginning of data collection. To secure the optic probe and ensure that data were captured from a uniform area, a 3D printed fixture (Fig. 14.4) was designed to focus the optic light for a circular area with a diameter of 1 cm. The fixture was complete with a slip fit insertion point and a hard stop angled at 45 degrees. A crosshair was extruded on the top of the optic fixture block to align the optic measurement area at the center of the color card or given object. After placing the optic fixture on top of a physical object, data collection was initiated and allowed to continue for a set integration time. At the completion of the integration time, the Spectrasuite software conducted preset algorithms to generate metrics quantifying the measured color in terms of CIELAB values, assuming a CIE D65 standard illuminant and a CIE standard 10 degrees observer. The intensity versus wavelength graphs from the spectrometer were also recorded. The CIELAB values for each color card were recorded and used as inputs for the ΔE^*_{94} color difference calculations.

2.2.2 Scanner color measurements

With color cards placed flat on a table surface, the Artec scanner was passed over the length of card slowly to permit maximum texture acquisition through the Artec Studio software. Real time feedback of scan progression was provided on the visual display of the laptop computer connected to the scanner, to ensure that the area of interest was being recorded. The scanner is equipped with its own intrinsic light source used for scan acquisition. For all scanning procedures, the room lights were turned off and window blinds were closed to minimize extrinsic light illuminating the target object. Although the scanner is also capable of scanning in well-lit environments, it was decided that possible variability associated with incandescent and florescent lighting could be minimized through implementation of a no light condition.

To quantify the color associated with the digital texture file generated by the scanner, the exported texture file of the 3D model was uploaded into Adobe Photoshop 2017 for analysis. The area selection tool was utilized to highlight the area of specific color blocks in the color card image. Using the color histogram tool, mean RGB values were obtained and recorded for each of the digital scans of the color cards. Because RGB values encoded in the images are independent of ICC (International Color Consortium) color profiles, no specific ICC profile was set for the image analysis. To compare the digital colors of the texture file with the spectrophotometric measurements of the color cards, the RGB values had to be converted to CIELAB values. To execute this conversion, a MATLAB code was generated using D65 illuminant and standard White Point values in line with CIE standards.

2.2.3 Color comparison

With both scanned digital and physical cards in CIELAB data format, it was possible to calculate a color difference (ΔE_{94}* value) for the test colors, to measure the color error introduced by the scanning process. The CIELAB values obtained from the scanned digital images and from the spectrometer were compared via the ΔE_{94}* color difference equation. The resulting ΔE_{94}* values constitute the color error for the scanned digital representation of the physical objects captured by the Artec scanner. The measured CIELAB values for each test color were also converted to RGB for the creation of graphs to provide a visual representation of the potential offsets in the RGB color space.

2.2.4 Scanner color calibration

A color calibration procedure was developed to minimize the color difference between the digital image and the physical object. This was accomplished within Adobe Photoshop 2017. Color cards for two colors (white and a dark shade of gray), whose CIELAB values were already known from the spectrometer measurements, were included within the scanned images of 25 color cards to serve as internal color controls. The scanned texture files were calibrated by opening the file in Adobe Photoshop 2017, choosing two points corresponding to the locations of the known colors, and assigning to those locations the spectrophotometer-based values for those colors. The software can automatically adjust the entire image file once the known values are assigned. Once the image was calibrated in this manner, the mean RGB values for each color were recorded as described previously, converted to CIELAB values in MATLAB, and compared with their respective spectrometer CIELAB values using the ΔE_{94}* color difference metric.

2.2.5 Effects of lighting conditions on scanner performance

The color accuracy of the Artec scanner was also evaluated under different lighting conditions as a means to establish the relationship between environmental lighting conditions and the color accuracy of the 3D scan. Scans of 11 arbitrary color cards were collected sequentially under 3 different lighting conditions. The lighting conditions used to test the scanner were selected because they are representative of lighting conditions that may be found at long-term care facilities, hospitals, or other locations where patient scans of pressure injuries may be collected in the future. The first scan was conducted during midday with the window shades open and overhead incandescent lights turned on. The second scan was conducted in the same way but with the window shades open and overhead incandescent lights turned off. The final scan was collected with natural light diffused through the closed window shades and overhead incandescent lights turned off. Appropriate time was permitted between scans to allow the scanner to remain at optimal scan temperature. It should be noted that time elapsed during this time was not enough to cause significant error due to environmental lighting changes. The ΔE_{94}* values of the 11 colors were collected individually from the respective texture files for each lighting condition and assessed.

2.3 Printer color accuracy assessment

2.3.1 Creating test color blocks

To assess the color accuracy of the Mcor Iris color 3D printer, the color difference between printed physical objects and the objects' digital color values had to be evaluated. To evaluate this color difference, a set of colored blocks (cubes with sides lengths of 1 inch) were generated to evaluate the physical color values from printed objects compared with the digital color values used to create those objects. To accomplish this, a range of test colors across the color spectrum were systematically selected using the online Adobe Color Selector in Adobe Color CC 2017, as shown in Fig. 14.5.

Within this program, the contrasting tool's provided color chart was utilized to draw a diametrical line from which colors would be selected. On each line, six colors were selected with brightness being randomly adjusted to 0%, 25%, 50%, or 100%, respectively. On acquisition of RGB values for six colors spanning a given line, the line was rotated 45 degrees to permit selection of another color set. This process was repeated 4 times until a sample of 100 colors was obtained. The RGB values from each color were used to generate digital 3D models of 1 inch × 1 inch × 1 inch (w × l × h) blocks with a different

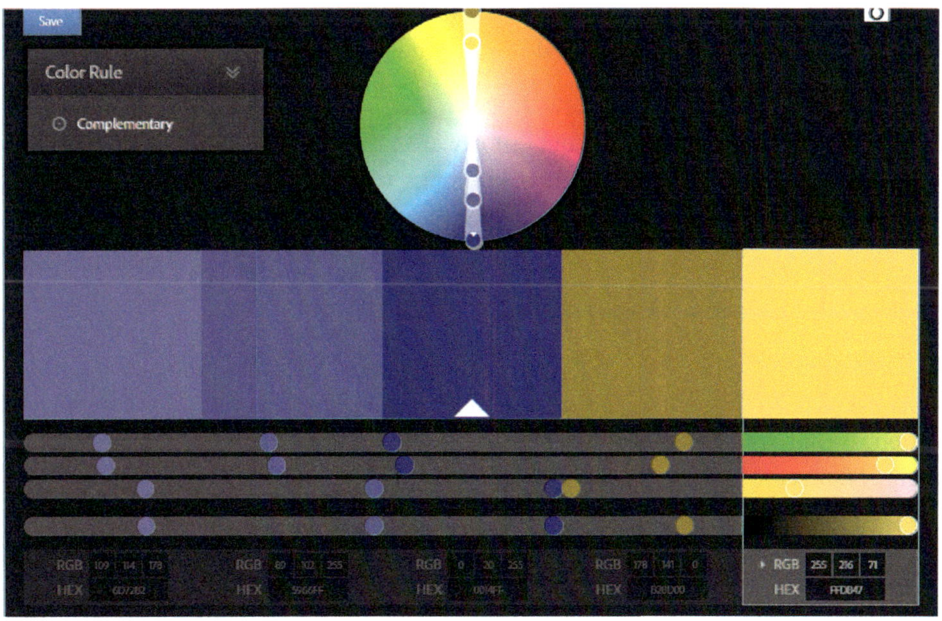

Figure 14.5
Color selection on Adobe color wheel.

Figure 14.6
Set of 25 color cubes used for printer color accuracy assessment.

color on each of the four sides and two opposing sides left blank (Fig. 14.6). Within Netfabb CAD software (Autodesk, San Rafael, CA, USA), the RGB values for each selected color were input into the color-creating feature and combined with the target face of the extruded block. These blocks were subsequently made watertight and exported as .wrl files for Mcor Iris preprint procedures in Mcor's SliceIT software.

When setting up the print in the preview space in the printer software interface, the blocks were oriented such that the sides without color were on the top and bottom of the print. This ensured that the colored sides of the test blocks would be formed from the cut edges of the paper used to create the print, rather than the flat face of a single page, so that potential color errors associated with the print process would be incorporated in the prints.

2.3.2 Spectrometer color measurements with integrating sphere

For the 3D printed color blocks, a new experimental setup for color reflectance measurements was developed by adding an integrating sphere (Fig. 14.7) (Newport Corporation, Irvine, CA, USA) in place of the 3D printed fixture shown in Fig. 14.4. Integrating spheres are typically used to diffuse light, so that it is uniform throughout the sphere, which aids in generating more accurate and consistent reflectance measurements. A Fianium supercontinuum laser (Fianium; Southampton, UK), with output between 450 and 900 nm, was directed into the integrating sphere. The color block was positioned at a different port of the integrating sphere. Incident light reflected off of the color block and inside of the sphere and ultimately passed through the fiberoptic cable into the Ocean

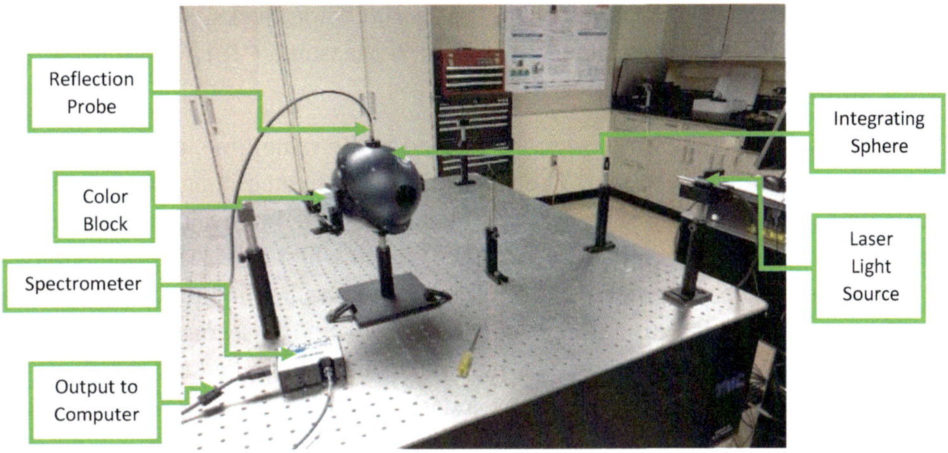

Figure 14.7
Color measurement setup for thee-dimensional printed color blocks.

Optics spectrophotometer with Spectrasuite software. CIELAB values were generated by the software as described previously, and the intensity versus wavelength graphs were recorded for each test surface.

2.3.3 Color comparison

The RGB values for each test color (Figs. 14.5, 14.6) were converted to CIELAB values via a MATLAB script. The ΔE_{94}* color difference metric was utilized to quantify the color difference between the digital image sent to print and the physical object that was produced. The resulting ΔE_{94}* constitute the color error for the digital 3D model submitted to the printer and the physical object generated by the 3D printer. The measured CIELAB values for each test color were also converted to RGB for the creation of graphs that would give a visual representation of the potential offsets in the RGB color space.

2.4 Developing the transfer function

2.4.1 Creating test color blocks for six grayscale colors

A transfer function was derived to perform the required texture file modifications to compensate for color errors introduced via the 3D printing process. To develop the transfer function, grayscale colors were used. Grayscales are colors with equal R, G, and B values, which makes it possible to only create one transfer function that could be applied simultaneously to R, G, and B values within a given texture file. The chosen grayscale colors spanned the range from RGB of 0–255 at regular intervals of 51. Two 1-inch cube blocks of three grayscale colors each were created using the previously mentioned procedures. The spectrophotometric measurements of CIELAB values for these grayscale blocks were obtained the same way as for the set of 100 color blocks, using the integrating sphere and the experimental setup shown in Fig. 14.7. Measured CIELAB values were converted to RGB values using a MATLAB script.

Three sets of these grayscale blocks were 3D printed at the same time, and each set received a different surface treatment. The first set was not treated in any way. In addition to noncoated blocks, two different types of coatings were tested to evaluate their effects on the color. The first coating was a mixture of Mcor C1430 white glue (Mcor Technologies; Dublin, Ireland) and water at a ratio of 1:4. This mixture was applied to the block using a small paintbrush covering it with a thin, evenly spread layer. The second type of coating that was chosen was a semigloss polycrylic material. This was applied the same way as the glue coating. These blocks were placed in a dark cabinet to dry for 1 day before color measurement.

2.4.2 Derivation of transfer function

The transfer function was developed in the RGB space because the inputs to the transfer function are from a digital source (the texture file from the scanner), and the outputs of the transfer function are sent as a digital 3D model to the 3D printer. Once the color measurements from across the RGB space were collected from the grayscale blocks, separate plots were made for each coating type (uncoated, glue coating, polycrylic coating). The input RGB value in Netfabb was used on the y-axis, and output RGB from the printer (using the spectrometer measurements of the 3D blocks) was used on the x-axis. A regression line was fit to each set of data. The regression line constitutes the transfer function for the given coating type. When using the transfer function to perform texture modification to compensate for color error associated with the 3D printing process, the desired RGB values served as the input to the function and the output of the function was the required input for the 3D printer.

A custom MATLAB script was developed to convert the RGB values of each pixel in a texture file with respect to the transfer function. The texture file from the scan served as the input of the program, and the output file was a new texture file that would be wrapped around the object file so that it was "corrected" before the print.

2.4.3 Testing transfer function effectiveness on selected target colors

Once the transfer function was computed, a new set of blocks were made to test its effectiveness in a preliminary trial. The test blocks were made by choosing eight color targets including four grayscales and four colors of varying RGB values (a green, a red, a blue, and a purple). The RGB values for the selected colors were run through the transfer function, to calculate the required input RGB values to submit to the printer to obtain the desired colors in the final printed object. The transfer function—corrected values (the required inputs calculated in the previous step) were then applied to blocks in Netfabb in the same way as described previously. Following this step, the "precorrected" blocks were 3D printed, coated with glue, then measured using the integrating sphere/spectrophotometer setup. The CIELAB values of the prints were converted to RGB and then compared with the desired RGB values using graphical methods. ΔE_{94}* color difference calculations were also performed. The graphical method was used to help visualize the potential correspondence between the desired RGB values of the printed objects and the actual measured values obtained from the prints.

2.4.4 Testing transfer function effectiveness on von Luschan skin colors

Based on the outcome of the preliminary evaluation of the transfer function's performance described in Section 2.4.3, the effectiveness of the transfer function was next evaluated more comprehensively for a color set with more relevance for human skin. CIELAB values of varying skin tones were obtained from the von Luschan scale (Table 14.1), which is a series of 36 opaque glass tiles used historically to characterize skin colors worldwide based on visual evaluation. CIELAB values for all 36 von Luschan tiles were previously reported by Ref. [26], and these reported values were used in the present study. ΔE_{94}* values were calculated to compare each tile color with its immediate neighbors, which revealed that the first 14 tiles were nearly indistinguishable; hence, tiles 1 to 13 were excluded from analysis. Tiles 22, 24, 31, and 32 were excluded for the same reason. Lastly, it was noted that tiles 34–36 fell outside the range of the transfer function because they are darker than the darkest color value that can be generated by the 3D printer used in this study (for colored surfaces formed from the cut edges of the paper used to create the print). Applying the transfer function to these colors would generate negative RGB values, which are invalid and cannot be interpreted by digital devices. The three colors that were not in the range of the function were still printed without applying the transfer function correction, to show to what extent the uncorrected blocks deviated from the optimal case.

The CIELAB values from the remaining von Luschan colors were converted into RGB using a conversion script in MATLAB. Then the RGB values were run through the transfer

Table 14.1: von Luschan Skin Chart.

	1	10			19	28	
	2	11			20	29	
	3	12			21	30	
	4	13			22	31	
	5	14			23	32	
	6	15			24	33	
	7	16			25	34	
	8	17			26	35	
	9	18			27	36	

function, and a new set of corrected RGB inputs was obtained. Two distinct sets of von Luschan color blocks were then printed: one with and one without the transfer function correction applied to the RGB values submitted to the 3D printer, so that the effect of the transfer function could be examined. After printing, the color of each block was measured using the integrating sphere as described previously.

2.4.5 Feasibility study for human subject scans

A feasibility study was conducted to demonstrate the application of the color correction algorithm to 3D printed skin models generated from 3D color scans of two healthy human subjects of varying skin tones. The human subject research protocol was approved by the Institutional Review Board of Miami University. Color 3D scans were collected of the forearm skin near the elbow, using the Artec scanner. Markers with known color values were included in the scans to permit image calibration as described in Section 2.2.4. After image calibration, the texture files generated by the scanner were transformed via the transfer function algorithm. Two distinct 3D models were then printed for each subject, one with transfer function correction and one without, to show the effect of the color correction technique on the printed skin model.

3. Results

3.1 Scanner color accuracy results

3.1.1 Color comparison

Figs. 14.8–14.10 show three separate plots (for R, G, and B channels, respectively) comparing the color values obtained from the scanned images of the Pantone color cards

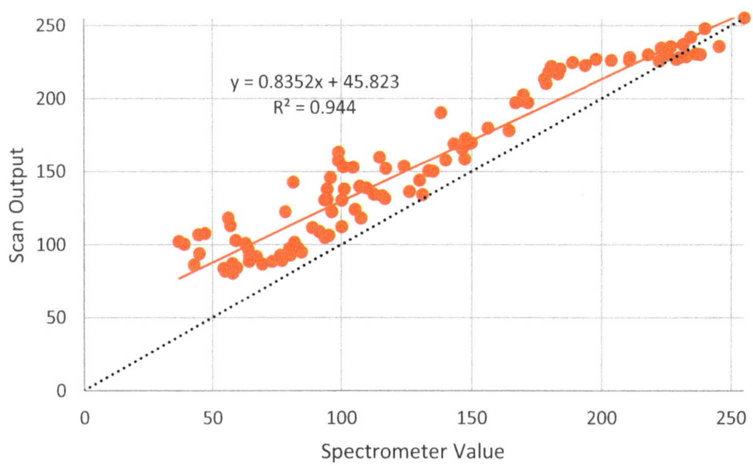

Figure 14.8

Scanner color accuracy assessment: scanner-based values for R channel versus spectrometer values.

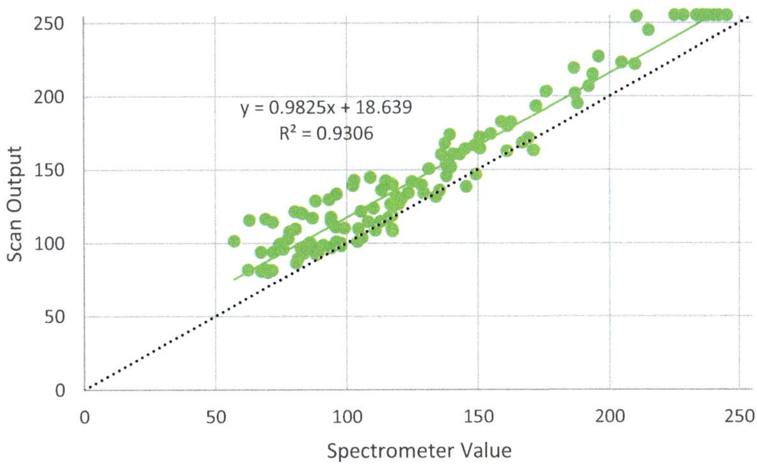

Figure 14.9

Scanner color accuracy assessment: scanner-based values for G channel versus spectrometer values.

(y-axis) versus the color values from the spectrometer measurements of the physical color cards themselves (x-axis). The dashed black line in each image represents the ideal case, shown as a line with a slope of unity indicating perfect correspondence where input is equal to output. The red and green channels showed reasonable agreement overall, with slight to moderate offsets in the lower end of the range. The blue channel showed more obvious nonlinearity at the low and high ends of the color range.

3.1.2 Effect of color calibration technique

ΔE_{94}* values were calculated for the set of 25 colors analyzed with and without the color calibration step in Adobe Photoshop 2017. Results are shown in Table 14.2. Without calibration, the average ΔE_{94}* was 12.4, which is somewhat high in relation to an approximate benchmark value of 2 being regarded as visually undetectable (see Fig. 14.2 for context). Using the color calibration step in Photoshop increased the color accuracy of the scan, improving the average ΔE_{94}* value from 12.4 to 9.28. The calibration process

Table 14.2: ΔE_{94}* values: scanner accuracy before and after Photoshop calibration.

	ΔE_{94}*	
	Without calibration	**With calibration**
Mean	12.4	9.18
Median	10.8	8.74
Minimum	5.59	1.86
Maximum	32.6	16.8
Standard Deviation	6.10	3.86
n	25	25

also reduced the maximum $\Delta E_{94}*$ value from 32.6 to 16.8. Although the color accuracy of the scanned images was not as high as anticipated, the improvement due to the color calibration step was noticeable and improved the color accuracy of the texture file as a whole.

3.1.3 Effects of lighting conditions on scanner performance

Ambient room lighting appeared to have almost no effect on the scans obtained with the Artec scanner. The color values obtained in the scans under different lighting conditions were almost identical, with a maximum color difference between lighting conditions of 0.233 (Table 14.3).

3.2 Printer color accuracy results

3.2.1 Color comparison

Figs. 14.11–14.13 show separate R, G, and B plots comparing the color values obtained from spectrometer measurements of 100 printed color blocks (y-axis) versus the known color values used as inputs to the 3D printing process (x-axis). Once again, the dashed black line represents the ideal case where input equals output. Overall, there were large differences between the input color value and the actual printed color value, for all three color channels. The green channel showed the least spread, whereas the blue channel showed the greatest variability. All three channels showed some cases with highly variable outputs for the same input.

The average $\Delta E_{94}*$ value associated with the printer, 18.56, showed that significant color inaccuracies were introduced by the printing process (Table 14.4). The $\Delta E_{94}*$ values also varied widely within the data set (Fig. 14.14). These spread from 3.38 to 41.12, with a peak frequency in the range of $\Delta E_{94}* = 10–15$.

Table 14.3: $\Delta E_{94}*$ values for varying lighting conditions during scan.

	$\Delta E_{94}*$	
	Window plus incandescent lighting versus no lighting	Window lighting versus no lighting
Mean	0.074	0.085
Median	0.052	0.108
Minimum	0.024	0.009
Maximum	0.233	0.153
Standard Deviation	0.082	0.061
n	11	11

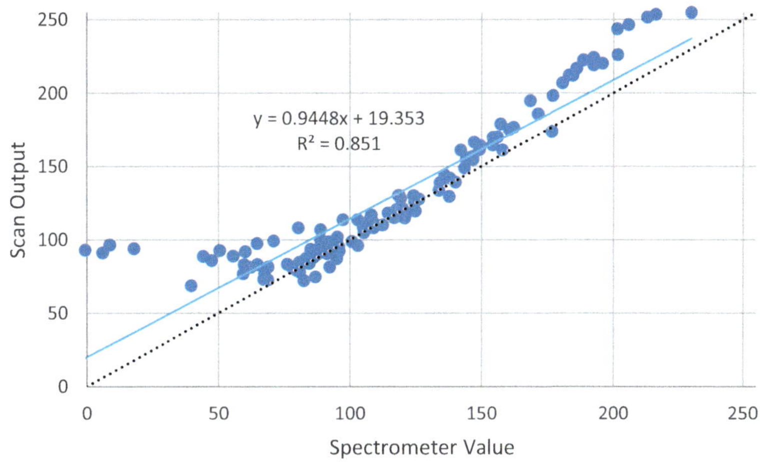

Figure 14.10

Scanner color accuracy assessment: scanner-based values for B channel versus spectrometer values.

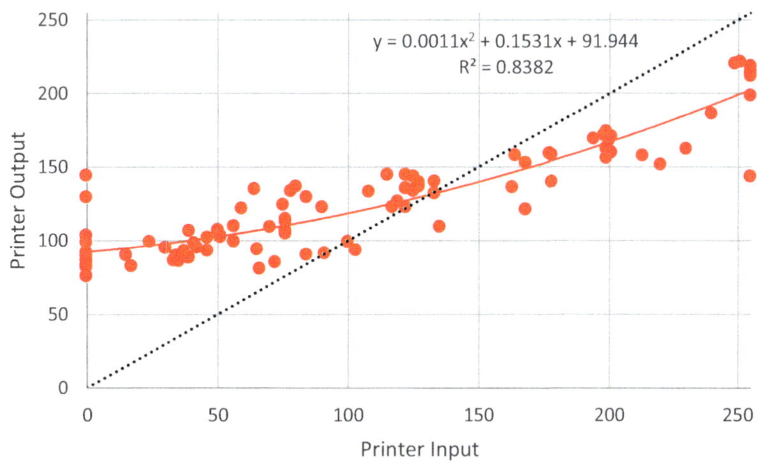

Figure 14.11

Printer output R values of 100 colors versus input.

3.3 Transfer function

Three transfer functions were developed, one for each of the three different surface treatments, to ascertain the best surface treatment for future prints. The visual effect of each surface treatment is illustrated in Fig. 14.15. The polycrylic coating produced a somewhat glossy surface. The glue coating appeared to increase the darkness of the printed color (a slight shift on the light/dark axis) with minimal effects on the other color axes. The effects of each coating are shown more quantitatively in the subsequent section.

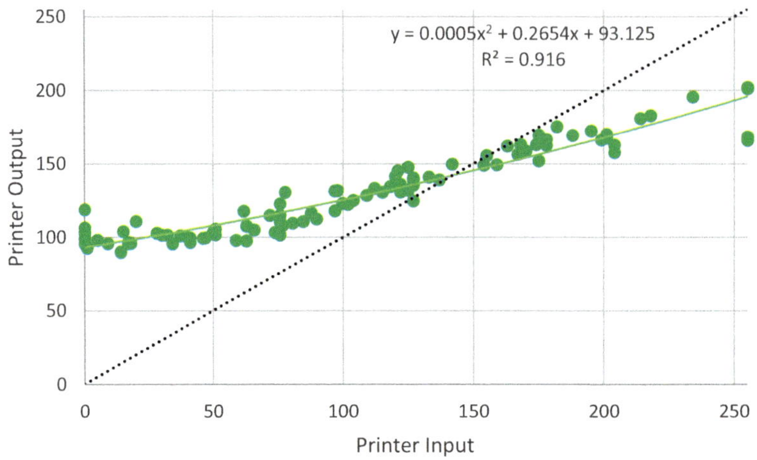

Figure 14.12
Printer output G values of 100 colors versus input.

Table 14.4: ΔE_{94}^* values of 100 three-dimensional printed color samples.

	ΔE_{94}^*
Mean	18.56
Median	15.71
Minimum	3.38
Maximum	41.12
Standard Deviation	8.77
n	100

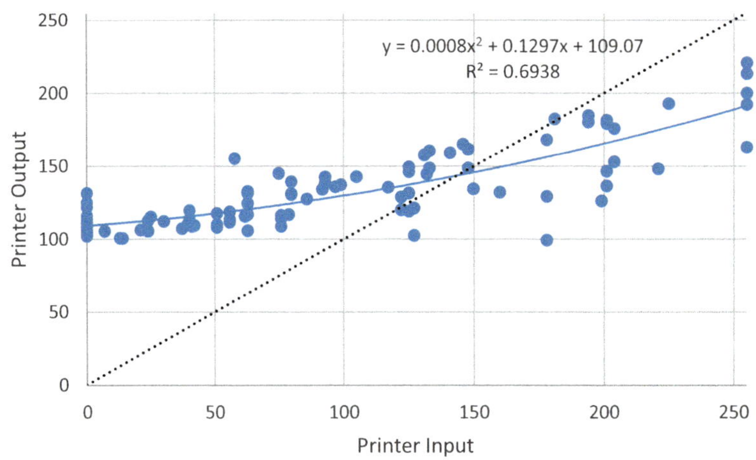

Figure 14.13
Printer output B values of 100 colors versus input.

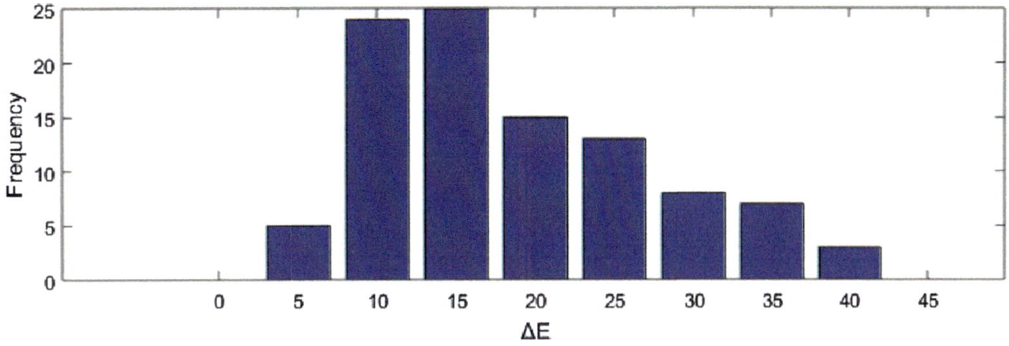

Figure 14.14
Distribution of ΔE_{94}* values for 100 three-dimensional printed colors.

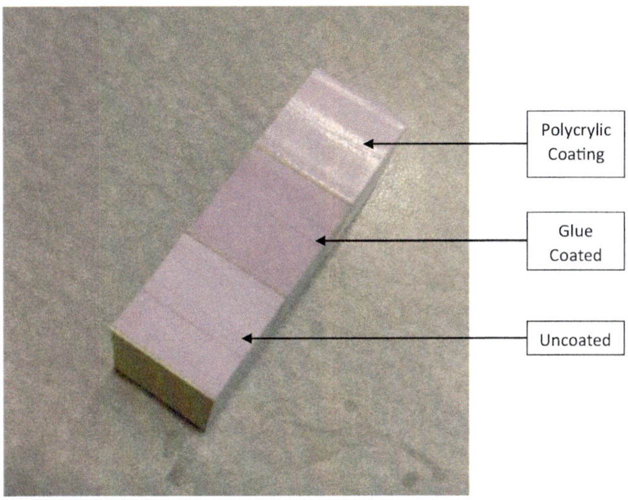

Figure 14.15
Color blocks with three different types of surface treatment methods.

The results for the 3D printed grayscale color blocks are shown in Fig. 14.16. Recall that a grayscale color has equal R, G, and B values. The digital R = G = B values used as inputs to the 3D printer are shown on the x-axis, and the measured colors for the printed blocks are shown on the y-axis. There was minimal difference between the three surface treatments, with the glue coating producing a slight downward shift of the printed colors, consistent with the darkening effect observed in Fig. 14.15. The semiglossy appearance of the polycrylic coating appeared to have minimal effect on the color of the prints.

Once the data were collected from the grayscale blocks for even increments across the RGB space (Fig. 14.16), separate curves were created for each surface treatment method (Figs. 14.17–14.19). The axes were reversed (with the printer RGB output on the x-axis

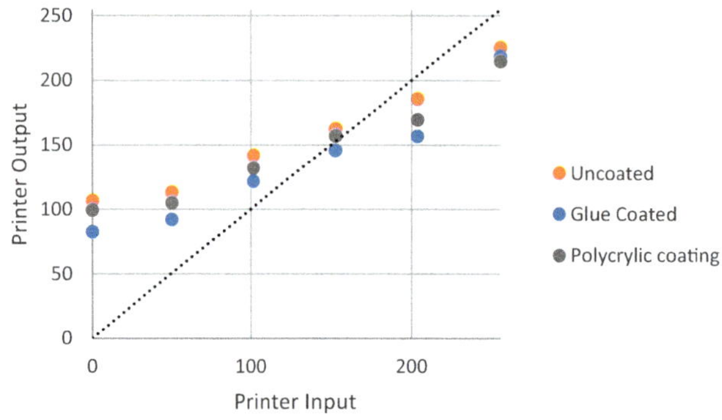

Figure 14.16
Printer output values of three types of coated grayscales versus input.

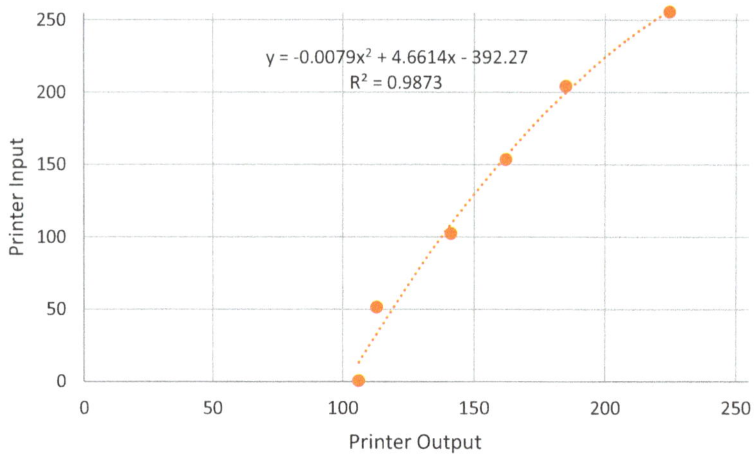

Figure 14.17
Graph of uncoated transfer function equation.

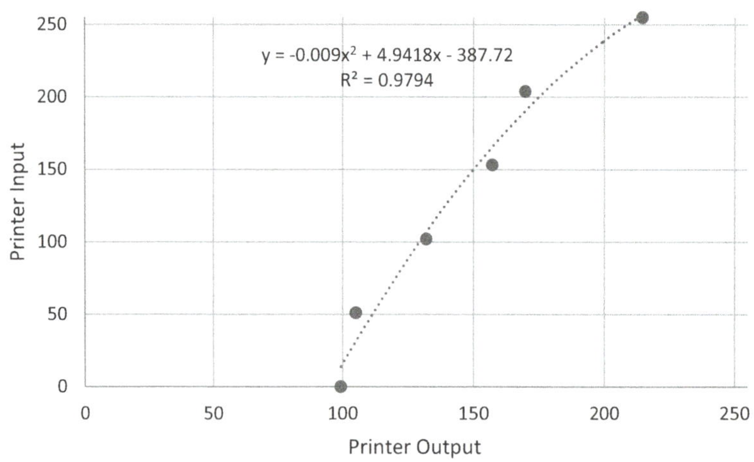

Figure 14.18
Graph of polycrylic coated transfer function equation.

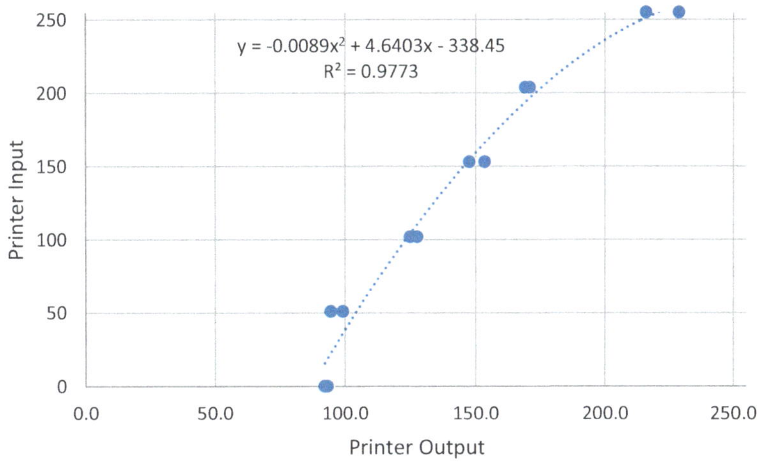

Figure 14.19
Graph of glue coated transfer function equation.

and printer RGB input on the y-axis) because the desired RGB outputs from the printer serve as the *inputs* to transfer function calculations. Quadratic best-fit lines were determined for each set of results. All R^2 values were above 0.97. The quadratic equation on each plot represents the calculated transfer function for each surface treatment.

The glue coating was selected as the optimal surface treatment for 3D printing skin tones, primarily based on the upper and lower bounds of the transfer function range for the glue coated blocks. It was found that none of the 36 von Luschan skin tones had an RGB value greater than 227, which is within the glue coating transfer function range. Furthermore, although none of the three transfer functions had lower range limits much below 100, the glue coating transfer function had the lowest limiting value of the three, consistent with the slight darkening effect of the glue coating itself. Thus, the transfer function associated with the glue coated grayscale blocks (Fig. 14.19) was selected for this study:

$$y = -0.0089x^2 + 4.6403x - 338.45 \quad (14.9)$$

This equation was implemented in the MATLAB code for texture modification as described, converting the RGB values of each pixel in an image with respect to the equation of best fit. However, as the equation only encompassed values down to 88 for each R, G, and B value, a further adjustment was needed. To address this problem, a linear equation was computed from (0,0) to the transfer function at x = 90, which had the form of

$$y = 0.0816x \text{ for RGB} < 90 \quad (14.10)$$

This change resulted in much more reasonable outputs of the code as it retained small gradient to the lower value colors, rather than turning all colors below the threshold black.

3.4 Transfer function effectiveness

3.4.1 Preliminary trial on selected colors

The preliminary trial of the transfer function showed a substantial improvement in color accuracy, having an average ΔE_{94*} value of 6.82 (Table 14.5). The effect of the transfer function can also be seen in the graphs (Figs. 14.20–14.22) showing the RGB values for prints generated with the transfer function correction, compared with the optimal line. Transfer function–corrected values generally lie close to the optimal line, and show an improved slope that better mimics the optimal case.

3.4.2 Results for von Luschan skin colors

Figs. 14.23–14.25 also show a substantial improvement in color accuracy for color blocks in von Luschan skin tones printed using the transfer function correction method (shown in red, green, and blue) versus those printed without the transfer function correction (shown in dark gray). As noted in previous data sets, there was less scatter in the green channel and slightly more scatter in the red and blue channels. For all three color channels, the single transfer function shifted the data toward the optimal line where input equals output, suggesting that the printed colors were closer to the desired colors than in the uncorrected case. This is also supported by the mean $\Delta E_{94}*$ values (Table 14.6), which improved from 15.19 for the uncorrected color blocks to 8.03 for the corrected color blocks.

The output colors of the von Luschan skin tone trials showed a high level of accuracy from the transfer function, as the output from the printer had minimal deviation from the desired outputs. The $\Delta E_{94}*$ values for transfer function–corrected blocks also had a smaller spread (Fig. 14.26), with the exception of one apparent outlier value of 19.12. The standard deviation, shown in black in Fig. 14.26, encompasses all values other than the outlier mentioned previously.

3.4.3 Feasibility study results for human subject scans

Forearm scan and print results from subject #1 are shown in Figs. 14.27 and 14.28. In Fig. 14.27 (left), the scan is shown before executing the color calibration step in Adobe

Table 14.5: $\Delta E_{94}*$ values of the corrected blocks compared to the wanted output color.

	$\Delta E_{94}*$
Mean	6.82
Median	6.31
Minimum	1.18
Maximum	11.02
Standard Deviation	3.64
n	8

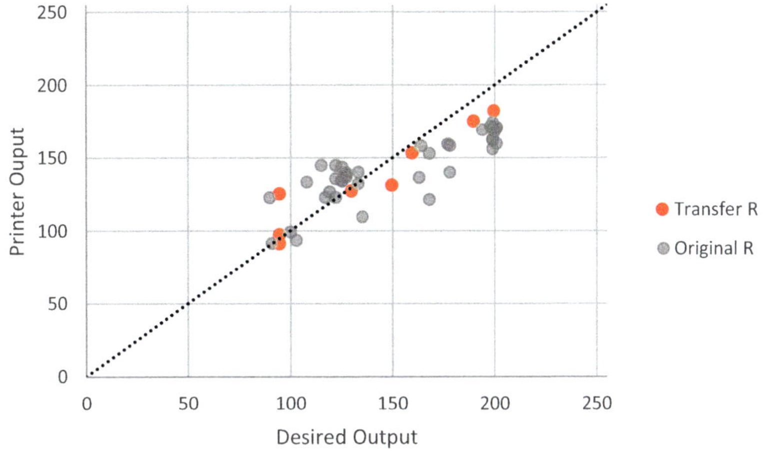

Figure 14.20
Printer output R values versus desired R values for initial test colors. Shaded data represent results from Fig. 14.8 but limited to the output range of the transfer function—corrected data (shown in *red*).

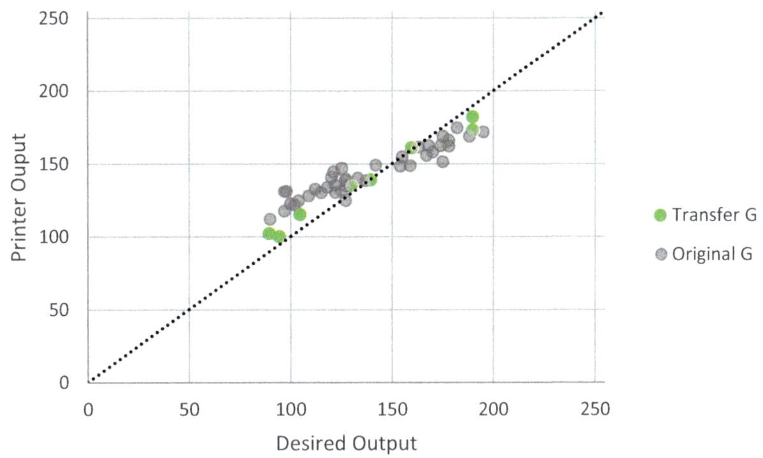

Figure 14.21
Printer output G values versus desired G values for initial test colors. Shaded data represent results from Fig. 14.9 but limited to the output range of the transfer function—corrected data (shown in *green*).

Photoshop. The uncalibrated scan was printed directly, without employing the transfer function correction technique. The resulting uncorrected print is shown in Fig. 14.27 (right). The uncalibrated scan appeared darker than the subject's actual skin color. The printed version of the scan seemed darker than the subject's actual skin as well, while also looking more faded.

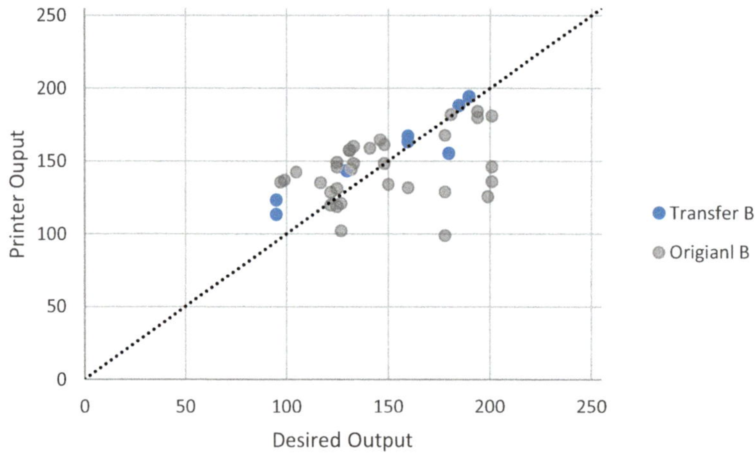

Figure 14.22
Printer output B values versus desired B values for initial test colors. Shaded data represent results from Fig. 14.10 but limited to the output range of the transfer function–corrected data (shown in *blue*).

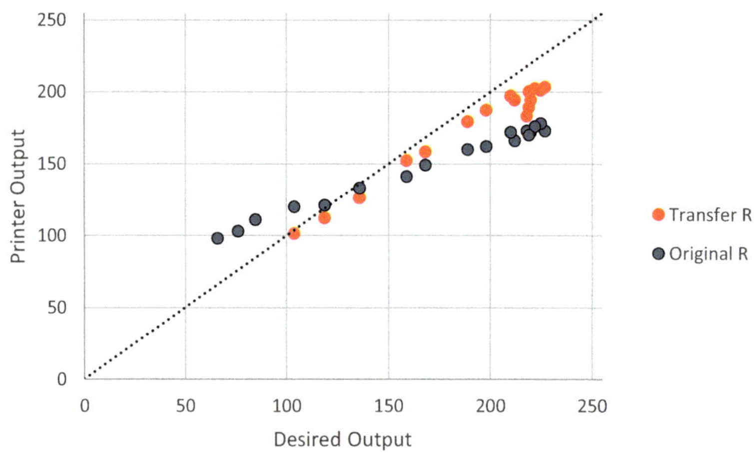

Figure 14.23
Printer output R values versus desired R values for von Luschan colors.

Next, the scan was subjected to the color calibration process in Adobe Photoshop as described, and then the calibrated texture file was sent through the transfer function for texture modification. The corrected texture file was mapped to the 3D geometry of the original scan and then sent to the 3D printer. The result at each step of that sequence is shown in Fig. 14.28. The color calibrated scan, and the final print, looked more like the subject's actual skin color. Also, the vein on the subject's forearm was more visible and less faded in the transfer function–corrected print than in the uncorrected print.

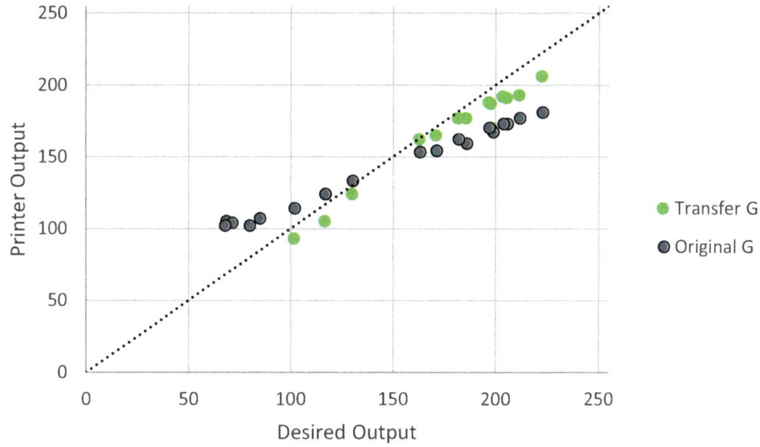

Figure 14.24
Printer output G values versus desired G values for von Luschan colors.

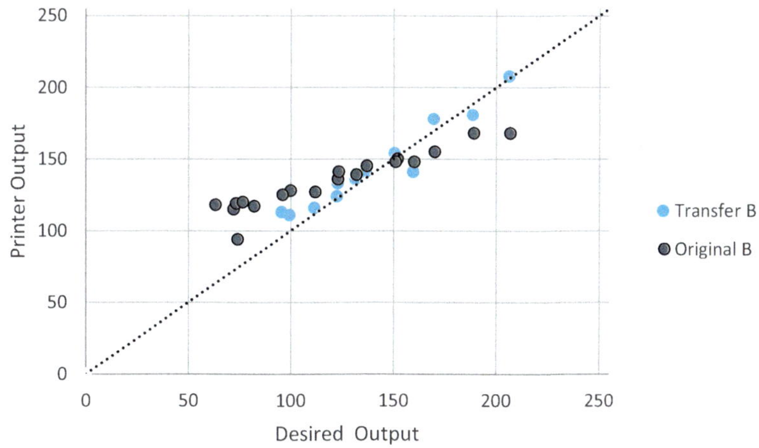

Figure 14.25
Printer output B values versus desired B values for von Luschan colors.

Table 14.6: $\Delta E_{94}{}^*$ values of von Luschan corrected and uncorrected color versus the wanted output color.

	$\Delta E_{94}{}^*$	
	Uncorrected colors	**Corrected colors**
Mean	15.19	8.03
Median	14.97	6.95
Minimum	13.46	3.25
Maximum	18.44	19.12
Standard Deviation	1.40	3.25
n	17	14

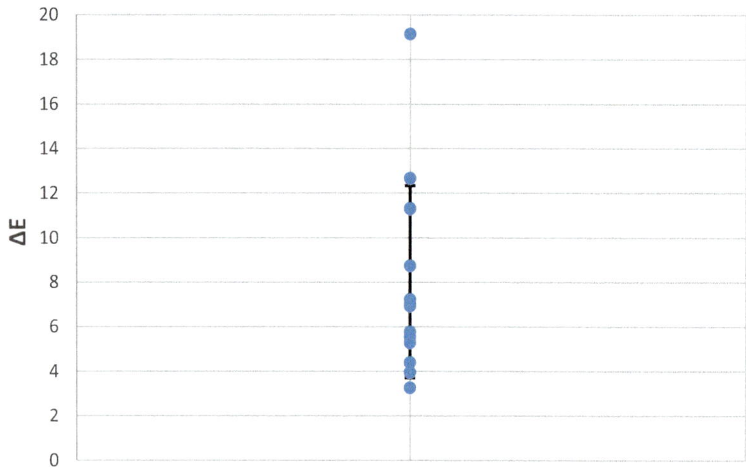

Figure 14.26
ΔE_{94}^* distribution for von Luschan colors.

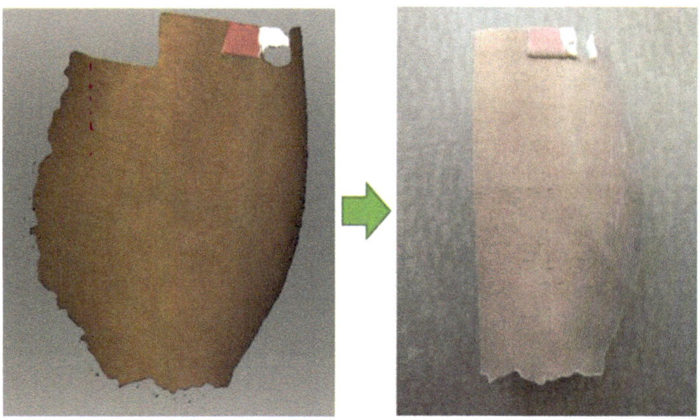

Figure 14.27
Subject #1 scan (left) that was printed directly, sanded, and glue coated (right).

Figure 14.28
Subject #1 scan (left) going through each step of the process, from left to right.

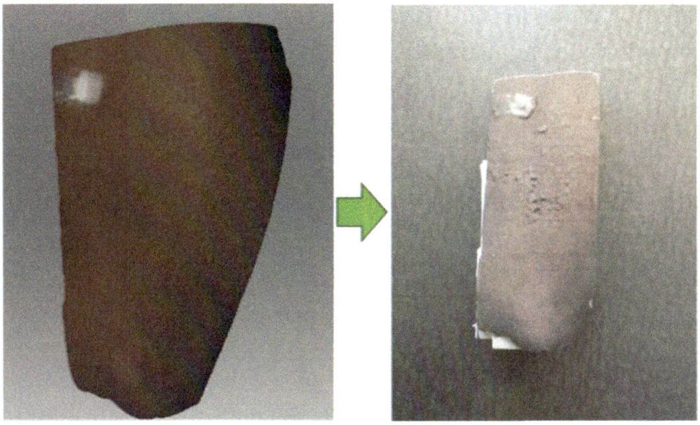

Figure 14.29
Subject #2 scan (left) that was printed directly, sanded, and glue coated (right).

Forearm scan and print results from subject #2 are shown in Figs. 14.29 and 14.30. Subject #2 had a darker skin tone than subject #1. Fig. 14.29 shows the scan before the color calibration in Adobe Photoshop and the resulting print created without the transfer function correction method. As with subject #1, the scan of subject #2 looked darker than the subject's actual skin color (Fig. 14.29). The uncorrected print (Fig. 14.29) was much lighter than the subject's skin and also looked as faded as the corresponding print from subject #1. A noticeable feature of the print was bumps throughout the vertical center. These are artifacts from the scanner software and are believed to be generated when image data from multiple angles are combined inaccurately. The bumps were much more evident in the print compared with the scan.

In Fig. 14.30, the original scan from subject #2 was subjected to color calibration in Adobe Photoshop, and then the calibrated texture file was sent through the transfer function algorithm before printing as described above. While the corrected print for

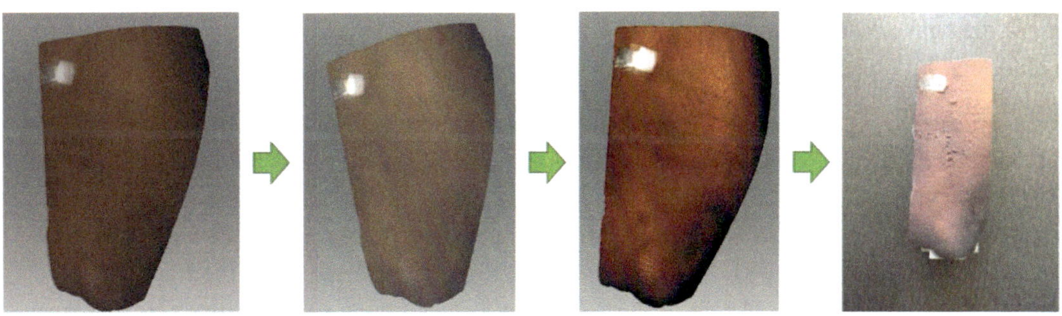

Figure 14.30
Subject #2 scan (left) going through each step of the process from left to right.

subject #2 looked more similar to the subject's actual skin color than the uncorrected print, the color on the printed model appeared to be redder than the subject's skin.

4. Discussion

The key experiments underlying our color correction protocol were scanner color accuracy tests with Pantone color cards, printer color accuracy tests with 3D printed color blocks, and derivation of the algorithm to correct for color errors in 3D printing based on color measurements of 3D printed glue coated grayscales.

When looking at the ΔE_{94}* values obtained for the scanner color accuracy assessment, much lower ΔE_{94}* values were expected from the scanner (Table 14.2 and Figs. 14.8–14.10). These value differences could have occurred as the light coming from the spectrometer (Fig. 14.4) was absorbed by the black probe fixture. However, the scan color accuracy improved after calibration to an internal reference color. In the feasibility study, the human skin scans stood out especially in this regard, looking rather dark without the correction and closer to the skin that was scanned with the changes applied (Figs. 14.28 and 14.30).

The ΔE_{94}* values for the Mcor 3D printer were almost double that of the scanner (Tables 14.2 and 14.4). This can be seen with both the high ΔE_{94}* values (Fig. 14.14) and the deviation from optimal observed in the RGB graphs (Figs. 14.11–14.13). The most notable finding of the study was that the transfer function derived in this work was able to increase the color accuracy of the prints, by decreasing the average ΔE_{94}* from 18.56 to 6.82. The performance of the transfer function was further evaluated on von Luschan skin tones, for which the ΔE_{94}* decreased from 15.19 to 8.03. It should be noted that the range of RGB values that can be generated by the printer was 106–255. The range was expanded to 88–255 after the application of the glue coating. As the RGB value range of the von Luschan skin chart extends from 63.3–227.2 the majority of von Luschan skin tones were encompassed by our transfer function (Fig. 14.31). However, as the appearance of early stage pressure injury in darkly pigmented skin can be more difficult to detect, it is important to work to expand the color correction methods developed in this work to include the full spectrum of skin tones to better meet clinical needs.

For both subject #1 and subject #2, the improvements wrought by the color correction methods were visible. This was most notable for subject #1, where the color of the printed model was visually very similar to that of the actual skin (Fig. 14.28). The model from subject #2 also improved in appearance after the color correction methods were applied, but not as much as subject #1. For that reason, further improvements to the transfer function will be necessary. An example would be to make an additional function for the

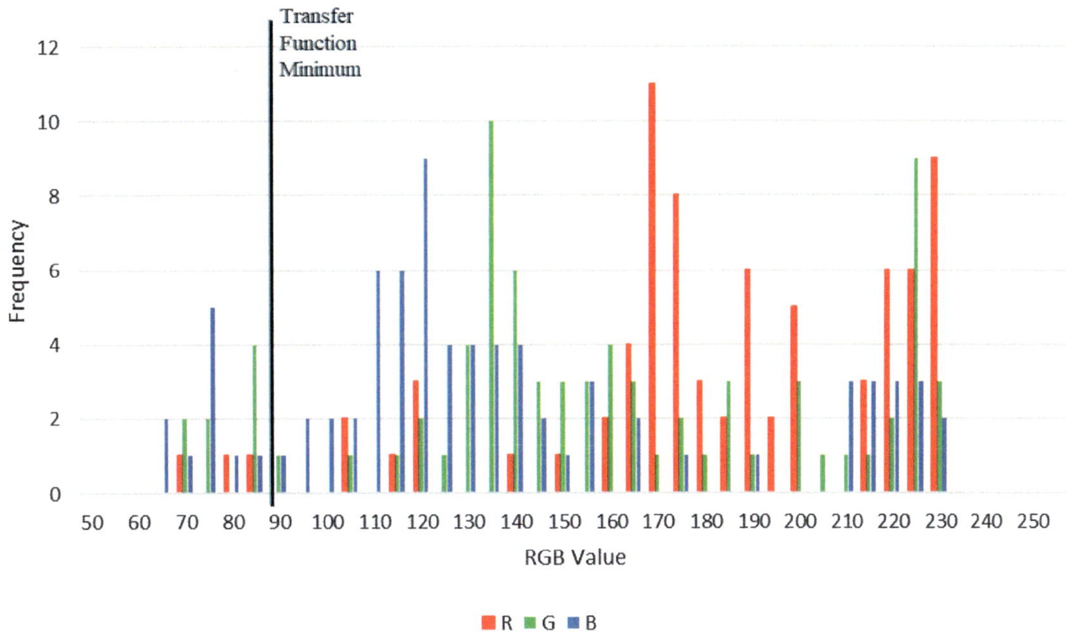

Figure 14.31
RGB value color frequency for measured skin tones from dermatology research and standardized charts (n = 78).

darker region of the printed colors, which would range between 88 and 130. This would increase the accuracy of the darker colors.

5. Conclusion

The research developed methods to compensate for color errors introduced via the processes of 3D color scanning and 3D color printing. The color correction methods described here produced measureable improvements in color accuracy, as assessed through the ΔE_{94}* color difference metric. The feasibility study demonstrated the applicability of the color correction methods to create more color-accurate 3D printed models of human skin. However, further work is needed to improve the printed skin models, particularly for subjects with darker skin tones. In future, after further refinement, these methods could be applied to create more realistic color 3D models based on scans from subjects with pressure ulcers, for training of health-care workers. The 3D models from patient scans could be either digital or physical representations of pressure ulcer appearance in subjects of varying skin tones; in general, the 3D representation of such injuries could advance the realism of training aids used in this field.

References

[1] Dziedzic M. Fast facts about pressure ulcer care for nurses: how to prevent, detect, and resolve them in a nutshell. In: Howse L, et al., editors. Wound practice & research. New York, NY: Springer Publishing Company; 2014.

[2] Williams L, Hopper P. Pressure ulcers. Understanding medical surgical nursing. 2015. Retrieved from: https://www.r2library.com/Resource/Title/0803640684/ch0054s3222.

[3] Edlich R, et al. Pressure ulcer prevention. J Long Term Eff Med Implants 2004;14(4):285−304.

[4] Guadagnin RV, Santana LA, Neves RS. Inference of pressure ulcer stage and texture on an image training set. Pattern Recognit Image Anal 2016;26(3):558−64. https://doi.org/10.1134/S1054661816030056.

[5] Allman RM. Pressure ulcer prevalence, incidence, risk factors, and impact. Clin Geriatr Med 1997;13(3):421−36.

[6] Black J, et al. National pressure ulcer advisory panel's updated pressure ulcer staging system. Adv Skin Wound Care May 2007;20(5):269−74.

[7] NDNQI − pressure injury training module III. Press Ganey Associated, Inc.

[8] Keelaghan E, Margolis D, Zhan M, Baumgarten M. Prevalence of pressure ulcers on hospital admission among nursing home residents transferred to the hospital. Wound repair and regeneration, vol. 16. Official Publication of the Wound Healing Society [and] the European Tissue Repair Society; 2008. p. 331−6 (3), http://doi.org/10.1111/j.1524-475X.2008.00373.x.

[9] Preventing Pressure Ulcers in Hospitals. Content last reviewed October 2014. Agency for Healthcare Research and Quality, Rockville, MD. https://www.ahrq.gov/professionals/systems/hospital/pressureulcertoolkit/index.html

[10] Russo C, Steiner C, Spector W. Hospitalizations related to pressure ulcers among adults 18 years and older. Agency for Healthcare Research and Quality; 2008.

[11] Harold B, et al. High cost of stage IV pressure ulcers. Am J Surg 2009;200(4):473−7.

[12] Clark M. Skin assessment in dark pigmented skin: a challenge in pressure ulcer prevention. Nursing Times; 2010.

[13] Beekman D, Schoonhoven L, Boucqué H, Van Maele G, Defloor T. Pressure ulcers: e-learning to improve classification by nurses and nursing students. J Clin Nurs 2008 Jul;17(13):1697−707.

[14] Xiao K, Zardawi F, van Noort R, Yates JM. Color reproduction for advanced manufacture of soft tissue prostheses. J Dent 2013;41(5):e15−23. https://doi.org/10.1016/j.jdent.2013.04.008.

[15] Xiao K, Zardawi F, van Noort R, Yates JM. Developing a 3D colour image reproduction system for additive manufacturing of facial prostheses. Int J Adv Manuf Technol 2014;70:2043−9. https://doi.org/10.1007/s00170-013-5448-1.

[16] Giorgianni EJ, Madden TE. Digital color management: encoding solutions. Chichester, I.K: Safari Books Online; 2008. J. Wiley,c.

[17] Takiwaki H. Measurement of skin color: practical application and theoretical considerations. The University of Tokushima School of Medicine; 1998.

[18] Krantz D. Color measurement and color theory: I. Representation theorem for Grassmann structures. J Math Psychol August 1975;12(3):283−303.

[19] Weatherall I, Coombs B. Skin color measurements in terms of CIELAB color space values. J Investig Dermatol 1992;99(4):468−73.

[20] León K, Mery D, Pedreschi F, León J. Color measurement in L*a*b* units from RGB digital images. Food Res Int December 2006;39(10):1084−91.

[21] Godbole H. Characterization of color, gloss, and mechanical performance of 3D printed structures. Rochester Institute of Technology; 2012. Thesis.

[22] Hill B, Roger T, Vohagen FW. Comparative analysis of the quantization of color spaces on the basis of the CIELAB color-difference formula. ACM Trans Graph 1997;16(2):109−54. https://doi.org/10.1145/248210.248212.

[23] McLaren K. The development of the CIE 1976(L*a*b*) uniform colour space and colour difference formula. Wiley Online Library; 1976.

[24] Oulton D, Porat I, Boston C, Walsby R. Imagemaster: precision colour communication based on CIE calibrated monitor screens. International conference on high technology. 1996.

[25] Loo MR, Rigg B. BFD (l:c) color difference formula. Part 1: development of the formula. JSDC February 1987;103:126–32. Part 2: performance of the formula. JSDC 103 (March), 86–94.

[26] Dlugos J, Taylor J. Visible reflectance spectroscopy of human skin: the use of CIEL*a*b* color analysis for *in vivo* ethnic skin characterization. Waltham, MA: Perkin Elmer; 2012.

Further reading

[1] Berman B. 3-D printing: the new industrial revolution. Bus Horiz 2014;55(2):155–62. https://doi.org/10.1016/j.bushor.2011.11.003.

[2] Central Bureau of the CIE. Colorimetry. In: Colorimetric observers. 2nd ed. CIE Publications; 1986.

[3] McMenamin PG, Quayle MR, McHenry CR, Adams JW. The production of anatomical teaching resources using three-dimensional (3D) printing technology. Am Assoc Anat 2014;7:479–86. https://doi.org/10.1002/ase.1475.

[4] McQuiston BK, Styts MR, Whitestone JJ, Henderson R. Quantifying wound morphology and tissue energetics for diagnosis and treatment of chronic wounds. Med Imag 1996. https://doi.org/10.1117/12.237891.

[5] Wyman C, Sloan P, Shirley P. Simple analytic approximations to the CIE XYZ color matching functions. J Comput Graph Tech 2013;2(2):1–11.

CHAPTER 15

Effects of friction and pressure on skin in relation to pressure ulcer formation

Leigh Toni Fleming, Pak Hung Leung
Centre for Precision Technologies, University of Huddersfield, Huddersfield, United Kingdom

Chapter Outline

The following chapter explores the effect that friction and pressure have on skin in relation to pressure ulcer formation. Pressure ulcers are a serious condition typically caused by external pressure experienced by parts of the body for a period of time. Many people will have had experience of pressure ulcers either themselves or observing a relative or friend experiencing one. They are not exclusive to acute settings, pressure ulcers that can also occur in the home, or in other care settings. They can also be experienced by people who use medical devices and prosthetics.

Pressure is not the sole factor in the onset of what are designated as pressure ulcers. Friction and resulting shear are also known factors that contribute in different ways. To understand how friction and pressure interact to contribute to pressure ulcer formation, it is important to understand them as scientific concepts.

Innovations and Emerging Technologies in Wound Care. https://doi.org/10.1016/B978-0-12-815028-3.00015-8

Table 15.1: Parameters and corresponding units.

Parameters		Corresponding Units
Normal force	A force that acts perpendicularly to a surface.	Newton (N)
Pressure [1]	"Pressure is the force acting normally on unit area of a surface or the ratio of force to area."	Pascal (Pa), Newton per square meter (N/m^2), millimeter of mercury (mmHg)
Friction [2]	"Friction is the force that resists the motion of one surface relative to another with which it is in contact."	Newton (N)
Shear [3]	"Shearing force is a force that acts parallel to a plane, as with a tensile or compressive force."	Newton (N)

Friction and pressure are physical phenomenon that occurs when any two surfaces are in contact with each other and they are subjected to a force. To understand the effects of friction and pressure on skin in relation to pressure ulcer formation, it is first important to understand the mechanisms of pressure and friction and also how they relate to each other.

To explore pressure and friction in more detail, common scientific terminology must be understood; Table 15.1 shows simple explanations of the key concepts that are important in the understanding of pressure and friction.

1. Scientific definition of terminologies

1.1 Pressure, friction, and shear

1.1.1 Pressure

If you apply the same force over areas of differing sizes, the surface with larger area would experience less damage compared with the smaller surface. The magnitude of pressure is a scientific explanation to this phenomenon. Eq. (15.1) is the equation for calculating the magnitude of pressure. "Force" and "area" are the magnitude of force measured in Newton, which is acting vertically to the area of the surface that is subjected to the vertical force, respectively. The unit of pressure is Newton per square meter (N/m^2).

$$\text{Pressure} = \frac{\text{Force}}{\text{Area}} \tag{15.1}$$

1.1.2 Friction

Friction is a force that acts in the opposite direction to motion in between two surfaces that are in contact. Friction is measured in Newton (N). The magnitudes of friction can be calculated with Eq. (15.2), in which F_f is the magnitude of friction, μ is the coefficient of friction, which is dependent on the material properties of the surfaces in contact, and N is the normal force acting on the surfaces in contact. The energy converts from kinetic energy to thermal energy in most cases. To reduce the magnitude of friction, the magnitude of normal force can be reduced. However, it is often the case that the magnitude of the normal force is constant or uncontrollable. Thus, the coefficient friction is reduced by different means.

$$F_f = \mu N \tag{15.2}$$

For most situations, friction is undesirable and engineers try to reduce the friction as much as possible. For instance, lubricants are introduced into the cylinder of automotive engines and gearboxes to reduce the magnitude of friction by reducing the value of the coefficient of friction. However, friction is necessary in some cases, for example, in the tire of the automotive and the bottom of the shoes. The presence of the friction acts as a driving force of the motion, which allows the automotive to roll and human to walk.

1.1.3 Shear

Shear force can be understood as forces that push parts of a subject in different directions, which often causes delamination and/or splitting. Shearing can be the consequence of the presence of friction. A daily example of the shearing action is cutting. A pair of scissors cut the paper easily because two opposite force is applied to the paper by two blades, which shear the paper resulting in cutting the paper apart.

When skin is subjected to this physical phenomenon under certain conditions, it can be very damaging and can compromise the integrity of skin, allowing for infection or other conditions that can detrimentally affect patient's quality of life.

1.1.4 Pressure and skin

As previously stated, pressure is defined as force per unit area. The atmospheric pressure of the earth is about 101 kN/m^2 at sea level. Skin is good at handling atmospheric pressure, for instance, skin is perfectly capable of withstanding pressure experienced when fully submerged under water. The hydrostatic pressure experienced by the new scuba diving world record holder, Ahmed Gabr [4], was 3440.86 kN/m^2 at 332.35 m under the sea level in 2014. His skin was intact and withstood the demands placed on it due to the fact that hydrostatic pressure was uniformly distributed, whereas when there is localized pressure, it is a different story: studies have shown that as little as 6 kN/m^2 of sustained pressure is sufficient to cause a pressure ulcer under certain conditions [5].

Localized pressure occurs in patients where there is either applied pressure from an external source, such as a medical device, or where patients are immobile and their inactivity results in areas of the body being subjected to pressure due to the weight of the patient resting on certain areas.

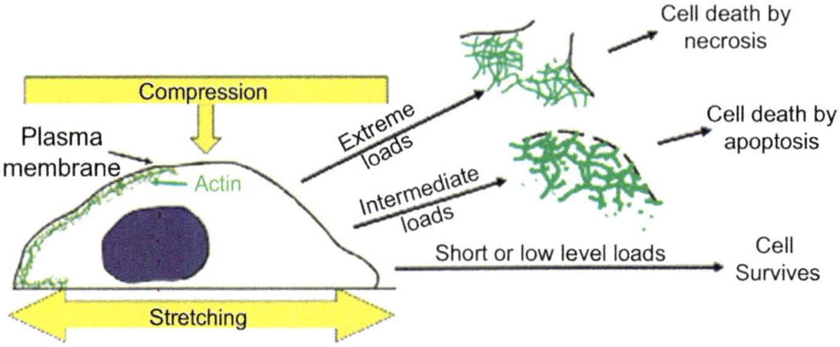

Figure 15.1
Cells deaths caused by pressure [7].

One of the main sources of loading is pressure arising from the weight of the patients. As the subcutaneous tissue, muscle, and other soft tissue have a cushioning effect that spreads the load across a bigger contact area with the supporting surface, pressure ulcers usually occur at bony prominences where there is little soft tissue. Typical anatomical locations include heel, elbow, back of the skull, and the sacrum [6].

There are typically two mechanisms or etiologies by which pressure ulcers occur. These are excessive shear and distortion of tissues causing deep tissue injury and ischemia. The magnitudes of the pressure contributing to the two etiologies are different. Higher magnitudes of pressure are typically responsible for causing pressure ulcers by excessive shear and distortion, while lower magnitude of pressure are capable of causing damage due to ischemia. The presence of pressure can lead to cell death by necrosis and apoptosis at the tissue scale [7] and distortion of internal soft tissues at bony prominence at skeletal muscle and bone level [8] as shown in Fig. 15.1.

The effects of magnitude of the pressure acting on skin is also time depending [9]; the time dependency is based on a number of different factors and can be affected by the condition of the skin due to nutrition, hydration, moisture, or other factors that affect it. The higher magnitudes of pressure experienced through self-weight over bony prominence often cause deep tissue injuries that appear as localized area of maroon or purple discoloration of intact skin or a blood-filled blister. This can be further exacerbated by the presence of friction not only causing damage directly to the skin but also inducing shear forces when combined with pressure.

During ischemia, the cutaneous blood vessels are occluded resulting in restriction in blood flow, accumulation of metabolic waste, changes in pH value, and often cell death [10]. This ultimately can result in pressure ulcers through pressure alone; if you also introduce the phenomenon of friction, further damage can be sustained, as with the deep tissue injury, this is further exacerbated by shear that can also be introduced. Ischemia is commonly the cause of pressure ulcers where there is interaction between the skin and a

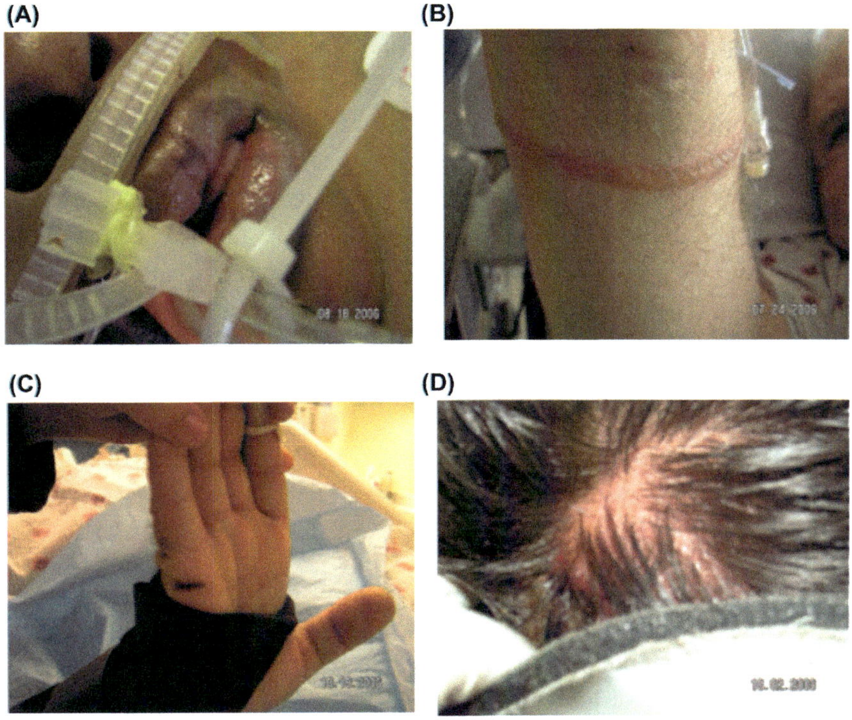

Figure 15.2
Pictures of medical device–related pressure ulcers [13].

medical device; this is often referred to as medical device–related pressure ulcer. The common medical devices found to cause pressure ulcers are cervical neck collar, O_2 saturation probe, noninvasive ventilation mask, among others, but there is a potential for a medical device–related pressure injury when there is any contact between a device and skin where there is a force applied either externally or through loading of some sort [11,12]. Medical device–related pressure ulcers were shown to contribute 33% to the total number of pressure ulcers, and in one study conducted by Black in 2010, they were mostly diagnosed to be stage I and II [13]. These medical devices are often attached to the patients' bodies to monitor their physiological conditions; however, the fixation methods (acting like a tourniquet) or the weight of the devices reduce the cutaneous blood flow. Fig. 15.2 shows the pressure ulcers caused by medical devices [13]. Fig. 15.2A–D are the pressure ulcers caused by endotracheal tube, arterial line tubing, wrist splint, and cervical collar, respectively.

1.1.5 Friction and skin

Friction presents where two surfaces are in contact, of interest to the situation of pressure ulcer formation includes anywhere there is contact with skin and another surface where there is likely to be movement and/or pressure. Examples of this are where skin is in contact with clothing, bed sheets, pillow covers, and pressure redistribution devices. It is unlikely that friction can be eliminated completely from these situations; thus, there are studies that show projects that try to reduce the magnitude of friction by reducing the friction coefficient, shown in Eq. (15.2), to have an impact on the overall magnitude of friction. Vilhena from the University of Coimbra conducted a study to investigate the coefficient of medical use fabrics including polyamide, polyester, silk, cotton, and wool against human skin in 2016 [14]. The coefficient ranged from approximately 0.1 to 1.1. The wide range of the coefficient indicates that carefully designed medical devices can reduce the physiological effects of friction to patients. Dinsdale conducted a study to investigate the magnitude of pressure required to initiate a pressure ulcer with and without pressure, which were found to be 5999.5 N/m^2 (45 mmHg) and 38,663.5 N/m^2 (290 mmHg), respectively, in 1974 [5]. Although it was an animal experiment, it still showed the importance of controlling the magnitude of friction for reducing the prevalence of pressure ulcers.

As friction has drawn more attention in recent years in the field of pressure ulcers prevention, it is more popular to see in the commercial market nowadays that different types of medical devices, including mattresses, mattress covers, patient transporting equipment, etc., claim to have low friction properties. However, to minimize the friction that acts on the patients' bodies, the whole system of the patient skin, device, clothing, and any other surface that interfaces directly with the patient should be considered. It is therefore vitally important to consider the whole system including the patient's own clothing as well as the implementation of devices that report to reduce friction. An example of this would be where a mattress has a low friction cover on; if a sheet and nightwear is between that low friction cover and the patient's skin, the benefits of the low friction cover are potentially negated by the other layers in the system.

2. Assessment of the system

To fully evaluate the system, it is important to be able to measure pressure, friction, and shear to establish good systems to reduce the risk of pressure ulcers. In the field of engineering, different methods have been developed over the years for measuring pressure and friction. These methods have been utilized in the field of skin integrity.

2.1 Measurement of pressure

Pressure can be measured simply by establishing the load applied and the area that it is applied over. Many devices are commercially available to enable this, most of which use an array of pressure sensors in a mat or similar to "map" peak pressure over a loaded area.

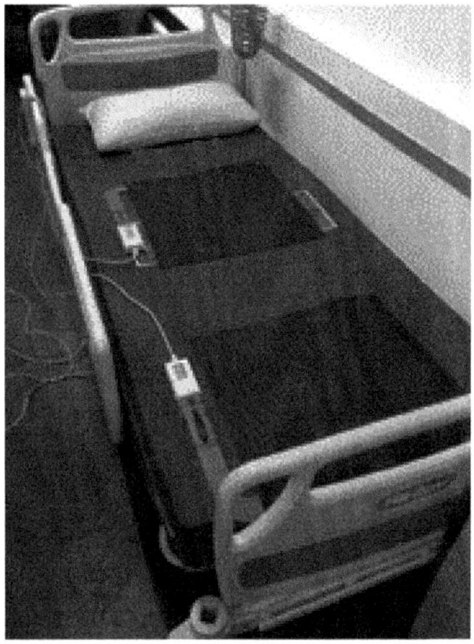

Figure 15.3
Pressure mapping equipment.

Fig. 15.3 is an experimental setup of such a system at the University of Huddersfield, which is used to establish the performance of the mattress by assessing the peak pressure over time.

Some research found out that the interface pressure may not be significantly important to the formation of deep tissue injury as the interface pressure could remain practically unchanged while the internal soft tissue is experiencing 1.5-fold or even higher stress and distortion [15,16]. However, for initial testing of mattress and the clinical practice, pressure mapping is an effective tool to evaluate the interface pressure experienced by patients at least for the prevention of the superficial pressure ulcers formation. Pressure mapping measurement can also be utilized to evaluate the interface pressure experienced by the patients caused by different medical devices to prevent the formation of medical device–related pressure ulcers.

2.2 Measurement of friction

Friction is also a phenomenon that can be quantified; it is complex to assess the friction of materials against skin; however, a number of studies have achieved this. Vilhena measured the friction of skin by utilizing a multicomponent force sensor (2-axis load cell) with a probe that was used to investigate a number of different materials used in medical applications. Both the normal and tangential forces were measured by load cells, and the coefficients of friction between the skin and the specific medical-use fabric were calculated using Eq. (15.2) [14]. Derler conducted an experiment measuring the friction between the human skin and "smooth" and "rough" glass [17]. The principle of the experiment employed a quartz 3-component dynamometer that was attached to a glass plate. The volunteers' hands were rubbed against the plate and the frictional and normal forces were measured; thus, the coefficient of friction could be deduced. The only difference between the two measurements was that the force sensors were attached to the stationary and moving components for the skin–glass and skin–fabric measurement, respectively.

3. Pressure ulcer classification

According to the guidelines provided by the European Pressure Advisory Panel [20], pressure ulcers can be categorized into stages:

- Category/Stage 1: Nonblanchable erythema of intact skin
- Category/Stage 2: Partial-thickness skin loss with exposed dermis
- Category/Stage 3: Full-thickness skin loss
- Category/Stage 4: Full-thickness skin and tissue loss

There are a further two categories, typically used when the skin is unbroken to begin with, but there is damage to underlying tissues without breakthrough of the skin, which often presents as discoloration of the skin.

- Unstageable
- Deep tissue pressure injury

4. Relieving pressure

The only way to completely remove pressure from an area of skin is to ensure that skin is not in contact with another surface or device. In practice, this is very difficult, so the ideal is to spread loading across as large an area as possible to redistribute and reduce pressure on the skin. From Eq. (15.1), it is evident that if the area of loading A is increased, and the force F remains the same, the pressure will be reduced.

Where a pressure ulcer has already begun to form, often the only option is to offload the area completely to allow for healing; it is important to ensure that the issue is not moved to the area of redistributed pressure.

There are a number of devices that are designed to redistribute pressure over larger areas, which can be used in conjunction with good clinical practice to seek to reduce the potential of pressure ulcers occurring in patients who demonstrate risk factors.

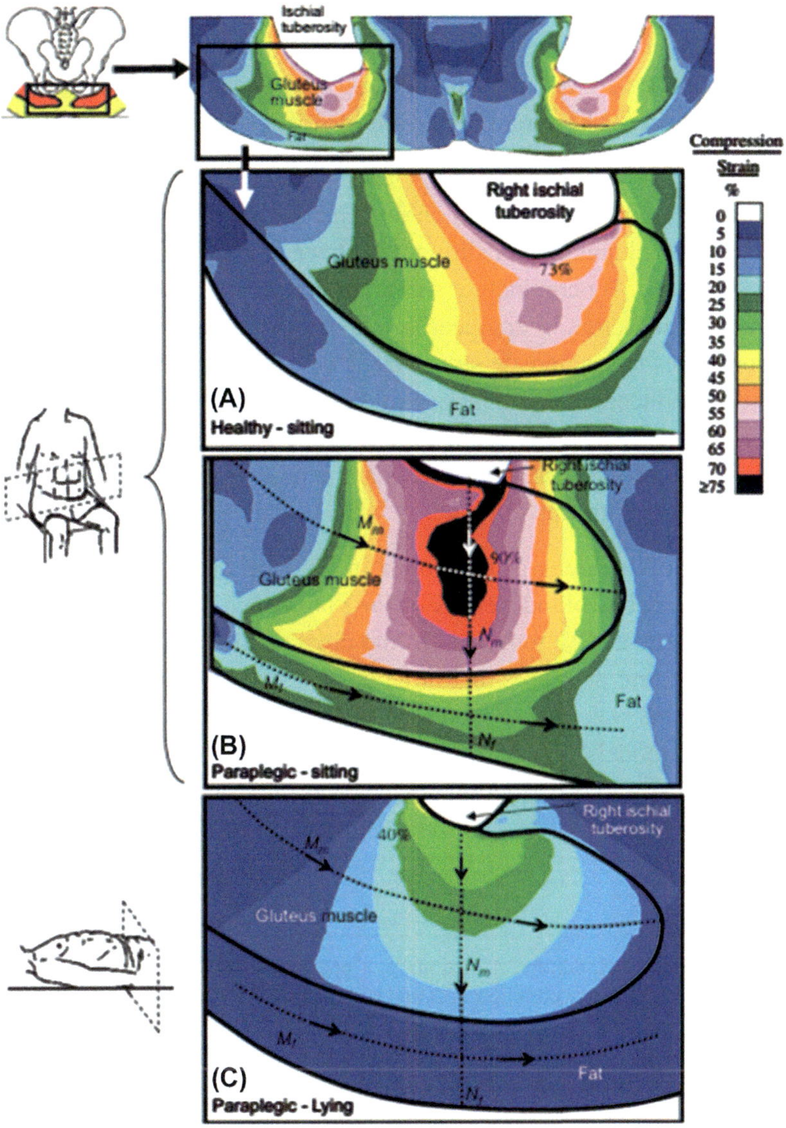

Figure 15.4
Graphical representation of internal loading mechanism of healthy and paraplegic volunteers in sitting posture. [18]

4.1 The role of modeling

Computational modeling is a reliable tool in engineering to predict and demonstrate the deformation or behavior of material under specific conditions. Finite element of analysis (FEA) and computational fluid dynamics are two typical types of computational modeling techniques that divide the designed computational models into many small pieces and simulate the behaviors of the materials by the means of mathematical calculations. A lot of computational models were developed over the years, including the ones with the purposes of demonstrating the internal distortion of soft tissue, reduction in the cross-sectional area of cutaneous blood vessels, as well as many others.

Computational models were developed to understand the internal strain experienced by healthy and paraplegic volunteers and sitting on rubber tires and semirigid flat foam by Linder-Ganz in 2008 [18]. By conducting simulations of the computational modeling (FEA), all the important factors in engineering perspective including compressive strain, tensile strain, von Mises stress, tensile stress, etc., for both the gluteus muscle and enveloping fat were deduced. Besides numerical data, the internal loading mechanism was illustrated by different graphical representation in Fig. 15.4. The computational model demonstrated that most of the aforementioned parameters were higher in paraplegic groups than the healthy volunteers groups [18].

A three-dimensional FEA model was developed in 2017 by Leung for demonstrating the deformation of the cutaneous blood vessels of the volunteers in sitting posture [19]. The model showed there to be a 19% reduction in the cross-sectional area of cutaneous blood vessels when typical sitting pressures measured by pressure mapping equipment were applied to the model, this was further increased when friction was also considered in the model. Computational models such as these offer a useful tool for device developers in evaluating possible pressure redistribution devices and investigating their effects on pressure, friction, and shear.

References

[1] Pressure. In: Law J, Rennie R, editors. A dictionary of physics. Oxford University Press; 2015. Retrieved from: http://www.oxfordreference.com.libaccess.hud.ac.uk/view/10.1093/acref/9780198714743.001.0001/acref-9780198714743-e-2414.

[2] Friction. In: Law J, Rennie R, editors. A dictionary of physics. Oxford University Press; 2015. Retrieved from: http://www.oxfordreference.com.libaccess.hud.ac.uk/view/10.1093/acref/9780198714743.001.0001/acref-9780198714743-e-1196.

[3] Shearing force. In: Law J, Rennie R, editors. A dictionary of physics. Oxford University Press; 2015. Retrieved from: http://www.oxfordreference.com.libaccess.hud.ac.uk/view/10.1093/acref/9780198714743.001.0001/acref-9780198714743-e-2760.

[4] Guinness World Records. Deepest scuba dive (male). 2014. Retrieved from: http://www.guinnessworldrecords.com/world-records/deepest-scuba-dive-(male.

[5] Dinsdale SM. Decubitus ulcers: role of pressure and friction in causation. Arch Phys Med Rehabil 1974;55(4):147–52.

[6] Agam L, Gefen A. Pressure ulcers and deep tissue injury: a bioengineering perspective. J Wound Care, 2007;16(8):336–42.

[7] Gefen A, Weihs D. Cytoskeleton and plasma-membrane damage resulting from exposure to sustained deformations: a review of the mechanobiology of chronic wounds. Med Eng Phys 2016;38(9):828–33. https://doi.org/10.1016/j.medengphy.2016.05.014.

[8] Gefen A. Deep tissue injury from a bioengineering point of view. Ostomy/Wound Manag 2009;55(4):26–36.

[9] Linder-Ganz E, Engelberg S, Scheinowitz M, Gefen A. Pressure–time cell death threshold for albino rat skeletal muscles as related to pressure sore biomechanics. J Biomech 2006;39(14):2725–32. https://doi.org/10.1016/j.jbiomech.2005.08.010.

[10] Liao F, Burns S, Jan Y. Skin blood flow dynamics and its role in pressure ulcers. J Tissue Viability 2013;22(2):25–36. https://doi.org/10.1016/j.jtv.2013.03.001.

[11] Davis JW, Parks SN, Detlefs CL, Williams GG, Williams JL, Smith RW. Clearing the cervical spine in obtunded patients: the use of dynamic fluoroscopy. J Trauma Inj Infect Crit Care 1995;39(3):435–8. https://doi.org/10.1097/00005373-199509000-00006.

[12] Murray JS, Noonan C, Quigley S, Curley MAQ. Medical device-related hospital-acquired pressure ulcers in children: an integrative review. J Pediatr Nurs 2013;28(6):585–95. https://doi.org/10.1016/j.pedn.2013.05.004.

[13] Black JM, Cuddigan JE, Walko MA, Didier LA, Lander MJ, Kelpe MR. Medical device related pressure ulcers in hospitalized patients. Int Wound J 2010;7(5):358–65. https://doi.org/10.1111/j.1742-481X.2010.00699.x.

[14] Vilhena L, Ramalho A. Friction of human skin against different fabrics for medical use. Lubricants 2016;4(1):6.

[15] Elsner JJ, Gefen A. Is obesity a risk factor for deep tissue injury in patients with spinal cord injury? J Biomech 2008;41(16):3322–31. https://doi.org/10.1016/j.jbiomech.2008.09.036.

[16] Linder-Ganz E, Shabshin N, Itzchak Y, Gefen A. Assessment of mechanical conditions in sub-dermal tissues during sitting: a combined experimental-MRI and finite element approach. J Biomech 2006;2007;40(7):1443–54. https://doi.org/10.1016/j.jbiomech.2006.06.020.

[17] Derler S, Gerhardt L-C, Lenz A, Bertaux E, Hadad M. Friction of human skin against smooth and rough glass as a function of the contact pressure. Tribol Int 2009;42(11):1565–74. https://doi.org/10.1016/j.triboint.2008.11.009.

[18] Linder-Ganz E, Shabshin N, Itzchak Y, Yizhar Z, Siev-Ner I, Gefen A. Strains and stresses in sub-dermal tissues of the buttocks are greater in paraplegics than in healthy during sitting. J Biomech 2007;2008;41(3):567–80. https://doi.org/10.1016/j.jbiomech.2007.10.011.

[19] Leung IPH, Fleming L, Walton K, Barrans S, Ousey K. Development of a model to demonstrate the effects of friction and pressure on skin in relation to pressure ulcer formation. Wear 2017;376–377:266–71. https://doi.org/10.1016/j.wear.2016.11.026.

[20] National Pressure Ulcer Advisory Panel. In: Haesler E, editor. European pressure ulcer advisory panel and pan pacific pressure injury alliance. Prevention and treatment of pressure ulcers: quick reference guide. Osborne Park, Australia: Cambridge Media; 2014.

CHAPTER 16

Silver—enzyme hybrids as wide-spectrum antimicrobial agents

Y. Dror, C. Ophir, A. Freeman

Department of Molecular Microbiology and Biotechnology, Faculty of Life Sciences, Tel Aviv University, Tel Aviv, Israel

Chapter Outline

1. Introduction

Chronic wounds such as pressure ulcers, diabetic foot ulcers, and venous leg ulcers often do not heal and become "stalled" due to prolonged inflammatory state initiated by exudates accompanied by high density of mixed population of contaminating microorganisms embedded in necrotic tissues and biofilms. Prevention of wound contamination, removal of necrotic tissue by debridement, and biofilm eradication by wide-spectrum antimicrobial agents are therefore essential actions required for shifting from the "stalled" state into an active healing process [1,2]. Topical antimicrobial treatments were hence recommended [3], mostly applied by wound dressings, which according to recent FDA classification comprise three products groups: "solid wound dressings" "gels, creams, and ointments" and "wound wash solutions" Silver is the most

Innovations and Emerging Technologies in Wound Care. https://doi.org/10.1016/B978-0-12-815028-3.00016-X

frequently wide-spectrum antimicrobial agent included in most products affiliated with all abovementioned dressing categories (FDA classification of wound dressings, 2016) [4].

The use of silver has gradually evolved throughout the past two decades from the use of ionic silver, e.g., as silver nitrate solution, silver sulfadiazine, or silver alginate, to the use of metallic silver, e.g., dressing embedded metallic silver nanoparticles [1,5].

The efficacy and safety of wound dressings impregnated with silver nanoparticles for the enhancement of wound healing were questioned, however, in a series of recent critical reviews, providing data obtained from either case studies or controlled clinical trials [5–9]. Although few reports on silver resistance and allergy concerns were not considered as a substantial obstacle to the extensive use of silver [9], its efficacy was questioned [5–8]. It appeared that because of heterogeneity of wound size and status, wide-range variations of admitted patients' age and wound condition, and accompanied medical issues, e.g., dialysis, clearly defined conclusions about the efficacy of silver-based treatments could not be made. At this time, it appears that silver-containing dressings were found effective as wide-range antimicrobial wound treatments in some cases while failed to be effective in others [5,7].

Furthermore, in an attempt to explain the gap between results obtained from in vitro studies describing high efficacy of ionic silver as effective wide-spectrum antimicrobial agent and results reported for metallic silver nanoparticles impregnated in wound dressings, thorough analysis of the molecular mechanisms underlying the antimicrobial activity of ionic versus metallic silver was carried out [10–14]. It appeared that ionic silver is responsible for the antimicrobial activity of silver, while pure metallic silver is practically inert until it is ionized.

The antibacterial activity of ionic silver is affected by a series of intracellular activities including inhibition of essential enzymes involved in electron transport throughout the respiratory chain, interaction with DNA, and rupture of prokaryotic cell membrane [10]. It appears, however, that while the mechanisms involved with the antibacterial activity of ionic silver were elucidated, data available on mechanisms involved in its antimicrobial activity on eukaryotic cells, e.g., yeast and fungi, are much more limited.

The ionization of metallic silver present on silver nanoparticles' surface strongly depends on environmental conditions, such as humidity and chemical oxidation, and may be hampered by particle's surface coating with adsorbed proteins present in wound's exudates [13]. The inherent nonhomogeneous environmental conditions prevailing in infected chronic wounds may thus result in large variations in effective silver ions release and triggering of their antimicrobial activity, especially from nanoparticles embedded in solid wound dressings.

Furthermore, it also appeared that soluble silver ions exhibit very poor migration capability in physiological medium, resulting from precipitation following interaction with chloride or phosphate ions as well as from adsorption onto the surface of negatively charged proteins and other biopolymers. The poor bioavailability of silver ions therefore calls for controllable continuous generation and supply of silver ions within the close vicinity of the targeted cells for effective antimicrobial activity. This prerequisite becomes critical in cases where the targeted microbial cells are embedded within necrotic tissues as biofilms, while silver is provided as metallic nanoparticles embedded within another phase, e.g., an outer phase of solid wound dressings.

In view of these observations, it appears that two critical steps are involved in enabling effective use of silver nanoparticles as a wide-spectrum antimicrobial agent for wound management: bringing the silver nanoparticles to close vicinity—preferably into contact—with the targeted microbial cell population and gaining control of (I) release of silver ions from the particles' surface and (II) enabling effective migration of the silver ions released to the targeted cells' surface [12,13]. Efficient control of the two processes will enable efficient antimicrobial activity of silver nanoparticles administered by all abovementioned three wound dressing categories for either infection prevention or eradication of microbial biofilms from contaminated wounds [12,13,15].

In view of the abovementioned prerequisites and mechanistic aspects, it appears that improvement of the efficacy of the next generation of silver nanoparticle–based dressings calls for *redesign* of their current structure, characteristics, and mode of administration. A key parameter in such redesign process appears to be a shift to *use of soluble, nanosize metallic silver nanoparticles, equipped with controllable leaching of the antimicrobial silver ions into the close vicinity of the targeted cells.*

Here, we describe design and characterization of such a new type of silver nanoparticles, which demonstrates redesign aiming at coping with the abovementioned prerequisites.

2. Design and fabrication of novel enzyme–silver hybrids as an antimicrobial agent

In view of the potential inherent in protein-mediated biotemplating, enabling directed deposition of metals onto the surface of protein molecules [16], we propose a "core and shell" approach for the fabrication of soluble silver nanoparticles comprising core made of biologically active oxidoreductase enzyme molecule, coated with a thin, porous metallic silver layer, affected by directed electroless silver deposition.

Molecules of the enzyme glucose oxidase (Fig. 16.1A), presenting a "peanut shaped," negatively charged dimer, 5×8 nm in size, were selected as the core for the fabrication of the proposed silver–enzyme hybrid.

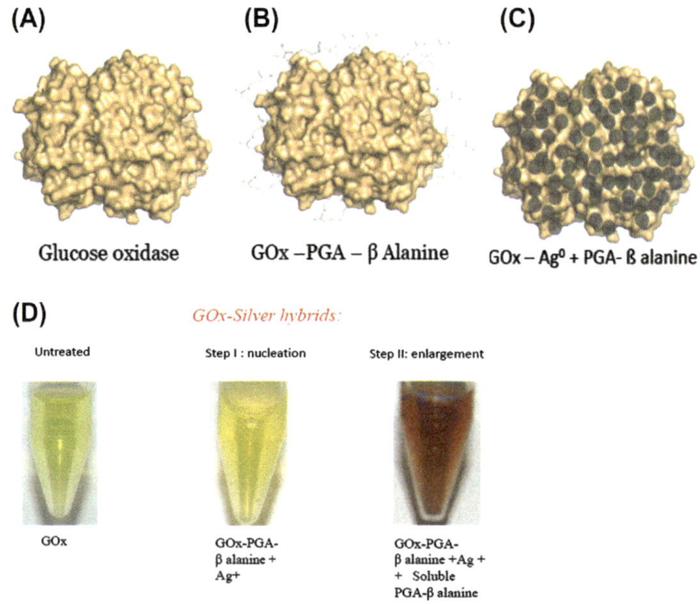

Figure 16.1
The main three steps in the fabrication of glucose oxidase–silver hybrids (A–C) and their physical appearance (D).

Glucose oxidase (GOx) is an analytical enzyme catalyzing the reaction:

$$\underset{\text{Glucose}}{C_6H_{12}O_6} + 2H_2O + O_2 \xrightarrow{GOX} \underset{\text{Gluconic acid}}{C_6H_{12}O_7} + 2H_2O_2$$

This enzyme is routinely used for the monitoring of blood glucose levels of diabetic patients [17].

A two-step mechanism of electroless silver deposition onto GOx molecules' surface was enabled by displaying on its surface polymeric chains of polyglutaraldehyde (PGA)–β-alanin conjugates (see Fig. 16.1B), capable of binding and reducing silver ions, thus providing a controllable nucleation step. Subsequent enlargement was carried out by the addition of soluble silver ions and polymeric reducer (Fig. 16.1C). A series of enzyme–silver hybrids could be thus prepared (as described in Fig. 16.1A–D and in Ref. [16] [and references cited therein]) with controllable metallic silver content of the hybrid.

The glucose oxidase–silver hybrids thus obtained displayed a porous metallic silver outer layer ~1.5 nm thick and retained the core enzyme's solubility and ~60% of its initial enzymatic activity.

As hydrogen peroxide is formed throughout the enzymatic oxidation of glucose by the active core comprising an enzyme molecule, a subsequent oxidation and leaching of silver

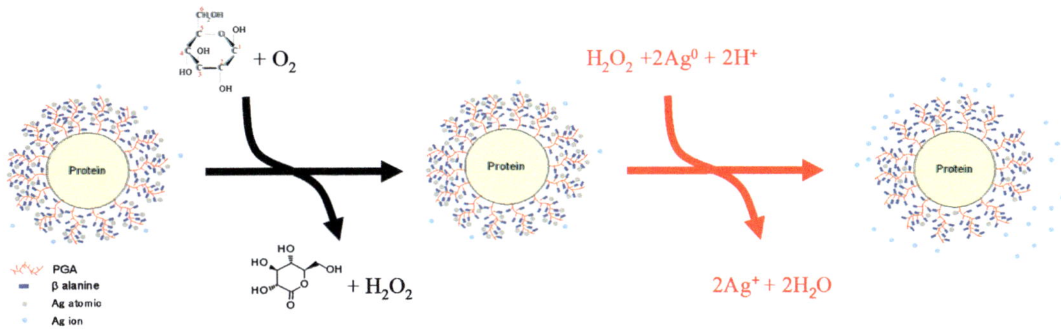

Figure 16.2
Schematic description of enzymatically attenuated silver ions release from glucose oxidase–silver hybrid.

ions from the metallic coating is enabled, providing a new mechanism of "enzymatically attenuated" silver ions release into the close vicinity of the hybrid, as schematically described in Fig. 16.2.

The soluble hybrids thus obtained could be readily formulated as solutions, lotions, or crèmes by the addition of pharmaceutical grade thickener, e.g., hydroxyethyl cellulose, to their buffered aqueous solutions.

2.1 Characterization of storage conditions, storage stability, and resistance to proteolytic digestion of silver–enzyme hybrid

As the enzyme core molecule may be susceptible to loss of enzymatic activity due to thermal denaturation or proteolytic digestion, e.g., by exposure to wound exudates containing proteolytic enzymes, a preliminary storage stability and resistance to proteolytic digestion study were carried out. A 0.04 mg/mL buffered silver–enzyme hybrid solution, thickened by 5% hydroxyethyl cellulose, was incubated at 4°C, room temperature (25°C), and 40°C for 10 days. Results (Fig. 16.3) clearly indicated that prolonged storage should be explored at 4°C.

Full retention of hybrid's initial activity was recorded in a subsequent prolonged storage study extending over 6 months at 4°C (Fig. 16.4).

Furthermore, the addition of other ingredients, e.g., glycerol added as a soothing agent, to the abovementioned 0.04 mg/mL buffered silver–enzyme hybrid solution thickened by 5% hydroxyethyl cellulose did not interfere with the full retention of stability recorded for 6 months storage at 4°C (Fig.16.5):

It appears, however, that addition of additives should be handled with care as in some cases, they may interfere with the abovementioned recorded storage stability, as illustrated below in the case of mannitol addition (Fig. 16.6).

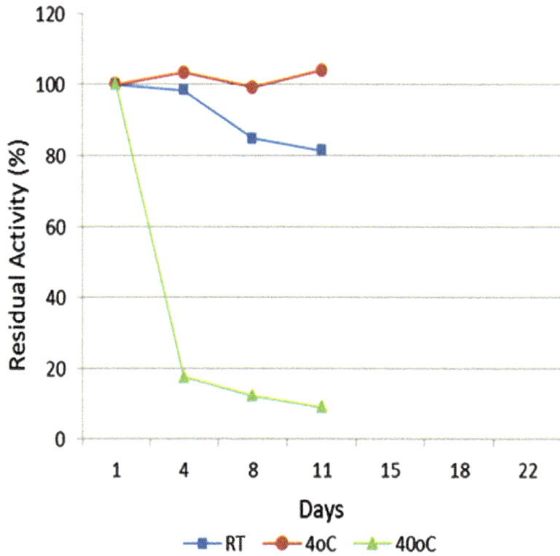

Figure 16.3
Preliminary study of glucose oxidase—silver hybrid thermal stability.

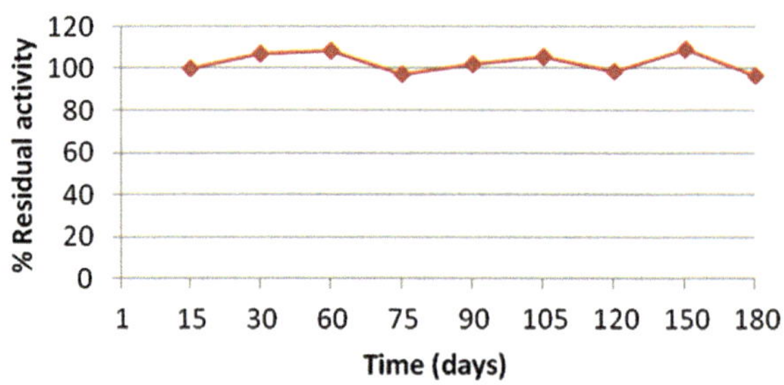

Figure 16.4
Extended study on storage stability at 4°C of buffered 0.04 mg/mL glucose oxidase—silver hybrid solution thickened with 5% hydroxyethyl cellulose.

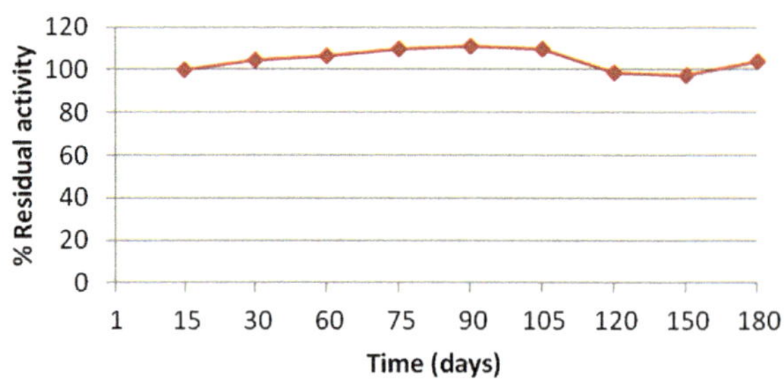

Figure 16.5
Extended study on storage stability at 4°C of buffered 0.04 mg/mL glucose oxidase—silver hybrid solution thickened with 5% hydroxyethyl cellulose containing 10% (V/V) glycerol.

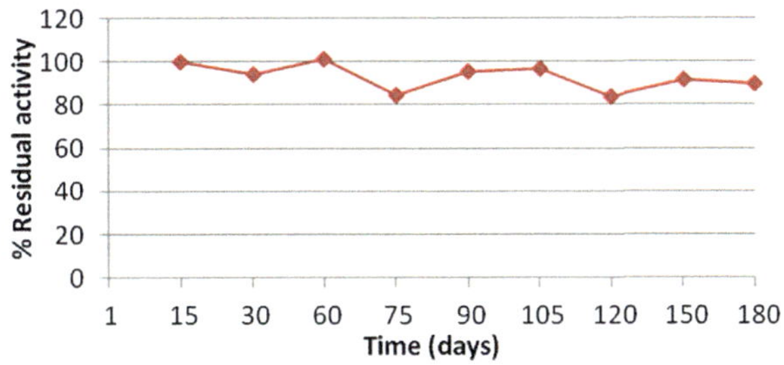

Figure 16.6
Extended study on storage stability at 4°C of buffered 0.04 mg/mL glucose oxidase–silver hybrid solution thickened with 5% hydroxyethyl cellulose containing 10% (V/V) D-mannitol.

The chemical modification of the core enzyme molecules by binding of polymeric chains of silver ions reducer PGA–β-alanine conjugates and the subsequently deposited metallic silver layer may also provide the hybrid with protection and improved resistance to proteolytic digestion that may be exerted by wound-secreted proteases. Comparison of the resistance of the hybrid to proteolytic digestion by the nonspecific protease Proteinase K to that of the native enzyme is shown in Fig.16.7.

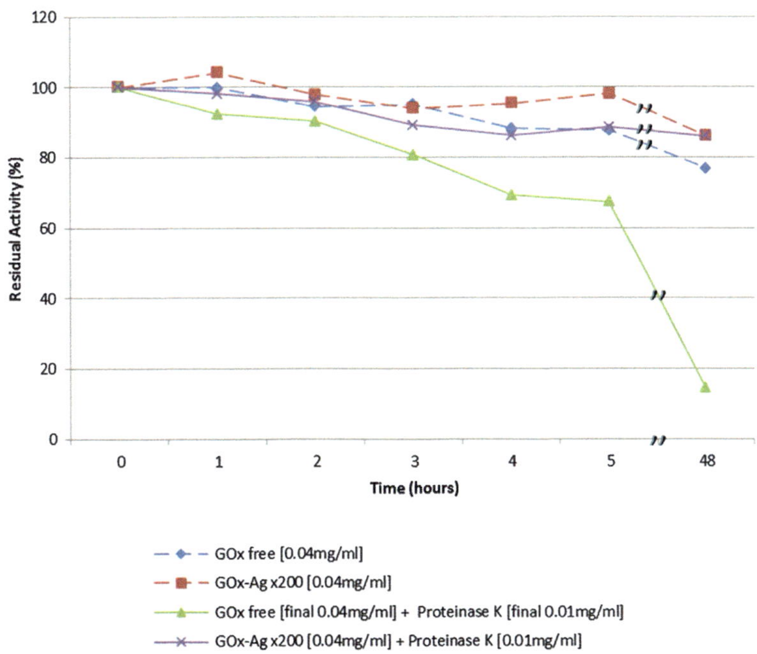

Figure 16.7
Resistance of buffered 0.04 mg/mL glucose oxidase–silver hybrid solution to proteolytic digestion by 0.01 mg/mL of the nonspecific protease, proteinase K, at 30°C.

Results obtained indicate significantly improved resistance of the hybrid to proteolytic digestion throughout the tested period (48 h incubation at 30°C).

3. Characterization of antimicrobial activity of the silver–enzyme hybrid by Zone of Inhibition tests on microbial prokaryotic and eukaryotic representative strains

Comparative tests of the antimicrobial activity of lotion samples of 0.4% mg/mL (w/v) GOx–silver hybrid were carried out in vitro on lab strains of microbial prokaryotic and eukaryotic representative strains, including the prokaryotic bacterial strains of *Escherichia coli* and *Staphylococcus aureus* and eukaryotic strains of the yeast *Candida albicans* and the filamentous fungi *Microsporum canis* and *Trichophyton rubrum*. An agar overlay version of the standard zone of inhibition (ZOI) assay [18,19] and references cited therein, was used, reflecting simultaneously antimicrobial activity and diffusional capabilities of the hybrid. Samples of 50 μL hybrid lotion were transferred to a well-carved agar plate comprising basal layer of 1.5% agar containing growth medium, topped with an overlay layer of 0.75% soft agar containing inoculum of the tested strain. Following incubation allowing cell growth throughout the overlay layer, the diameters of the resultant clear zone indicating the relative efficacy of the antimicrobial activity of the hybrid were recorded (Fig. 16.8).

The following ZOI diameter values (in mm) were recorded:

Bacterial strains: *E. coli*—13 mm; *S. aureus*—19 mm;
Yeast and fungi: *C. albicans*—14 mm; *T. rubrum*—64 mm; *M. canis*—51 mm.

It appears that the hybrid exhibited wide-spectrum antimicrobial activity, resulting in inhibition of biofilm formation by all strains tested, with exceptionally higher efficacy of growth inhibition of biofilms of filamentous fungi, tested here for two representative strains of medical importance: *T. rubrum*, affecting fungal skin and nail infections, and

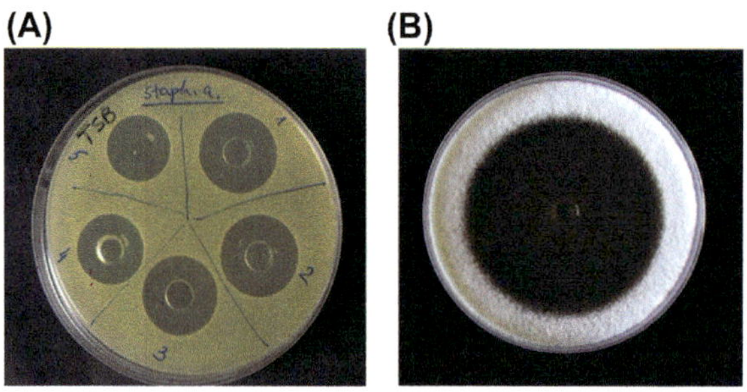

Figure 16.8
Illustration of zone of inhibition recorded for bacterial (A) and fungal (B) tested strains.

M. canis, affecting fungal hair infection. As eradication of such filamentous fungi biofilms involves a tedious and prolonged series of treatments, it appears that the prevention of growth and development of such contaminating biofilms may be enabled by pretreatment with the hybrid developed.

In attempt to characterize the molecular mechanism underlying the observed results obtained for the inhibition of filamentous fungi biofilms, we carried out an in vitro biochemical study aiming at the elucidation of the molecular mechanism involved in the observed highly effective prevention of filamentous fungi biofilm formation by silver ions released from the hybrid nanoparticles.

4. Elucidation of the molecular mechanism underlying the prevention of filamentous fungal biofilm formation by silver ions

As mentioned above, while the mechanisms involved with the antibacterial activity of ionic silver were elucidated, data available on mechanisms involved in its antimicrobial activity on eukaryotic cells, e.g., yeast and fungi, are much more limited. It appears that the mechanism involved in the antibacterial activity of micromolar silver ion concentrations on *E. coli* involves cell penetration followed by *multisite intracellular effects*, including shutting down of respiratory chain and loss of DNA replication capability [20].

As our findings indicate that filamentous fungi biofilm formation was relatively much more sensitive to low levels of silver ions than the bacterial strains tested, our working hypothesis was that the antifungal activity of silver ions results from an *effective inhibition of a specific intracellular enzymatic activity that is essential for cell growth.*

Such essential enzymatic activity is known and served as target for most commonly used antifungal competitive inhibitors, e.g., the antifungal azoles series: the enzyme sterol-14α-demethylase (also known as CYP51), a microsomal cytochrome P450 in eukaryotes, essential for the biosynthesis of sterols that are an essential component of fungal cell membrane [21—25].

Cytochrome P450 enzymes (CYPs) are heme-containing monooxygenases present in virtually all groups of living organisms. Although most bacteria investigated so far can survive without CYPs, microbial eukaryotes usually have at least one essential CYP, e.g., the fungal enzyme CYP51. Such fungal cytochrome P450s are monooxygenases (mixed function oxidases) catalyzing the incorporation of a single atom of molecular—O_2—oxygen into the substrate, with a concomitant reduction of the other oxygen atom into water by another enzyme, resulting in the overall reaction described in the following equation [24]:

$$RH + O_2 + NAD(P)H + H^+ \rightarrow ROH + H_2O + NAD(P)^+$$

This process is performed by CYPs with the active involvement of three different cofactors (non–amino acids) molecules, affecting (1) specific binding of the substrate and binding and splitting of the molecular oxygen, both taking place in the crucial *heme*-containing binding site and (2) donation of electrons and hydrogen atoms from a *reducer* (NADPH) for the reduction of the second oxygen atom into water by electrons and hydrogen transfer from NADPH to FAD molecule, from the FAD to FMN molecule, and finally from the FMN to the heme. These three functions are often naturally located within two or three different adjacent proteins [25].

The biotechnological synthetic potential inherent in such CYPs for the production of pharmaceuticals has also attracted much attention [24]. Within this context, a self-sufficient CYP, namely integrating all abovementioned functions into a fused single-chain protein molecule that may be readily expressed and produced in large quantities by molecular biology, was sought.

Cytochrome P450 BM3 (CYP 102A1) from *Bacillus megaterium* (ATCC 14581) is a bifunctional polypeptide comprising "P-450 analogue" heme—containing domain, linked to a P-450 reductase analogue FAD—and FMN—containing domains, thus providing a catalytically self-sufficient single polypeptide chain serving as effective bio-catalytic agent as well as a representative model for understanding the molecular aspects of eukaryotic (class II) P450 functions [26,27].

We have therefore decided to conduct our biochemical inhibition study using the Cyt P450 BM3 enzyme as a model enzyme representing the wide group of fungal CYPs, including the CYP51 enzyme.

The catalytic cycle of Cyt P450 BM3 involves several consecutive steps taking place between its complementary single chain domains: specific binding of a substrate, i.e., long chain fatty acid, and cosubstrate (molecular oxygen) at the heme-containing binding site, of NADPH oxidation at the reductase domain (specifically at the subdomain of FAD) and electrons and hydrogens transfer—from NADPH to FAD, from FAD to FMN, and from FMN—through subdomain movement—to the heme for oxygen atom reduction and generation of water [26].

Following the expression, production, and purification of Cyt P450 BM3, we conducted a series of in vitro biochemical kinetic studies on the three steps involved in its catalysis of stereospecific hydroxylation, carried out by the purified enzyme.

All studies were carried out using buffered reaction mediums devoid of salts or other reagents, which may interfere or modify the effective silver ions input concentrations.

The *overall* effect of the addition of micromolar concentrations of ionic silver to enzyme solution was investigated first in this study. Results obtained from measuring the effect of

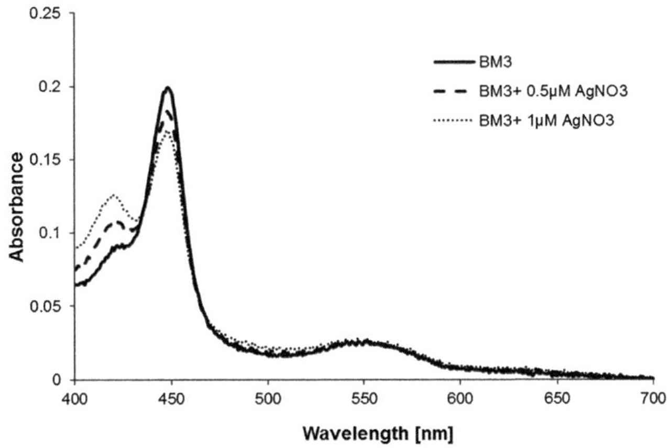

Figure 16.9
CO spectrum of cytochrome P450 BM3 (0.5 mg/mL) treated with 0.5 and 1 μM $AgNO_3$ in 0.1 M PB pH 7.5.

silver ions addition on cytochrome P450 BM3 monooxygenase *overall activity* measured at increasing silver ions micromolar concentrations (0.1–2 μM) resulted in the observation that 80% of the total enzymatic activity was lost already at a molar ratio of 5:1 silver ions to heme ratio, indicating a preferred, strong, site-specific interaction.

These results were accompanied by changes recorded in the enzyme's CO difference spectrum, which switched from the active peak of 450 nm (no silver ions added) to the enzyme's inactive peak of 420 nm on the addition of 0.5–1.0 μM silver ions (see Fig. 16.9).

As interaction with –SH cysteine residues appears to be one of the main mechanisms generally involved with silver ions interference with various enzymatic activities [28], we looked at –SH cysteine residues availability on the surface of Cyt P450 BM3. It appears that BM3 sequence contains nine cysteine groups in its primary amino acid sequence, three of them located at the monooxygenase domain and the rest at the reductase domain. Two cysteine residues, specifically Cys^{400}, located at the heme domain and Cys^{999} located at the reductase domain are known to play major roles in the catalytic activity of each domain, respectively [26,29]. Chemical assay of the amount of accessible –SH groups displayed on the surface of Cyt P450 BM3 using Ellman's reagent method yielded a ratio of five such sulfhydryl groups per heme-containing protein molecule. The observation that the BM3 enzyme lost 80% of its activity already at silver ion to enzyme molar ratio of 5:1, a "one-on-one" Ag^+-accessible cysteine interactions, appears to play a major role in Cyt P450 BM3 inhibition by silver ions.

Furthermore, the inactivation of Cyt P450 BM3 by silver ions could be partially reversed by the addition of substantial molar excess of cysteine, affected following exposure of the enzyme to *low* silver ions concentration, which was, however, ineffective following preexposure of Cyt P450 BM3 to *high* concentrations of Ag^+ ions. These results indicate that silver ions interacted first with some or all accessible SH groups and subsequently—when large silver ions excess became available—also affected other targeted groups, e.g., negatively charged residues, affecting combined and irreversible structural impact.

Results obtained from measuring the impact of added silver ions on substrate-specific binding of the values of K_m and K_m^{app}, and V_{max} and V_{max}^{app} (calculated from Lineweaver–Burk plots) clearly indicated that the suggested silver ions–SH interaction did not interfere with substrate binding, leaving specific binding values unaffected, as reflected by recorded, preserved *Km* value, while the overall enzymatic activity of BM3 exhibited decreases in *Vmax* values. This result is in accordance with the known sequence of amino acids comprising the substrate-binding pocket—it does not contain cysteines [26,30].

Results obtained from CO difference spectrum of Cyt P450 BM3 treated with 0.5 or 1.0 μM silver ions reflected a structural change interfering with the oxygen cosubstrate binding to the heme.

The fifth coordination of the protoporphyrin ring of heme is highly conserved as residue of cysteine throughout the cytochrome P450 family. This cysteine residue (Cys^{400} in Cyt P450 BM3) plays a key role in the enzyme's catalytic ability, enhancing heme iron electron density and promoting the splitting of oxygen [26]. Ag^+ ions may affect Cyt P450 BM3 enzymatic activity either by its direct interaction with this cysteine residue or through a combined effect with their additional interaction with other residues, causing subtle changes in the enzyme conformation that will change the positioning of the cysteine as the fifth ligand.

Cytochrome P450 is unique in the class of iron protoporphyrin IX containing heme proteins, in its complex with CO, which has a Soret absorbance band maximum at about 450 nm, while the ferrous–carbonyl complexes of the other members of this class absorb at about 420 nm. Spectral analysis of cytochrome P450 on denaturation, affected by either thermal or pH changes, showed P450 to P420 absorption conversion, indicating a slight protein unfolding. Accordingly, the addition of silver ions to cytochrome P450 BM3, which resulted in the observed decrease in the absorbance at 450 nm, which was accompanied by a shift of the absorbance to 420 nm in the CO spectrum, in a concentration-dependent manner, indicates that silver ions affected inhibition of cytochrome P450 BM3 activity by affecting first the positioning of Cys^{400}, resulting in *local conformational change in the heme environment,* which *interfered with oxygen binding and splitting*. These changes were subsequently followed by larger, irreversible conformational changes in the protein, affected by the higher silver ions concentrations, leading to irreversible loss of activity.

Kinetic analysis of silver ions impact on reductase activity of BM3, i.e., electrons and hydrogen transfer from NADPH to FAD and from FAD to FMN, was carried out by monitoring cytochrome *c* reduction by BM3 following its preincubation with micromolar concentrations of silver ions, followed by the addition of NADPH. The binding of cytochrome *c* as substrate to the reductase domain of BM3 is expected to minimize the distance between the two redox centers—FMN and heme—enabling maximal contact area between the two molecules; it also allows NADPH to enter/exit freely, while the FMN and FAD are "buried" within the complex formed. It should be mentioned within this context that Cys^{570} is positioned in close proximity to the binding site of cytochrome *c*.

To test the impact of the addition of a series of micromolar range concentrations of silver ions on the reductase domain, ionic silver (0.02–0.4 μM) was added to reaction mixture containing 0.01 mg/mL purified cytochrome P450 BM3, and its impact on its reductase activity was investigated, using silver ion to enzyme molar ratios of 0.5:1–10:1. Ag^+ ions exhibited an inhibitory effect of cytochrome P450 BM3 reductase activity in a concentration-dependent manner. The residual activity measured as initial rate of reductase activity showed that the highest degree of loss of activity—about 85%—was observed *already at silver ions to enzyme molar ratio of 5:1.*

At low concentrations of silver ions, increased K_m values were found for cytochrome *c* binding to the "P-450 reductase analogue" FAD, and FMN-containing domain, without affecting the overall reaction rate, indicating a change in the binding affinity. It thus appears that interaction of silver ions with residues at the interface of FMN domain, e.g., Cys^{570}, decreased the affinity of the positively charged cytochrome c to its docking area.

Another potential site for the silver ions binding in the reductase domain is an essential cysteine residue highly conserved in the diflavin reductase family, at or near the NADPH-binding site. Mutation in Cyt^{999} in the reductase domain of BM3 was reported to cause a significant decrease in the rate of hydride transfer from NADPH to FAD, indicating its role in increasing the rate of hydride transfer to the FAD cofactor and in stabilizing the complex $FADH_2$–$NADP^+$ Ref. [29].

5. Summary of main findings of kinetic study on the inhibition of Cyt P450 BM3 by ionic silver

Substrate (Lauric acid) binding: *Not affected* (preserved Km; reduced Vmax);

Cosubstrate (oxygen) binding and splitting: *Strongly inhibited* due to distortion of heme positioning by silver ions binding to –SH groups (loss of 80% of initial activity at 5:1 molecular ratio of Ag^+ to exposed –SH groups and shift of CO absorption spectra (from P450 nm to 420 nm);

Oxygen reduction by electron transfer from NADPH: *Inhibited* due to Ag^+ binding and interference with the interaction between FMN subdomain and the heme monooxygenase domain.

6. Conclusion

An overview of the molecular mechanisms involved in the antimicrobial activity clearly shows that the inhibition activity of silver is affected by intracellular submicromolar silver ions concentrations, following their penetration into the targeted cells. While the antibacterial activity is affected by *multisite intracellular effects,* including shutting down of respiratory chain and loss of DNA replication capability [20], results from our abovementioned study indicates that its antifungal activity is affected primarily by *specific Cyt P450 inhibition.*

Combining the accumulating data with the observations that (1) silver antimicrobial activity is affected by silver ions and (2) currently used silver dressings are based on metallic silver nanoparticles, it is evident that there are two major factors affecting silver nanoparticles' antimicrobial efficacy: leaching of silver ions from metallic nanoparticles surface and effective uninhibited migration of the silver ions into contact and penetration into the targeted cells.

A structural redesign of the currently used metallic silver nanoparticles to cope with the criticized poor efficacy of many currently used silver dressing is therefore required.

It is within this context that our abovementioned novel enzyme—silver hybrids demonstrate that such redesign is feasible, hopefully inspiring the design of additional novel and effective silver nanoparticles, offering improved and effective wide-spectrum antimicrobial treatments for chronic wounds.

References

[1] Percival SL, Finnegan S, Donelli G, Vuotto C, Rimmer S, Lipsky BA. Antiseptics for treating infected wounds: efficacy on biofilms and effect of pH. Crit Rev Microbiol 2016;42:293—309.
[2] Metalf DG, Bowler PG. Biofilm delays wound healing: a review of the evidence. Burns Trauma 2013;1:5—12.
[3] Lipsky BA, Hoey C. Topical antimicrobial therapy for treating chronic wounds. Clin Infect Dis 2009;49:1541—9.
[4] Executive Summary FDA. Classification of wound dressings combined with drugs. 2016. p. 11—3.
[5] Sood A, Granick MS, Tomaselli NL. Wound dressings and comparative effectiveness data. Adv Wound Care 2014;3:511—29.
[6] Ruttermann M, Maier-Hasselmann A, Nink-Grebe, Burckhardt M. Local treatments of chronic wounds. Dtsc Arztebl Int 2013;110:25—31.
[7] Greer N, Foman NA, MacDonald R, Dorrian J, Fitzgerald P, Rutks I, Wilt MD. Advanced wound care therapies for nonhealing diabetic, venous and arterial ulcers. Ann Intern Med 2013;159:532—42.

[8] Wilkinson LJ, White RJ, Chipman JK. Silver and nanoparticles of silver in wound dressings: a review of efficacy and safety. J Wound Care 2011;20:543–9.
[9] Sterling JP. Silver resistance, allergy and blue skin: truth or urban legend? Burns 2014;40S:S19–23.
[10] Marks DE, Barillo DJ. Silver in medicine: the basic science. Burns 2014;40S:S9–18.
[11] Maillard JY, Hartemann P. Silver as an antimicrobial: facts and gaps in knowledge. Crit Rev Microbiol 2013;39:373–83.
[12] Reidy B, Haase A, Luch A, Dawson KA, Lynch I. Mechanisms of silver nanoparticles release, transformation and toxicity: a critical review of current knowledge and recommendations for future studies. Materials 2013;6:2295–350.
[13] Le Ouay B, Stellacci F. Antibacterial activity of silver nanoparticles: a surface science insight. Nano Today 2015;10:339–54.
[14] Duran N, Duran M, Bispo de Jesus M, Seabara AB, Favaro WJ, Nakazato GA. Silver nanoparticles: a new view on mechanistic aspects on antimicrobial activity. Nanomed Nanotechnol Biol Med 2016;12:789–99.
[15] Abboud EC, Settle JC, Legare TB, Marcet JE, Barillo DJ, Sanchez JE. Silver-based dressings for the reduction of surgical site infection: review of current experience and recommendation for future studies. Burns 2014;40S:S30–9.
[16] Freeman A. Protein–mediated biotemplating on the nanoscale. Biomimetics 2017;2:14–29.
[17] Heller A, Feldman B. Electrochemistry in diabetes management. Accounts Chem Res 2010;43:963–73.
[18] Castellano JJ, Shafii SM, Ko F, Donate G, Wright TE, Mannari RJ, Payne WG, Smith DJ, Robson MC. Comparative evaluation of silver-containing antimicrobial dressings and drugs. Int Wound J 2007;4:114–22.
[19] Balouri M, Sadiki M, Ibnsouda SK. Methods for *in vitro* evaluating antimicrobial activity: a review. J Pharmaceut Anal 2016;6:71–9.
[20] Holt KB, Bard AJ. Interaction of silver ions with the respiratory chain of *E. coli*: an electrochemical and scanning electrochemical microscopy study of the antimicrobial mechanism of micromolar Ag^{+}. Biochemistry 2005;44:13214–23.
[21] Lepesheva GI, Waterman MR. Structural basis for conservation in the CYP51 family. Biochim Biophys Acta 2011;1814:88–93.
[22] Kathiravan MK, Salake AB, Chothe AS, Dudhe PB, Watode RP, Mukta MS, Gadhwe S. The biology and chemistry of antifungal agents: a review. Bioorg Med Chem 2012;20:5678–98.
[23] Bossche HV, Ausma J, Bohets H, Vermuyten K, Willemsens G, Marichal P, Meerpoel L, Odds F, Borgers M. The novel azole R126638 is a selective inhibitor of ergosterol synthsis in *Candida albicans, Trichophyton* spp and *Microsporum canis*. Antimicrob Agents Chemother 2004;48:3272–8.
[24] Berhardt R. Cytochromes P450 as versatile biocatalysts. J Biotechnol 2006;124:128–45.
[25] Cresnar B, Petric S. Cytochrom P450 enzymes in the fungal kingdom. Biochim Biophys Acta 2011;1814:29–35.
[26] Munro AW, Leys DG, McLean KJ, Marshall KR, Ost TWB, Daff S, Miles CS, Chapman SK, Lysek DA, Moser CC, Page CC, Dutton PL. P450 BM3: the very model of a modern flavocytochrome. Trends Biochem Sci 2002;27:250–6.
[27] Munro AW, Girvan HM, McLean KJ. Cytochrome P450-redox partner fusion enzymes. Biochim Biophys Acta 2007;1770:345–59.
[28] Lemire JA, Harrison JJ, Turner RJ. Antimicrobial activity of metals: mechanisms, molecular targets and applications. Nat Rev Microbiol 2013;11:371–84.
[29] Warman AJ, Roitel O, Girvan HM, Seward HE, Murray SA, McLean KJ, Joice MG, Toogood H, Holt RA, Scrutton NS, Munro AW. Flavocytochrome P450 BM3: an update on structure and mechanism of a biotechnologically important emzyme. Biochem Soc Trans 2005;33:747–53.
[30] Sevrioukova LH, Zhang H, Peterson JA, Poulos TL. Structure of a Cytochrome P450-redox partner electron-transfer complex. Proc Natl Acad Sci USA 1999;96:1863–8.

CHAPTER 17

Biomechanical aspects of skin aging—the risk of skin breakdown under shear loading increases with age

Daniel Garcia-Martinez[1], Maria Fabiola Leyva-Mendivil[1,2], Amit Gefen[3], Georges Limbert[1,2,4]

[1]*National Centre for Advanced Tribology at Southampton (nCATS), Department of Mechanical Engineering, Faculty of Engineering and Physical Sciences, University of Southampton, Southampton, United Kingdom;* [2]*Bioengineering Science Research Group, Department of Mechanical Engineering, Faculty of Engineering and Physical Sciences, University of Southampton, Southampton, United Kingdom;* [3]*Department of Biomedical Engineering, Faculty of Engineering, Tel Aviv University, Tel Aviv, Israel;* [4]*Biomechanics and Mechanobiology Laboratory, Biomedical Engineering Division, Department of Human Biology, Faculty of Health Sciences, University of Cape Town, South Africa*

Chapter Outline

Innovations and Emerging Technologies in Wound Care. https://doi.org/10.1016/B978-0-12-815028-3.00017-1

1. Introduction

The human skin, being the largest organ of the human body [1], acts as a multifunctional interface to the external environment and, as such, is involved in a wide range of mechanical contact interactions [2–4] with clothing fabrics, footwear, seating and bedding surfaces, sports equipment, consumer products, and medical devices. These contact interactions are mainly dominated by mechanical aspects [5–7]; they also feature other types of physical processes such as heat transfer [8] or water-induced chemical degradation of mechanical properties [9,10]. The mechanical behavior of the skin at macroscopic and microscopic levels is a key in conditioning its tribological response through scale-bridging mechanisms [11–13]. Understanding the interplay between the material and structural properties of the skin under physiological and pathophysiological conditions is therefore pivotal in gaining a mechanistic insight into its tribology and internal tissue loads under contact.

As a macroscopic biological structure, skin possesses strong anisotropic, inhomogeneous and nonlinear mechanical properties that vary according to body site, individuals, age, health status, and presence of a disease (e.g., obesity or diabetes), as well as ethnicity and exposure to specific environmental conditions [1,14,15]. In the skin, like in most biological tissues, there is an intimate relationship between structure and function [16]. Given the multiple functions of the skin, it is not surprising to observe that the skin, much like other tissue structures, features a complex multiscale hierarchical structure [1,17,18]. Skin is generally considered as a multilayer assembly made up of three main distinct structures: the epidermis, dermis, and hypodermis (Fig. 17.1). The epidermis—which is avascular—can be further divided into the *stratum corneum* and living epidermis (also called viable epidermis), which contains living basal cells—mainly corneocytes—which undergo mitosis and continuously migrate from the *stratum basale* to the *stratum corneum* [18].

The living epidermis is connected to the underlying dermis through a three-dimensional (3D) interlocking wavy interface called the dermal–epidermal junction (DEJ) or *basal lamina*, which is effectively a 0.5–1 μm-thick basement membrane [19]. Papillae are the protrusions of the papillary dermis into the dermis (Fig. 17.1). These finger-like structures increase the contact surface area between the reticular dermis and the living epidermis and are thus believed to favor biochemical mass exchanges between these layers, e.g., transport of oxygen or nutrients [19–21].

The DEJ controls the transit of biomolecules between the dermis and epidermis according to their dimension and charge. It allows the passage of migrating and invading cells under normal (i.e., melanocytes and Langerhans cells) or pathological (i.e., lymphocytes and tumor cells) conditions [20]. The behavior of keratinocytes is influenced by the DEJ via modulation of cell polarity, proliferation, migration, and differentiation. The defective

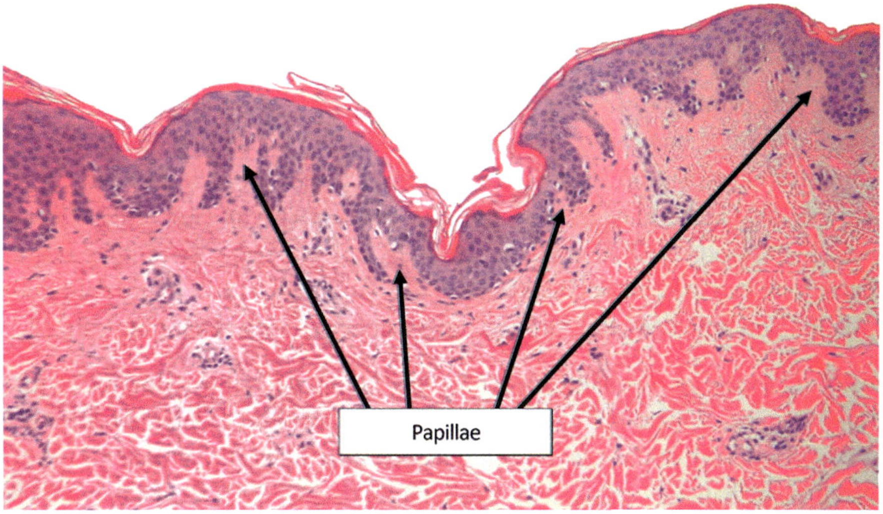

Figure 17.1
Histological section of human skin highlighting the two-dimensional topography of the dermal–epidermal junction and individual papillae. A fresh midback skin sample was obtained from a 30-year-old healthy white Caucasian female patient following biopsy. The specimen was processed, sliced, and stained with hematoxylin and eosin. Arrows indicate papillae protruding from the dermis into the viable epidermis.

biochemical functions of the basement membrane zone (BMZ) of skin are critical in certain genetic or autoimmune diseases linked to alterations in the production of key BMZ proteins. For example, in a condition called *epidermolysis bullosa*, either a defective laminin-5, a defective BP180, or a defective β_4 integrin results in abnormal skin fragility and chronic subepidermal blister formation ([19] and references therein). This strongly indicates that the BMZ plays an important mechanical role in maintaining skin integrity in response to external mechanical loads.

The pattern of transfer of mechanical loads between the skin surface (i.e., the *stratum corneum*) and its deeper strata critically depends on the loading mode (tension, compression, shearing, or any compound loading configuration), the contact interface (i.e., the coefficient of friction of the loading object against the *stratum corneum* that is environment-dependent, e.g., with regard to wetness), and the structural and mechanical properties of each constituents of skin within the influence zone (i.e., where the mechanical load is influential).

The BMZ features very complex arrangement of hemidesmosomes, anchoring filaments, anchoring fibrils, lamina lucida, and lamina densa [17,20,22], and it is clear that the mechanical characteristics of this multicomponent and multiscale structure are intrinsically linked to the mechanical properties of each of its constituents and how they structurally interact. As pointed out earlier in this section, alterations of proteins of the BMZ lead to abnormal loss of skin integrity [19].

If one considers the next upper level of the hierarchical scale of the BMZ, i.e., the BMZ as a continuum structure, it is intriguing to question whether the purely geometrical characteristics, for example, the 3D wavy topography disregarding the mechanical properties of the BMZ, may play an important role in the ability of skin to resist certain types of mechanical loads, such as shear. If the BMZ loses its integrity through degeneration of its ultrastructure, can the contact between the papillary dermis and living epidermis be somehow maintained kinematically through the existence of geometric constraints?

Geometrically interlocking mechanisms connecting components of multiphasic structure are a common feature of many natural biological materials (i.e., bone and nacre) [23,24]. Analytical and numerical studies have been conducted to assess the effects of interface geometry on the mechanical behavior of idealized joints [23–25]. It was demonstrated that specific interlocking geometries provide structural stiffness, strength, and tolerance to repeated (cyclic) loading.

In chronologically and photoaged skin, the DEJ tends to flatten out [26–29]. This reduction in the amplitude of the papillae is also accompanied by a decrease in their density (unit per surface area) [26,29]. The flattening of the DEJ leads to a reduction of up to 20% in epidermis thickness [30]. Aging of the skin has other consequences as well, such as increased interlinking of collagen fibrils, overall decreased elastin and collagen contents, rise in transepidermal water loss, reduction in skin moisture content, diminished sebum production, arteriosclerotic changes in the small and large vessels, thinning of vessel walls [31], reduction in mast cells, melanocytes, Langerhans cells, Meissner's cells, Merkel cells, and Pacinian corpuscules [31,32], and increase in skin surface pH [33]. It has been suggested that the flattening of the DEJ may facilitate the formation of skin wrinkles by compromising the structural integrity of the basement membrane, which links the dermis to the epidermis [34], thereby increasing the likelihood of relative motions between these two layers. Besides these important structural effects, the flattening of the DEJ apparently has important metabolic consequences as it reduces the surface area for nutritional exchange and waste clearance between the dermis and epidermis. As a result, epidermal cell turnover is slowed down and free radicals accumulate. It has also been shown through image analysis that intrinsically aged skin features the same density of veins and arteries as young skin, but their diameters are reduced. In photoaged skin, the dermal vasculature is progressively lost and the diameter of veins and arteries is also reduced [35].

To date, it appears that no study has examined the role of the DEJ topography characteristics on the mechanical response of skin when subjected to surface loads, particularly shear. In situations where patients with a fragile skin, especially geriatric and malnourished patients, or patients with incontinence problems are being moved or are moving in bed or on a chair, clinical evidence is well established that the skin may break down [36]. Consider, for example, an elderly patient at a nursing home who is being repositioned in bed every several hours, and at each such repositioning maneuver, the skin is being sheared intensively against the clothing and bedsheets, possibly under high friction contact if there is urine leakage or perspiration that is wetting the skin. Skin lesions, from irritation to serious pressure ulcers, are common in these facilities but still insufficiently understood.

Here, it is proposed to explore this clinically relevant topic using an idealized, yet sophisticated, parametric multilayer 3D finite element (FE) model of human skin as it ages.

The structural behaviors of five age-dependent model configurations of the DEJs (distinguished by the wavelength and amplitude of the DEJs) are quantitatively compared, by determining load transmission pathways as well as calculating local stress distributions. The focus is placed on the effects of the age-dependent geometrical characteristics of the DEJ and the associated changes in the thickness of the epidermis on the mechanical performance and tolerance of the aging skin to shear loads, which indicates the risk for skin breakdown in real-world scenarios in care facilities such as described above.

2. Materials and methods

To evaluate the influence of the age-dependent geometry of the DEJ in combination with variations in the mechanical properties of the different skin layers, a series of computer-aided design (CAD) solid and FE models of a volume of interest in a 3D skin structure was devised. Five model configurations incorporating the geometries of the multilayer skin structure were considered, each corresponding to a different ages (30, 50, 70, 80, and 90 years).

2.1 Parametric four-layer model of skin and rigid indenter

A 3D nonuniform rational B-spline—based parametric solid model, representing the skin as an idealized composite structure containing the *stratum corneum*, living epidermis, papillary, and reticular dermis, was developed in the software package Rhino3D 5.0 (Robert McNeel & Associates, Seattle, WA, USA). Additionally, the truncated part of a 500 μm spherical indenter surface was built in Rhino3D 5.0. Similar to the indenter surface, each individual skin layer was then exported as a STEP file into ABAQUS/CAE 6.13 (Simulia, Dassault Systèmes, Providence, RI, USA) for FE preprocessing.

The focus of this study was on the role of the age-related geometry of the DEJs and its relation to the transmission of loads applied to the external surface of the *stratum corneum*. From this observation and to minimize the final size of the composite FE model of skin, it was decided not to include the basal lamina and hypodermis and also to limit the thickness of the reticular dermis to 400 μm.

The *stratum corneum* and reticular dermis were modeled as regular cuboid shapes, whereas the papillary dermis and living epidermis layers were modeled as two cuboid shapes separated by a 3D wavy interface representing the DEJ junction (Fig. 17.2). The wavy interface of the DEJs was parameterized using the tensor product of two cosine functions:

$$z(x, y) = \mathscr{A} \cos(\mathscr{K}x)\cos(\mathscr{K}y) \tag{17.1}$$

where $\mathscr{A}$ is the amplitude, $\mathscr{K}$ is the in-plane wave number, and (x,y,z) are the spatial coordinates of the DEJ. Diridollou et al. [37] found that the skin thickness increases linearly with age up to 60 years, time by which the trend is reversed but still stays linear.

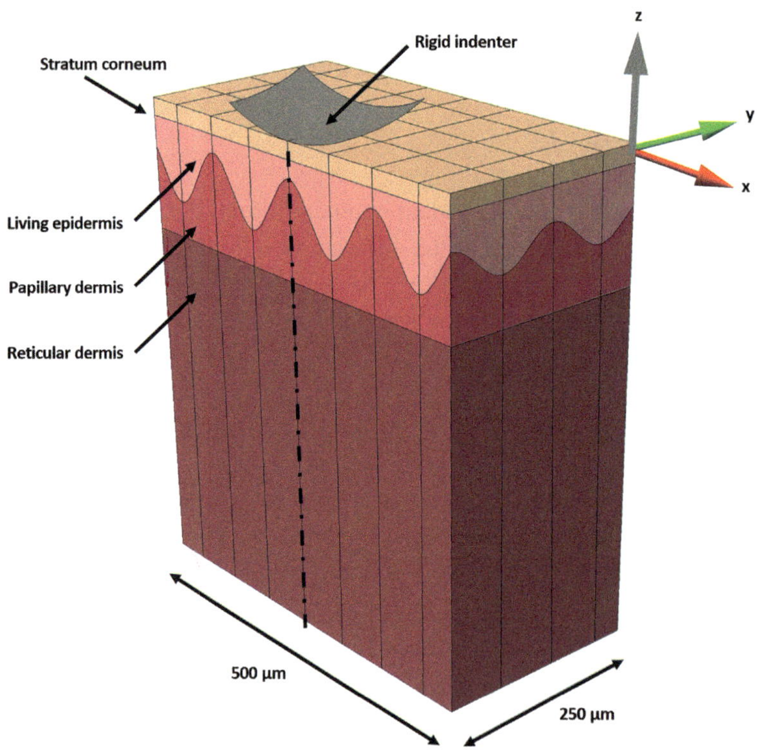

Figure 17.2
Solid representation of the idealized four-layer model of human skin and that of the truncated rigid spherical indenter of diameter 500 μm (the depicted parametric model corresponds to a 30-year-old individual). Using symmetry conditions, only half of the full three-dimensional geometry is necessary.

However, the same authors also found that the thickness of the epidermis decreases with age. In the present study, both the papillae height $\mathscr{A}$ and epidermal thickness t_e are assumed to be linear functions of age a:

$$\mathscr{A} = \begin{cases} 70 & \text{if } 20 \leq a \leq 60 \\ -2a + 190 & \text{if } a \geq 60 \end{cases} \tag{17.2}$$

$$t_e\left[\mu\text{m}\right] = 20 + \mathscr{A} = \begin{cases} 90 & \text{if } 20 \leq a \leq 60 \\ -2a + 210 & \text{if } a \geq 60 \end{cases} \tag{17.3}$$

To account for the age-dependent structural differences in the skin layers within the FE model and combine it with the effect of aging on the density of papillae, a simple empirical relationship was derived from the data of Refs. [29,37]. It provides the wave number of the DEJ topography as a function of age a:

$$\mathscr{K} = \mathscr{K}(a) = \begin{cases} 2\pi(-0.0643 \times a + 8.654) & \text{if } 20 < a \leq 60 \\ 2\pi(-0.0188 \times a + 6.382) & \text{if } 60 \leq a < 90 \end{cases} \tag{17.4}$$

Five FE models representing the skin at different ages (30, 50, 70, 80, and 90 years) were developed. The characteristics of the DEJ topography ($\mathscr{A}$ and $\mathscr{K}$) and thickness of the epidermis (t_e) for each of these models are summarized in Table 17.1 and depicted on Fig. 17.3.

The thickness of the *stratum corneum* was assumed to be independent of age, as demonstrated experimentally and reported by several authors [29,38,39]. The DEJ topography in each model was designed so that the wave amplitude in both in-plane directions (**x** and **y**) was maximal along the central vertical axis of the rigid indenter (z-direction) (Fig. 17.2). This was done to ensure consistency of the subsequent comparisons of results from the model configurations across the different ages. All the

Table 17.1: Geometrical characteristics of the DEJ and epidermal thickness t_e as a function of age. The thickness of the stratum corneum is fixed at 20 μm, so that the thickness of the living epidermis t_{le}[μm] is simply obtained as follows: $t_{le}[\mu m] = t_e[\mu m] - 20[\mu m]$. The thickness of the papillary dermis is 40 μm while that of the reticular dermis is 400 μm.

Age (years)	$\mathscr{A}[\mu m]$	$\mathscr{K}[\mu m^{-1}]$	$t_e[\mu m]$	Dermal-epidermal junction surface area (μm^2)
30	70	46.30	90	186,630.61
50	70	42.26	90	142,947.56
70	50	31.82	70	136,811.48
80	30	30.64	50	131,326.13
90	10	29.45	30	125,666.56

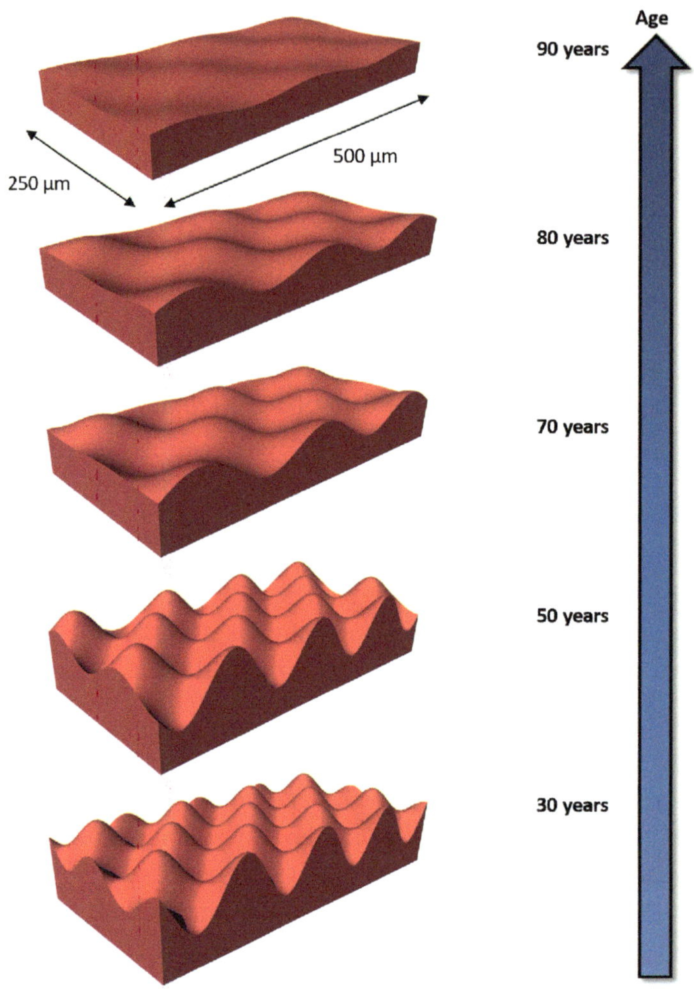

Figure 17.3
Solid representation of the parametric papillary dermis layer at various ages (30, 50, 70, and 90 years), highlighting the geometry of the dermal–epidermal junction.

skin layers were assumed to be perfectly fixed to each other, in lack of relevant (age-dependent) experimental data for human skin. This assumption will be elaborated upon later, in the **Discussion** section.

To get a sense of how age affects the contact area between the dermis and living epidermis in the multilayer skin finite element model, an age-normalized DEJ surface area metric was defined. This was accomplished by normalizing the real surface area of the DEJ by the real surface area of the DEJ at age 30 (Fig. 4A). The contact area of the DEJ at 90 years is about 67% of its value at 30 years. Similarly, another metric, the normalized projected surface area, was defined as the ratio between the real surface area of the DEJ

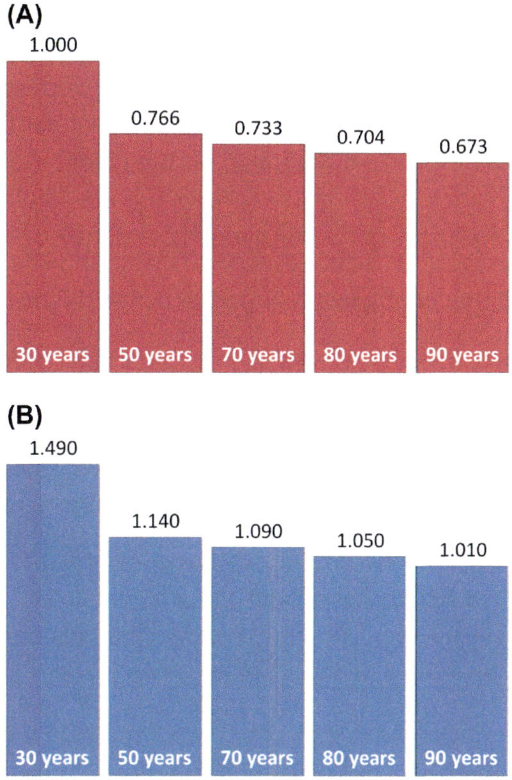

Figure 17.4

Geometrical characteristics of the dermal–epidermal junction (DEJ) in terms of surface area ratios. (A) Age-normalized surface area of the DEJ: the normalizing reference DEJ surface area is at 30 years; (B) ratio of the DEJ surface area over that of a flat surface featuring the same base dimensions (500×250 μm).

and that of the projected surface area (i.e., that of the base area, equal to 500×250 μm), which could also be called apparent surface area (Fig. 4B).

3. Mechanical properties of skin and its individual layers

As highlighted in a recent paper by Ref. [40] (see Table 17.1 of that paper), there is a significant variability of the mechanical properties of individual skin layers reported in the literature. This is mainly due to the intrinsic intra- and interindividual variability in skin biophysical characteristics, differences in tissue preservation methods, and experimental protocols (e.g., measurement techniques, spatial scales considered, and environmental conditions). As the focus of the present study is on the influence of the geometry of the DEJs on the mechanical response of the skin to surface shear loads, generic mechanical properties consistent with literature data are used here [40–42].

All the skin layers were assumed to obey a neo-Hookean constitutive formulation, which allows for a minimal number of constitutive parameters to be set while employing finite deformation theory. These constitutive parameters c_{10} and κ_0 correspond, respectively, to half the shear modulus and bulk modulus of an isotropic linear elastic material [43]. c_{10} and κ_0 can be expressed as functions of the initial elastic modulus E and Poisson's ratio ν:

$$c_{10} = \frac{E}{4(1+\nu)} \quad \text{and} \quad \kappa_0 = \frac{E}{4(1-2\nu)} = \frac{2}{D} \tag{17.5}$$

where D is the compressibility coefficient required to define the degree of compressibility in the FE code [44].

It is relevant to point out that, in the multilayer FE model of the skin, which has been introduced in the previous section, any structural effects of the DEJ can only be revealed if the viable epidermis and papillary dermis possess distinct mechanical properties as, otherwise, these two layers would form a homogeneous continuous medium.

For each age represented in the modeling (i.e., the model configurations of 30, 50, 70, 80, and 90 years), three series of FE model variants ("model variant A" "model variant B" and "model variant C") were considered. These three model variants differ by the choice of mechanical properties of the dermis and *stratum corneum* (Table 17.2).

Table 17.2: Material properties of individual skin layers used in the present computational studies. From the Young's modulus and Poisson's ratio, denoted respectively by E and ν, the equivalent shear and bulk moduli, μ andκ, are derived. The corresponding neo-Hookean parameters required by Abaqus are c_{10} and D.

		Constitutive parameters				
Variant	Skin layer	E (MPa)	ν	$c_{10} = \mu/2$ (MPa)	$D = 2/\kappa$ (MPa^{-1})	κ (MPa)
A	*Stratum corneum*	0.6	0.3	0.115385	4	0.5
	Viable epidermis	0.6	0.3	0.115385	4	0.5
	Papillary dermis	1	0.45	0.172414	0.6	3.33
	Reticular dermis	1	0.45	0.172414	0.6	3.33
B	*Stratum corneum*	0.6	0.3	0.115385	4	0.5
	Viable epidermis	0.6	0.3	0.115385	4	0.5
	Papillary dermis	2	0.45	0.344828	0.3	2
	Reticular dermis	2	0.45	0.344828	0.3	2
C	*Stratum corneum*	10	0.3	1.92308	0.24	8.33
	Viable epidermis	0.6	0.3	0.115385	4	0.5
	Papillary dermis	1	0.45	0.172414	0.6	3.33
	Reticular dermis	1	0.45	0.172414	0.6	3.33

It is well established that hydration-induced plasticization of the *stratum corneum* and the structural effects of drying stress [45,46] can lead to significant alteration of its elastic modulus, e.g., one can observe variations from about 5 to 1000 MPa in the in-plane direction (parallel to the skin surface) and from about 1 to 25 MPa in the out-of-plane direction. Wu et al. [10] reported a range of tensile elastic moduli from 3 MPa at 100% relative humidity (RH) to 370 MPa at 30% RH.

4. Boundary, loading, and contact conditions

For all the FE analyses considered in this paper, the following common boundary conditions were applied: the bottom face of the reticular dermis was fixed along the *z*-direction, the two external lateral faces of the composite skin in the *xz*-plane were fixed in the *y*-direction (symmetry conditions), and the two bottom edges of the external lateral faces of the reticular dermis layer in the *yz*-plane were fixed in the *x*-direction to prevent their collapse during deep indentation (Fig. 17.2).

An ABAQUS/standard finite sliding node-to-surface contact algorithm (2012) was used to model the frictional contact between the *stratum corneum* and the rigid indenter, which served the purpose of inflicting the worst-case scenario shear loads on skin, to facilitate the comparison of structural skin behaviors under these loads across the ages. To maximize the transmission of shear loads to the deeper strata of the skin, a coefficient of friction of 0.6 was chosen. It is consistent with values that have been experimentally measured [2,47–49]; however, it does not represent a highly wet skin that would increase the friction coefficient and exacerbate the exposure to internal tissue loads as shear is applied to the surface [50,51].

Application of a combined indentation and surface shear load to the *stratum corneum* was done by prescribing a displacement to the rigid indenter: this displacement consisted of a vertical displacement along the z-direction (−20 μm) and a horizontal displacement along the x-direction (30 μm) representing real-world interactions of the skin with objects, which always involve application of both pressure and shear simultaneously.

5. Protocol of simulations

5.1 Finite element analysis series 1—model variant A

This first series of FE analyses considered *the five model geometries* (corresponding to the five ages) and assumed that the *stratum corneum* and viable epidermis had the same mechanical properties [40,41] (see Table 17.1, variant A), while the two dermal layers featured identical properties. The elastic modulus of the *stratum corneum* was assumed to be 0.6 MPa.

5.2 Finite element analysis series 2—model variant B

To simulate the stiffening effect of the *stratum corneum* as a result of dryer environmental conditions, separately from the microanatomical changes, this series of FE analyses considered a higher elastic modulus (10 MPa) compared with that which was used in model variant A1; however, other mechanical properties remained unchanged.

5.3 Finite element analysis series 3—model variant C

Because intrinsic and extrinsic aging is often being associated with stiffening of the dermis, mostly due to increased collagen cross-linking, the FE analysis considering the 90-year-old geometry in series 1a was rerun by considering a 2 MPa elastic modulus for the dermal layers [15]. It was decided to consider only the model corresponding to the most-elder skin to identify and highlight any major differences in the structural behavior between a young and an old skin.

6. Results

Color plots of the normal and shear strains at the interface between the viable epidermis and papillary dermis (i.e., DEJ) are reported in Fig. 17.5 (σ_{xx}), Fig. 17.6 (σ_{yy}), Fig. 17.7 (σ_{xz}), and Fig. 17.8 (σ_{yz}) for the 30-, 50-, 70-, and 90-years-old DEJ topographies for variant FE models. Given that negligible differences in results between the models featuring a 70- and 80-years-old DEJ topography were found, the latter model was subsequently excluded from the analyses. Although the notion of stress is defined as the amount of force per unit area, a normal stress represents the vector component of that force normal to the material cross section on which it acts, per unit area. For example, σ_{xx} is defined as the projection of the force vector along the x-axis on an element of surface

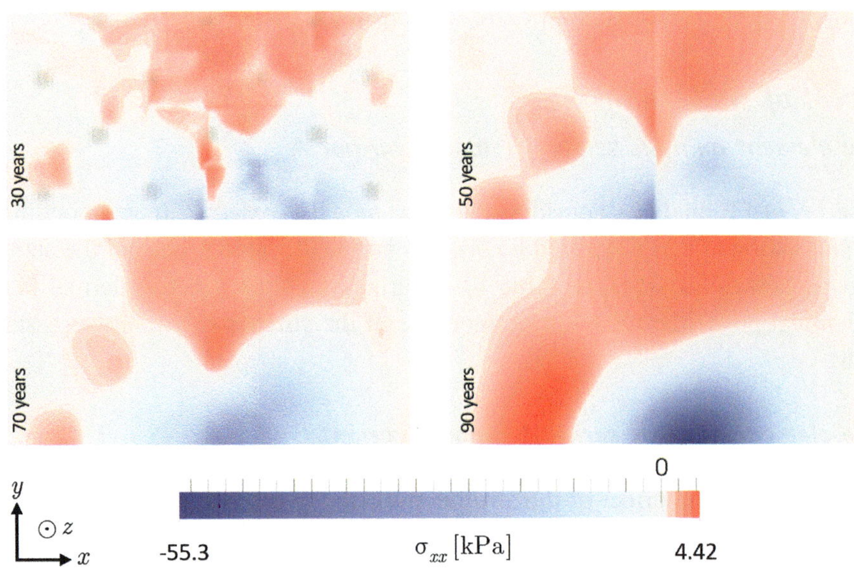

Figure 17.5
Color plots of the normal stress σ_{xx} at the top surface of the papillary dermis layer corresponding to each age-dependent multilayer skin model (30, 50, 70, and 90 years).

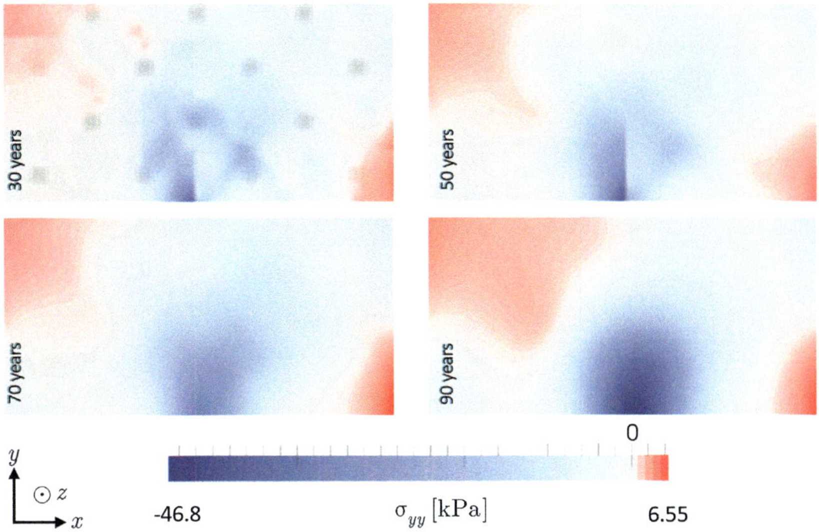

Figure 17.6
Color plots of the normal stress σ_{yy} at the top surface of the papillary dermis layer corresponding to each age-dependent multilayer skin model (30, 50, 70, and 90 years).

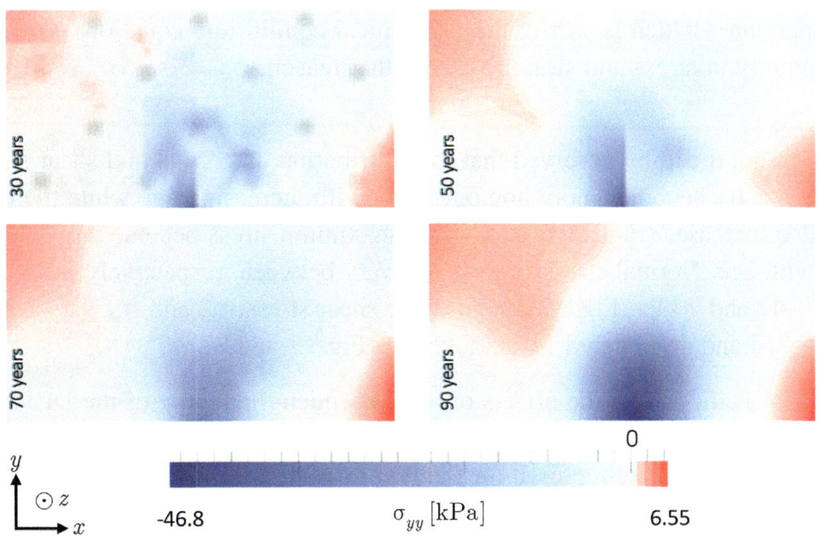

Figure 17.7
Color plots of the shear stress σ_{xz} at the top surface of the papillary dermis layer corresponding to each age-dependent multilayer skin model (30, 50, 70, and 90 years).

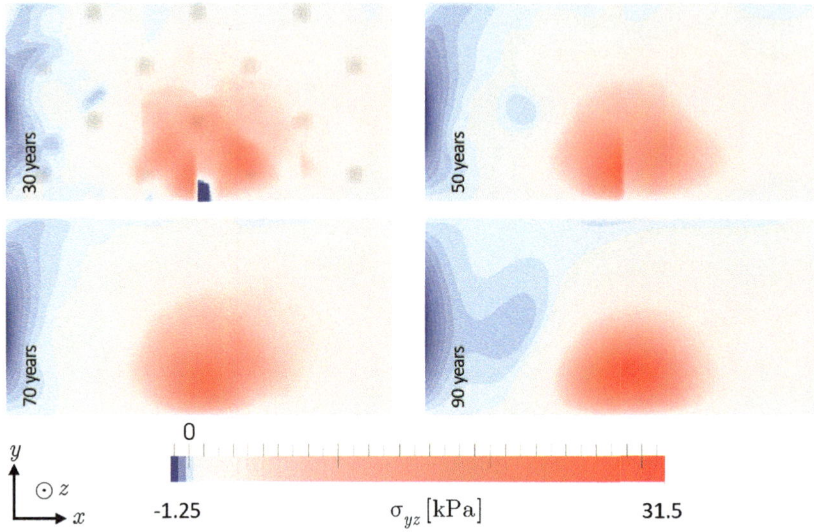

Figure 17.8
Color plots of the shear stress σ_{yz} at the top surface of the papillary dermis layer corresponding to each age-dependent multilayer skin model (30, 50, 70, and 90 years).

whose normal is aligned with the x-axis. A shear stress is defined as the projection of the force vector onto the plane coplanar with the surface on which it acts, per unit area. As an illustration, the shear stress σ_{xz} corresponds to the projection of the force vector acting on the surface with normal aligned with the x-axis along the z-axis. Because of the balance of angular momentum—which is part of the mechanical equilibrium equations—there is an inherent symmetry in stress and strain [52]. For that reason, $\sigma_{xz} = \sigma_{zx}$, $\sigma_{yz} = \sigma_{zy}$, and $\sigma_{xy} = \sigma_{yx}$.

As a general trend, it can be observed that the distribution of normal and shear stresses at the DEJ consistently becomes more homogeneous with increasing age while their magnitude also increases (Figs. 5–8). Zones of maximum stress become larger as the DEJ flattens out with age. Normal stress σ_{xx} and σ_{yy} vary between, respectively, -55 and 4.5 kPa and -47 and 7 kPa (Figs. 5 and 6) while shear stress σ_{xx} and σ_{yy} vary between, respectively, -14 and 46 kPa and -1 and 32 kPa (Figs. 7 and 8).

To focus more specifically on the effects of age-dependent flattening of the DEJ topography with variations in mechanical properties, values of normal and shear strain distributions at the DEJ were reported for the two extreme age cases (30 and 90 years) for model variants A, B, and C in Fig. 17.9 (σ_{xx}), Fig. 17.10 (σ_{yy}), Fig. 17.11 (σ_{xz}), and Fig. 17.12 (σ_{yz}). For both normal and shear stresses, it is clear from Figs. 9–12 that the region of the DEJ experiencing maximum intensity stress is much larger for the 90 rather than the 30-years-old DEJ topography.

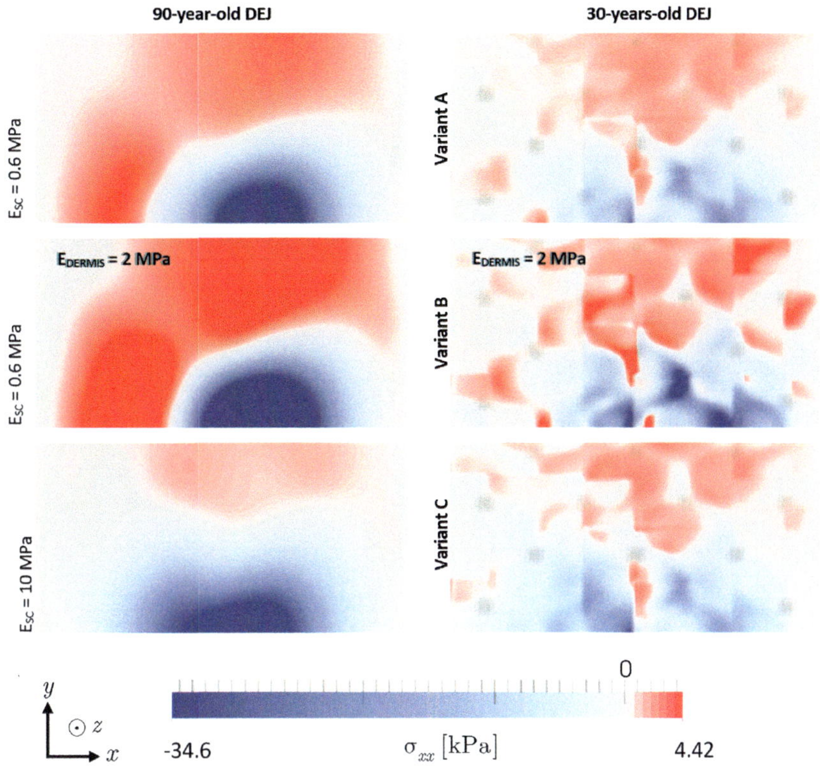

Figure 17.9

Color plots of the normal stress σ_{xx} at the top surface of the papillary dermis layer for the dermal–epidermal junction (DEJ) topographies corresponding to (left) 90-years-old and (right) 30-years-old subjects. The color plots from top to bottom correspond, respectively, to variants A, B, and C (see Table 17.2). Unless indicated (middle row) the Young's modulus of the epidermis is 1 MPa.

The magnitude of normal stresses σ_{xx} and σ_{yy} is very similar (Figs. 9–10), ranging from 38 kPa in compression to 5 kPa in tension. However, the distributions of normal stresses associated with, respectively, the x- and y-axes are noticeably different. For a given normal stress measure, there is also a drastically different stress distribution pattern when comparing the two extreme age DEJ topographies. As expected, stresses are more uniformly distributed for the 90-year-old case when compared with the youngest DEJ topography (30-year-old) so that continuous zones of high stress are much larger. Compared with the other model variants, model variant B (middle row of Fig. 17.9 and, to a lesser extent, Fig. 17.10) exhibits a striking stress pattern where the zone of maximum tensile stress is significantly larger.

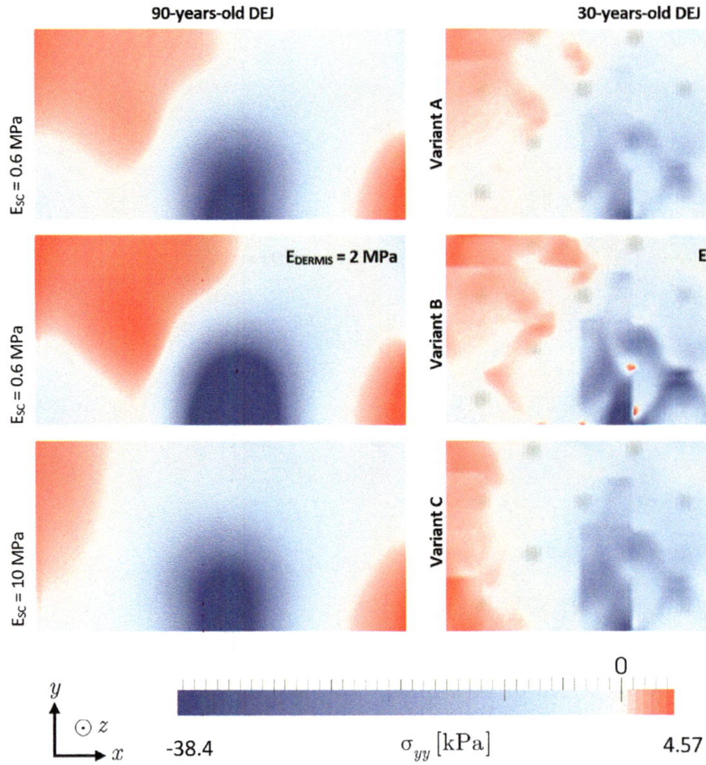

Figure 17.10

Color plots of the normal stress σ_{yy} at the top surface of the papillary dermis layer for the dermal–epidermal junction (DEJ) topographies corresponding to (left) 90-years-old and (right) 30-years-old subjects. The color plots from top to bottom correspond, respectively, to variants A, B, and C (see Table 17.2). Unless indicated (middle row) the Young's modulus of the epidermis is 1 MPa.

The distribution of shear stresses σ_{xz} (Fig. 17.11) and σ_{yz} (Fig. 17.12) is much more heterogeneous for the 30-year-old DEJ topography, with no large continuous regions of high stress like what is observed for the eldest DEJ case. It is clear that the DEJ of a young subject experiences much less shear stress than an older one. The magnitude of σ_{xz} and σ_{yz}, respectively, range from about 23 to 34 kPa and from 9 to 26 kPa. For σ_{xz}, there are clearly two main zones of high stress. They correspond to the back (negative shear stress, in dark blue) and front (positive shear stress, in intense red) of the indenter moving along the x-axis. Although, in an average sense, these zones also exist for the 30-year-old case, they are not so well defined and so large (Fig. 17.11). Like for normal stresses, it appears that for a stiffer dermis (E_{DERMIS} = 2 MPa instead of 1 MPa), the regions of maximum shear stress are larger (Figs. 11 and 12). As expected, for shear stress σ_{yz}, the response is more symmetric behind and in front of the indenter for the 90-year-old DEJ.

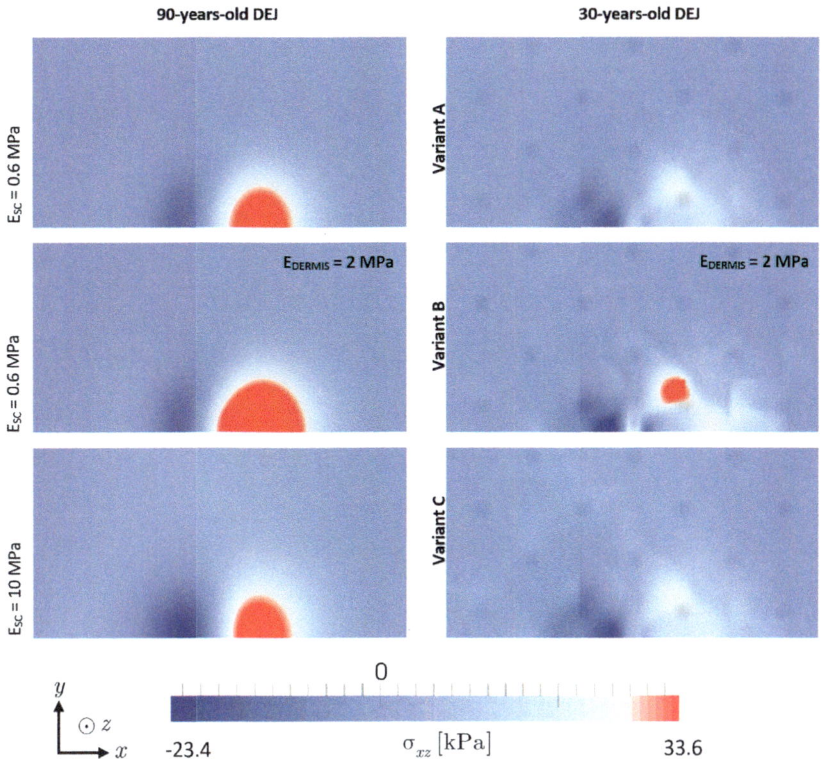

Figure 17.11

Color plots of the shear stress σ_{xz} at the top surface of the papillary dermis layer for the dermal–epidermal junction (DEJ) topographies corresponding to (left) 90-years-old and (right) 30-years-old subjects. The color plots from top to bottom correspond, respectively, to variants A, B, and C (see Table 17.2). Unless indicated (*middle row*) the Young's modulus of the epidermis is 1 MPa.

Box plots of shear stresses σ_{xy}, σ_{xz}, and σ_{yz} in the papillary dermis for the two extreme ages (30 and 90-year-old) of model variants A, B, and C are reported in Fig. 17.13. The focus of the computational study is on comparing extreme values of shear stress as a function of DEJ topography and mechanical properties as they are correlated with likelihood of skin damage. For this reason, the box plots in Fig. 17.13 considered a subset of all shear stress magnitude by filtering out values lower than 10 kPa.

For the three shear stress components (Fig. 17.13A–C), it is clear that interquartile and peak values are significantly higher for the case of a DEJ from a 90-year-old when compared with that of a 30-year-old subject. Doubling the Young's modulus of the dermis has also a significant effect on the peak and interquartile values of shear stress for both age-dependent DEJ topographies. They are both increased as a result of a 100% stiffening of the dermis. An increase of the elastic modulus of the stratum corneum from 0.6 to

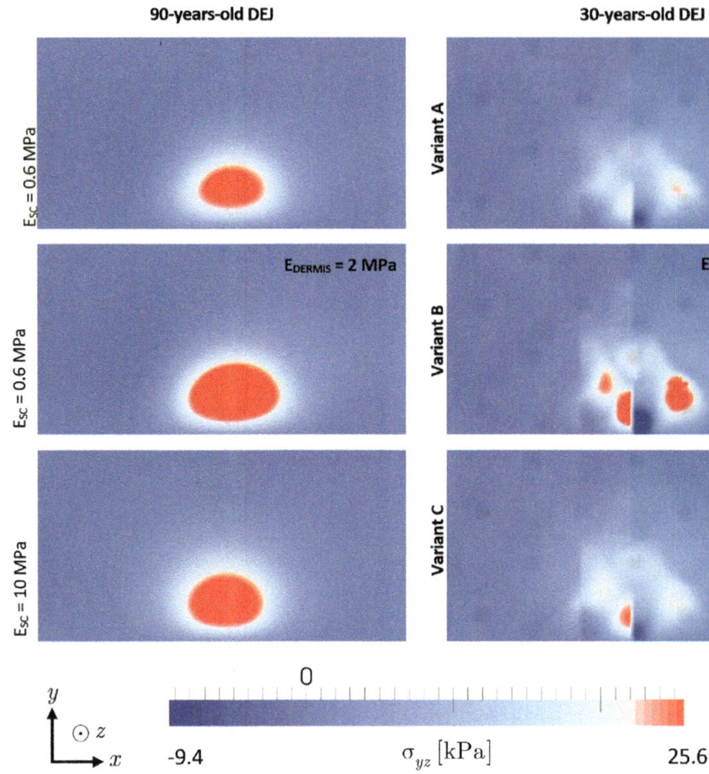

Figure 17.12

Color plots of the shear stress σ_{yz} at the top surface of the papillary dermis layer for the dermal–epidermal junction (DEJ) topographies corresponding to (left) 30-years-old and (right) 90-years-old subjects. The color plots from top to bottom correspond, respectively, to variants A, B, and C (see Table 17.2). Unless indicated (*middle row*) the Young's modulus of the epidermis is 1 MPa.

10 MPa does not appear to have a major effect on the interquartile and extreme values of shear stress at the exception of σ_{xy} (Fig. 17.13A).

7. Discussion

The results of this computational study with regard to the age-dependent structural characteristics of the DEJ have implications for a large number of biological, clinical, and engineering areas where the mechanics and tribology of the skin coupled with biological processes such as aging and interaction with the environment play a central role [53–55].

For a given imposed macroscopic deformation (e.g., sliding indentation) and coefficient of friction between the *stratum corneum* and a spherical rigid indenter, the flattening of

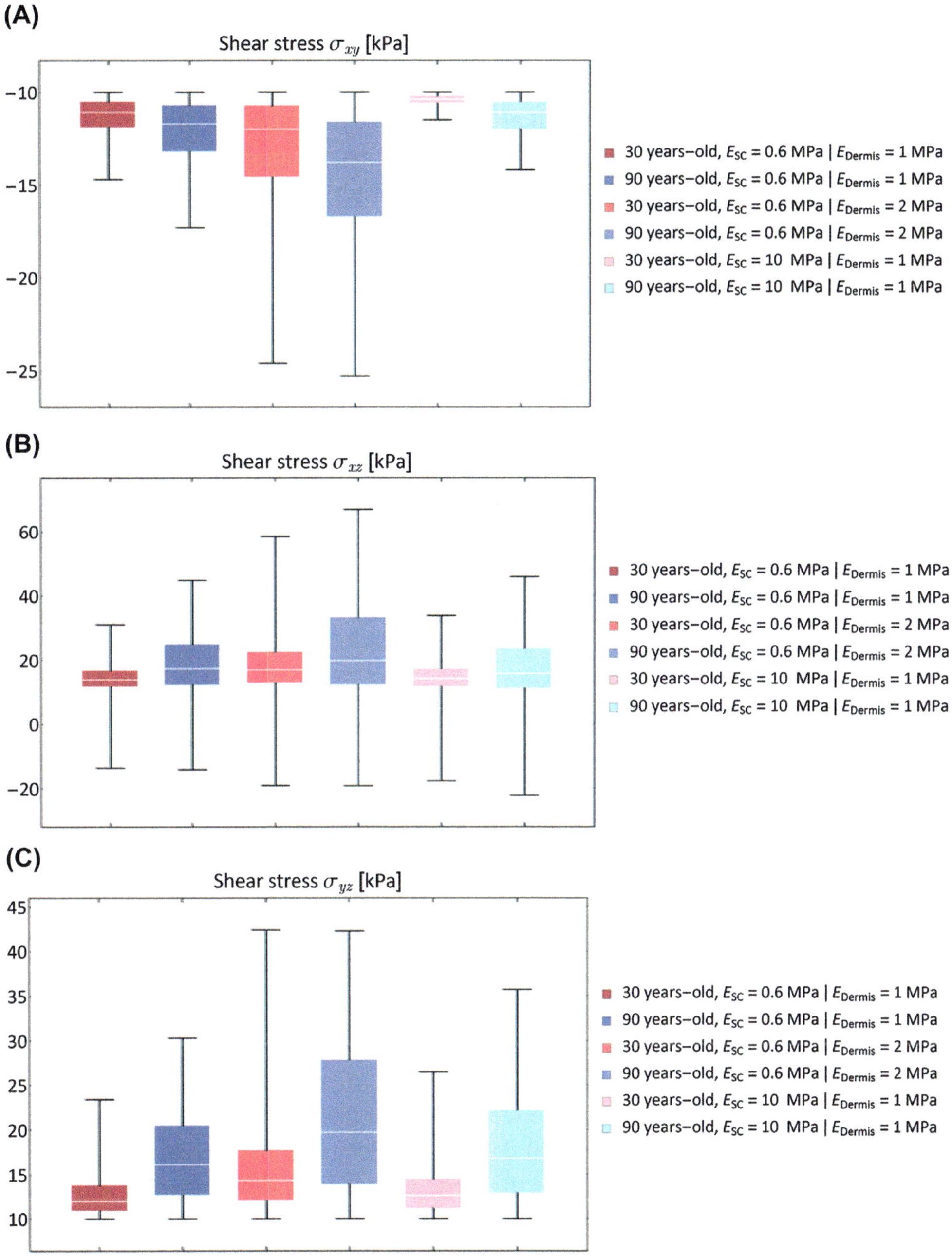

Figure 17.13

Box plots of shear stresses in the papillary dermis for the two extreme ages of model variants A, B, and C: 30- and 90-years-old subjects. Values of stress lower than 10 kPa were discarded to emphasize the significance of more extreme values between age groups. (A) σ_{xy}; (B) σ_{xz}; and (C) σ_{yz}.

the DEJ topography with age (Eq. (2)–(4))—in combination with age-independent mechanical properties—was shown to amplify the level of shear stresses at the DEJ as well having the effect of concentrating these high stresses more uniformly. This implies that shear stresses at the DEJ are more likely to reach critical (i.e., failure) values in an aged skin subjected to shear deformations compared with a younger one. It is important to point out that the potential alterations of mechanical properties of the skin layers with age were ignored in the present FE modeling. Nevertheless, as geometry in many cases dominates over mechanical properties in determining the mechanical state in biological structures, the value in demonstrating this vulnerability of the aged skin to shear loads is fundamentally important.

In numerous scenarios, the aged skin is being loaded in shear, and in the medical arena, this happens many times each day, e.g., when repositioning elderly patients in bed, when elevating the head of the bed to improve respiration, when preparing geriatric patients for washing or when changing diapers and dressings, and when people who use a wheelchair reposition themselves in the chair, to mention just a few such conditions. Each of these conditions involves a risk of formation of a skin lesion, either mild or severe, and the present modeling enlightens the mechanisms for such injuries as the skin ages. Additionally, there are medical devices that directly inflict shear on skin, particularly prosthetic sockets for lower and upper extremities, respiration masks, head and leg rests for surgical procedures, footwear and specialized socks for diabetic patients, again to mention a few of many examples. Currently, age and the degradation of skin tolerance with age is not a factor in the design of these medical devices, and in other words, there are no adaptations that are being made, e.g., in a socket of a prosthesis or in hospital clothing or bedsheets or diapers to accommodate to the fact that the skin is becoming more fragile as the patient becomes older. The fundamental understanding that the present modeling provides not only highlights such conditions but also prepares the ground for more application-specific modeling frameworks that could aid in the design of medical devices that contact the skin, while considering the age-dependent structure–function relationships that have been demonstrated here. In cases where the skin is moist or wet in such daily circumstances, e.g., due to perspiration (particularly when fevering) or an incontinence issue, friction coefficients between the skin and the clothing/bedsheets or between the skin and a contacting medical device will rise, and shear forces will rise accordingly, amplifying the shear loads at the DEJ. In future work, an attempt should be made to more specifically connect these clinical scenarios to the microscale interactions at the DEJ, which will likely require multiscale, multiphysics modeling to capture these phenomena at the different spatial scales, and represent the (thermodynamic) microclimate effects at the skin surface on the loads at the DEJ.

It is established that the skin in general, and the dermis in particular, tends to stiffen with intrinsic and extrinsic aging through biochemical and mechanobiological processes (e.g., photochemically induced enzymatic degradation of elastic fibers, abnormal collagen deposition/remodeling, or collagen cross-linking via glycation and/or ultraviolet irradiation) [56]. To simulate this effect, model variant B was devised, whereby the Young's modulus of the papillary and reticular dermis was doubled from 1 to 2 MPa (compared with model variant A). It was demonstrated that such stiffening of the dermal layer has a clearly visible effect on the magnitude and distribution of normal and shear stress at the DEJ: both magnitude and areas of maximum significantly stress increase. This demonstrates that alterations of the structural properties of the skin in combination with variations in mechanical properties have the potential to place patients at higher risk of sustaining shear-induced skin injuries.

There is evidence of ethnic-specific variations in biochemical and biophysical properties of the skin [57−59], which are further accentuated as manifestations of intrinsic and extrinsic aging [60]. A notable difference in intercellular cohesion properties was reported by Ref. [57]. In the light of the present computational study, it would be legitimate to wonder whether ethnic-specific mechanostructural properties of the skin might be correlated with ethnic-specific and age-dependent types of skin injury. However, this is way beyond the scope of the study reported in this paper.

The friction properties of the skin against a dry or lubricated solid medium are conditioned by small- to large-scale physical interactions including surface energy effects [61−65] and elastic/inelastic microscopic deformations [2,3,47,51,66] as well as thermomechanical phenomena [50] in combination with macroscopic deformations.

The friction response of soft materials involves the contribution of both an adhesion and a deformation component [67]. The adhesion component is connected to the notion of real area of contact (sum of microasperity contact areas), while the deformation component is associated with the geometry and deformations of asperities that resist the relative motion of the contacting surfaces. This implies that there is a strong coupling between adhesive forces and deformations. Because deformations are directly linked to the mechanical stress environment, it is straightforward to appreciate the link between material and structural properties of the skin and its friction behavior. The microscopic deformations of the skin and those of its respective layers are intrinsically determined by the material and structural properties of these individual layers and are therefore, through multiscale-bridging mechanisms, potential key contributing factors to macroscopic friction properties as recently evidenced by Ref. [62] by means of a modeling-based experiment. In another computational study that accounted for the real anatomical structure of the skin layers, Leyva-Mendivil et al. [40] showed that the geometrical characteristics of each skin layer, particularly those of the *stratum corneum*, play a significant role in modulating macroscopic strains so that local (or microscopic) strains are reduced or amplified. It is anticipated that these effects would have a potentially significant impact on the transmission of surface shear loads from the *stratum corneum* to the deeper tissue layers and the DEJ. Moreover, consideration should be given to diseases that change the structural properties of the skin, diabetes being a common example of this. In diabetes, chronic hyperglycation leads to thickening and increased interlinking of collagen, thereby affecting the structural behavior and stiffness properties of the skin layers, which should indirectly affect its frictional behavior as well. Again, these conditions warrant separate computational studies focusing on chronic skin conditions to investigate how the frictional behavior of skin may potentially relate to its pathophysiology.

In the present study, it was assumed that the *stratum corneum* had a simple cuboid shape while the DEJ was highly idealized under the form of a regular wavy surface described by the tensor product of cosine functions. Accounting for the real geometrical properties of these structural layers together with inhomogeneities in the distribution of mechanical properties, e.g., based on information extracted from optical coherence tomography or 3D digital reconstructions from histology, would likely affect the quantitative results of the current study and may highlight different/novel structural effects. However, it is felt that the representation of the waviness of the DEJs by means of closed-form mathematical functions, as was done in the current model, was a necessary step to isolate the age-related structural and material effects of a nonplanar DEJ topography and, by so doing, facilitate the interpretation of results particularly for comparison across age groups.

This study demonstrated that it is possible to use an idealized computational model of the skin grounded in simplifying assumptions and yet highlight the role that the age-dependent DEJ topography and mechanical properties of the skin layers can play on the mechanical response of the skin under surface shear loads. The use of an idealized geometry is an advantage as it allows systematic comparisons across age while filtering out any intra- and interindividual biological variability. Similar remarks apply to the choice of the constitutive model used in the FE simulations. The hypothesis-driven approach presented in this study is valuable to suggest biomechanically plausible scenarios explaining age-dependent sensitivity of the skin to shear loads and, ultimately, to develop a more complete mechanistic understanding of friction-induced injuries so that effective preventive or treatment solutions could be developed. In principles, it is conceivable to experimentally validate in vitro the model proposed in this study within the confine of a well-controlled environment offered by a tissue-engineered skin construct. However, this would likely be a very challenging endeavor.

An important biophysical aspect, not captured in this study, is the fact that the skin microrelief is correlated with age and extrinsic aging effects such as those induced by exposure to ultraviolet radiations or smoking [32]. Leyva-Mendvil and colleagues demonstrated that not only does skin microrelief play a role in modulating subsurface shear stress transmission [12] but also is a key contributor to deformation-induced macroscopic friction of the skin against a solid surface [13]. In reality, transmission of normal and shear stresses from the skin microrelief to the DEJ zone is likely to be significantly more heterogeneous than in the case of a perfectly flat skin surface. Moreover, the structural and material properties of the BMZ, yet to be characterized and quantified, were not considered in this study. It is believed that, for the simulated mechanical conditions, they would likely play a critical role in determining how stress and deformation are transmitted at the DEJ, from the epidermal to dermal layer.

Apart from a variation of the Young's modulus of the *stratum corneum*, no humidity-dependent properties were considered in our simulations. Skin microclimate (i.e., RH level and temperature) can modify the physicochemical properties of the contact interface and lead to higher friction force via alteration of the interfacial shear strength [68] rather than solely that of the contact area. Furthermore, water can be absorbed into the *stratum corneum* intercellular space, increasing the thickness of this outer layer up to three times for a 4-hour water exposure [69]. Through the development of an idealized parametric multilayer FE model of the skin, Sopher et al. [51] evaluated the combined effects of *stratum corneum* thickness, shear modulus, and coefficient of friction at the skin–support interface on the shear stress distribution within the skin layers.

In their study, they showed how these parameters can increase or reduce the risk of potential damage within the skin. Swelling of the *stratum corneum* in response to increased RH would modify the skin surface microstructure, smoothing out skin crests and superficial furrows, and closing up the deeper furrows. In turn, these geometrical alterations would likely affect stress transmissions from the outer skin surface to the DEJ. Furthermore, the thickness of this outer layer would modify the overall skin elastic behavior via structural effects [62]. So, it is likely that, if one takes into account the structural changes humidity causes in the *stratum corneum*, these changes could also significantly influence the global skin friction response, and therefore, the magnitude, distribution, and concentration of peak shear stresses at the DEJ. It can be conjectured that, compared with smooth skin, skin featuring a high degree of topographic roughness—as is the case in elderly subjects (e.g., deep wrinkles)—is particularly prone to generate higher macroscopic friction if the *stratum corneum* stiffness is minimal. Now, in reality, the water-induced volume changes of the *stratum corneum* layer coupled to an increase in adhesive forces at the local level would likely lead to even larger increase in microscopic friction. This could therefore be a plausible mechanism for the prevalence of skin tears in the elderly population [36].

The study has shown that the age-dependent topography of the DEJ coupled to variations in mechanical properties of the skin layers has major effects on normal and shear stress magnitude and distribution at the DEJ. This would suggest that many published computational models of the skin might need to be revisited in the light of our finding.

Future studies should include more realistic aspects of skin micromechanics such as skin collagen fiber architecture and anisotropic elasticity [1,70–72], which would likely produce more heterogeneous strain and stress distributions. However, for the loading and boundary conditions enforced, it is believed that the stretch-stiffening anisotropic characteristics provided by the dermis would be potentially not very significant. It is relevant to point out that collagen orientation is markedly different in the papillary and reticular dermis. The hierarchical structure of the basal lamina and its associated mechanical properties would also be an important element to integrate into future FE models but would require substantial efforts in terms of experimental characterization. From the modeling point of view, the disparity of spatial scales between the size of the BMZ and that of the epidermal and dermal layers would require a multiscale approach [73,74].

Besides the complex and multiple biochemical functions of the BMZ, and from an evolutionary mechanical point of view, it could be conjectured that the convoluted geometry of the DEJ acts as a stress diffusion mechanism avoiding very localized zones of high stress concentration while also providing a kinematic locking mechanism through geometric constraints resisting shear loading. With age, through the flattening of the DEJ, combined with alterations in mechanical properties of the skin layers, these capabilities are progressively reduced. This would suggest that age-related skin fragility could be explained by such a mechanism.

8. Conclusion

By means of a parametric idealized 3D multilayer FE model of the skin subjected to surface indentation and shear loads, the potentially significant role of the age-dependent topography of the DEJ in resisting and diffusing these loads was demonstrated. The flattening of the DEJ with age was shown to induce higher and more uniformly concentrated normal and shear stresses at the DEJ, which are correlated with higher likelihood of material damage. This study therefore suggests that the risk for skin breakdown under shear loading increases with age.

Acknowledgments

Financial support from the Engineering and Physical Sciences Research Council [Grant EP/F034296/1], the Royal Academy of Engineering, and the University of Southampton is gratefully acknowledged. Prof. Amit Gefen was awarded a Distinguished Visiting Fellowship from the Royal Academy of Engineering and was hosted at the University of Southampton. The discussions with Prof. Georges Limbert during his fellowship and visits to the United Kingdom in that framework contributed to this chapter. Daniel Garcia-Martinez would like to thank his parents for their financial support. Dr. Maria Fabiola Leyva-Mendivil would like to acknowledge the financial support of CONACyT through the award of a PhD scholarship.

References

[1] Lanir Y. Skin mechanics. In: Skalak R, Chien S, editors. Handbook of bioengineering. New York: McGraw-Hill; 1987.
[2] Derler S, Gerhardt LC. Tribology of skin: review and analysis of experimental results for the friction coefficient of human skin. Tribol Lett 2012;45:1–27.
[3] van Kuilenburg J, Masen MA, Groenendijk MNW, Bana V, van der Heide E. An experimental study on the relation between surface texture and tactile friction. Tribol Int 2012;48:15–21.
[4] van Kuilenburg J, Masen MA, van der Heide E. The role of the skin microrelief in the contact behaviour of human skin: contact between the human finger and regular surface textures. Tribol Int 2013c;65:81–90.
[5] Limbert G. Mathematical and computational modelling of skin biophysics-a review. P Roy Soc A-Math Phys 2017;473:1–39.
[6] Limbert G. Constitutive modelling of skin mechanics. In: Limbert G, editor. Skin biophysics-from experimental characterisation to advanced modelling. Hiedelberg: Springer; 2019a.
[7] Limbert G. Skin biophysics-from experimental characterisation to advanced modelling. In: Gefen A, editor. Studies in mechanobiology, tissue engineering and biomaterials. Heidelberg: Springer; 2019b.
[8] Xu F, Lu TJ, Seffen KA. Biothermomechanics of skin tissues. J Mech Phys Solids 2008;56:1852–84.
[9] Wildnauer RG, Bothwell JW, Douglas AB. Stratum corneum biomechanical properties. J Investig Dermatol 1975;56:72–8.
[10] Wu KS, van Osdol WW, Dauskardt RH. Mechanical properties of human stratum corneum: effects of temperature, hydration, and chemical treatment. Biomaterials 2006;27:785–95.
[11] Leyva-Mendivil MF, Lengiewicz J, Limbert G. Skin friction under pressure. The role of micromechanics. Surf Topogr Metrol Prop 2017a;6:014001.
[12] Leyva-Mendivil MF, Lengiewicz J, Page A, Bressloff NW, Limbert G. Implications of multi-asperity contact for shear stress distribution in the viable epidermis – an image-based finite element study. Biotribology 2017b;11:110–23.
[13] Leyva-Mendivil MF, Lengiewicz J, Page A, Bressloff NW, Limbert G. Skin microstructure is a key contributor to its friction behaviour. Tribol Lett 2017c;65:12.
[14] Kvistedal YA, Nielsen PMF. Estimating material parameters of human skin in vivo. Biomech Model Mechanobiol 2009;8:1–8.
[15] Silver FH, Freeman JW, DeVore D. Viscoelastic properties of human skin and processed dermis. Skin Res Technol 2001;7:18–23.
[16] Fung YC. Biomechanics: mechanical properties of living tissues. New York: Springer-Verlag; 1981.
[17] Burns T, Breathnach S, Cox N, Griffiths C. Rook's textbook of dermatology. 7th ed. ed. Oxford: Blackwell Science; 2004.
[18] Shimizu H. Shimizu's textbook of dermatology. Hokkaido University Press - Nakayama Shoten Publishers; 2007.
[19] Chan LS. Human skin basement membrane in health and autoimmune diseases. Front Biosci 1997;2:343–52.
[20] Burgeson RE, Christiano AM. The dermal-epidermal junction. Curr Opin Cell Biol 1997;9:651–8.
[21] Ciarletta P, Ben Amar M. Papillary networks in the dermal-epidermal junction of skin: a biomechanical model. Mech Res Commun 2012;42:68–76.
[22] Briggaman RA, Wheeler CE. The epidermal-dermal junction. J Investig Dermatol 1975;65:71–84.
[23] Li Y, Ortiz C, Boyce MC. Bioinspired, mechanical, deterministic fractal model for hierarchical suture joints. Phys Rev 2012;85:031901.
[24] Zhang Y, Yao H, Ortiz C, Xu J, Dao M. Bio-inspired interfacial strengthening strategy through geometrically interlocking designs. J Mech Behav Biomed Mater 2012;15:70–7.
[25] Jasinoski SC, Reddy BD, Louw KK, Chinsamy A. Mechanics of cranial sutures using the finite element method. J Biomech 2010;43:3104–11.

[26] Lagarrigue SG, George J, Questel E, Lauze C, Meyer N, Lagarde JM, Simon M, Schmitt AM, Serre G, Paul C. In vivo quantification of epidermis pigmentation and dermis papilla density with reflectance confocal microscopy: variations with age and skin phototype. Exp Dermatol 2012;21:281–6.

[27] Lavker RM. Structural alterations in exposed and unexposed aged skin. J Investig Dermatol 1979;73:59–66.

[28] Querleux B, Baldeweck T, Diridollou S, de Rigal J, Huguet E, Leroy F, Barbosa VH. Skin from various ethnic origins and aging: an in vivo cross-sectional multimodality imaging study. Skin Res Technol 2009;15:306–13.

[29] Sauermann K, Clemann S, Jaspers S, Gambichler T, Altmeyer P, Hoffmann K, Ennen J. Age related changes of human skin investigated with histometric measurements by confocal laser scanning microscopy in vivo. Skin Res Technol 2002;8:52–6.

[30] Marks R. Skin disease in old age. 2nd ed. London: Martin Dunitz; 1999.

[31] Gilchrest BA. Age-associated changes in the skin. J Am Geriatr Soc 1982;30:139–43.

[32] Gilchrest BA. Skin aging and photoaging: an overview. J Am Acad Dermatol 1989;21:610–3.

[33] Rippke F, Schreiner V, Schwantitz H. The acidic milieu of the horny layer: new findings on the physiology and pathophysiology of skin pH. Am J Clin Dermatol 2002;3:261–72.

[34] Grove GL. Physiological changes in older skin. Clin Geriatr Med 1989;5:115–25.

[35] Chung JH, Yano K, Lee MK, Youn CS, Seo JY, Kim KH, Cho KH, Eun HC, Detmar M. Differential effects of photoaging vs intrinsic aging on the vascularization of human skin. Arch Dermatol 2002;138:1437–42.

[36] LeBlanc K, Baranoski S. Skin tears: state of the science: consensus statements for the prevention, prediction, assessment, and treatment of skin tears. Adv Skin Wound Care 2011;24:2–15.

[37] Diridollou S, Vabre V, Berson M, Vaillant L, Black D, Lagarde JM, Gregoire JM, Gall Y, Patat FQ. Skin ageing: changes of physical properties of human skin in vivo. Int J Cosmet Sci 2001;23:353–62.

[38] Grove GL, Kligman AM. Age-associated changes in human epidermal cell renewal. J Gerontol 1983;38:137–42.

[39] Kligman LH. Photoaging. Manifestations, prevention and treatment (review). Dermatol Clin 1986;4:517–28.

[40] Leyva-Mendivil MF, Page A, Bressloff NW, Limbert G. A mechanistic insight into the mechanical role of the stratum corneum during stretching and compression of the skin. J Mech Behav Biomed Mater 2015;49:197–219.

[41] Geerligs M, van Breemen LCA, Peters GWM, Ackermans PAJ, Baaijens FPT, Oomens CWJ. In vitro indentation to determine the mechanical properties of epidermis. J Biomech 2011;44:1176–81.

[42] Tran HV, Charleux F, Rachik M, Ehrlacher A, Ho Ba Tho MC. In vivo characterization of the mechanical properties of human skin derived from MRI and indentation techniques. Comput Methods Biomech Biomed Eng 2007;10:401–7.

[43] Ogden RW. Non-linear elastic deformations. West Sussex, England: Ellis Horwood Ltd.; 1984.

[44] ABAQUS. Theory manual. Providence, RI, USA: Simulia, Dassault Systèmes; 2012.

[45] Levi K, Baxter J, Meldrum H, Misra M, Pashkovski E, Dauskardt RH. Effect of corneodesmosome degradation on the intercellular delamination of human stratum corneum. J Investig Dermatol 2008;128:2345–7.

[46] Levi K, Weber RJ, Do JQ, Dauskardt RH. Drying stress and damage processes in human stratum corneum. Int J Cosmet Sci 2010;32:276–93.

[47] van Kuilenburg J, Masen MA, van der Heide E. Contact modelling of human skin: what value to use for the modulus of elasticity? Proc Inst Mech Eng J J Eng Tribol 2013a;227:349–61.

[48] Veijgen NK, Masen MA, van der Heide E. A novel approach to measuring the frictional behaviour of human skin in vivo. Tribol Int 2012;54:38–41.

[49] Veijgen NK, Masen MA, van der Heide E. Variables influencing the frictional behaviour of in vivo human skin. J Mech Behav Biomed Mater 2013;28:448–61.

[50] Gefen A. How do microclimate factors affect the risk for superficial pressure ulcers: a mathematical modeling study. J Tissue Viability 2011;20:81−8.
[51] Sopher R, Gefen A. Effects of skin wrinkles, age and wetness on mechanical loads in the stratum corneum as related to skin lesions. Med Biol Eng Comput 2011;49:97−105.
[52] Holzapfel GA. Nonlinear solid mechanics. A continuum approach for engineering. Chichester, UK: John Wiley & Sons; 2000.
[53] Limbert G, Pond D, McBride A. Constitutive modelling of skin ageing. In: Limbert G, editor. Skin biophysics-from experimental characterisation to advanced modelling. Heidelberg: Springer; 2019.
[54] Limbert G, Simms C. Editorial: special issue on skin mechanobiology. J Mech Behav Biomed Mater 2013;28:395−6.
[55] Pond D, McBride A, Davids L, Reddy BD, Limbert G. Microstructurally-based constitutive modelling of the skin. Linking intrinsic ageing to microstructural parameters. J Theor Biol 2018;444:108−23.
[56] Sherratt MJ. Age-related tissue stiffening: cause and effect. Adv Wound Care 2013;2:11−7.
[57] Rawlings AV. Ethnic skin types: are there differences in skin structure and function?1. Int J Cosmet Sci 2006;28:79−93.
[58] Vashi NA, Buainain De Castro Maymone M, Kundu RV. Aging differences in ethnic skin. J Clin Aesthet Dermatol 2016;9:31−8.
[59] Wesley NO, Maibach HI. Racial (ethnic) differences in skin properties. Am J Clin Dermatol 2003;4:843−60.
[60] Vierkötter A, Krutmann J. Environmental influences on skin aging and ethnic-specific manifestations. Derm Endocrinol 2012;4:227−31.
[61] Israelachvili JN. Intermolecular and surface forces. 3rd ed. ed. Burlington, MA, USA: Academic Press; 2011.
[62] Leyva-Mendivil MF, Lengiewicz J, Page A, Bressloff NW, Limbert G. Skin microstructure is a key contributor to its friction behaviour. Tribology Letters 2017;65:12. https://doi.org/10.1007/s11249-016-0794-4.
[63] Pailler-Mattei C, Pavan S, Vargiolu R, Pirot F, Falson F, Zahouani H. Contribution of stratum corneum in determining bio-tribological properties of the human skin. Wear 2007;263:1038−43.
[64] Pailler-Mattei C, Vargiolu R, Zahouani H. Analysis of adhesion contact of human skin in vivo. Contact angle, wettability and adhesion. 4; 2006. p. 501−14.
[65] Pailler-Mattei C, Zahouani H. Study of adhesion forces and mechanical properties of human skin in vivo. J Adhes Sci Technol 2004;18:1739−58.
[66] van Kuilenburg J, Masen MA, van der Heide E. A mechanistic approach to predicting the friction behaviour of human skin. In: Proceedings of the asme/stle international joint tribology conference, ijtc 2012; 2013. p. 45−7.
[67] Persson BNJ. Sliding friction. Berlin, Heidelberg: Springer Berlin Heidelberg; 2000.
[68] Adams MJ, Briscoe BJ, Johnson SA. Friction and lubrication of human skin. Tribol Lett 2007;26:239−53.
[69] Warner RR, Stone KJ, Boissy YL. Hydration disrupts human stratum corneum ultrastructure. J Investig Dermatol 2003;120:275−84.
[70] Limbert G. A mesostructurally-based anisotropic continuum model for biological soft tissues—decoupled invariant formulation. J Mech Behav Biomed Mater 2011;4:1637−57.
[71] Ní Annaidh A, Bruyère K, Destrade M, Gilchrist MD, Maurini C, Otténio M, Saccomandi G. Automated estimation of collagen fibre dispersion in the dermis and its contribution to the anisotropic behaviour of skin. Ann Biomed Eng 2012;40:1666−78.
[72] Ní Annaidh A, Bruyère K, Destrade M, Gilchrist MD, Otténio M. Characterization of the anisotropic mechanical properties of excised human skin. J Mech Behav Biomed Mater 2011;5:139−48.
[73] Kouznetsova V, Brekelmans MWA, Baaijens TFP. An approach to micro-macro modeling of heterogeneous materials. Comput Mech 2001;27:37−48.
[74] Šolinc U, Korelc J. A simple way to improved formulation of FE^2 analysis. Comput Mech 2015;56:905−15.

CHAPTER 18

Developing standard test methods for assessment of medical devices in the fields of wound prevention and care

Evan Call[1], Adam Cheney[2]

[1]*Adjunct Faculty, Microbiology, Weber State University, Ogden, UT, United States;* [2]*Adjunct Faculty, Rhetoric, University of North Texas, Denton, TX, United States*

Chapter Outline

Innovations and Emerging Technologies in Wound Care. https://doi.org/10.1016/B978-0-12-815028-3.00018-3

1. Introduction

In the field of wound prevention and care, pressure injuries/ulcers represent a significant problem in terms of pain, disfigurement, suffering, and prolonging recovery, in addition to the monetary costs associated with each. To decrease the probability of pressure injuries/ulcers when possible and manage them when they occur, multiple approaches are required. First, caregivers must assume all patients are at risk. Second, identifying type of risk requires clinical judgment and trained, intuitive sense. Third, identification can be assisted, not supplanted, by the use of assessment tools [1]. A fourth step should include the use of best available support surfaces, cushions, and positioning devices, in which caregivers should be informed by the development of standards through rigorous testing and expert assessment.

First and foremost in wound prevention and care, the issue of pressure-induced tissue deformation must be addressed because pressure is a major causative factor in injury/ulcer formation [2,3]. Those factors can be further understood by breaking them down into related categories and examining the elements both independently and as affecting one another. Those categories are (*a*) intensity of pressure, (*b*) its duration, (*c*) the tolerance of tissue under pressure, and (d) force direction that defines the difference between pressure and shear. Tolerance of tissue is impacted by both intrinsic and extrinsic factors. Shear, fiction, heat, and moisture, or microclimate, are the extrinsic factors. These can be directly addressed during time of care by employing best practices when using appropriate support surfaces, positioning devices, and cushions to greatest combined benefit.

Support surfaces are not necessarily those available in a retail store. For these studies, a support surface is a specialized device for pressure redistribution designed for management of tissue loads, microclimate, and/or other therapeutic functions [4].

Identification, development, and application of appropriate support surfaces are being advanced by rigorous testing methods and scrupulous development of national and international standards. These measures narrow the information gap between manufacturers, rehabilitation technology suppliers, clinicians, and users.

2. An overview of standards organizations

2.1 American national standards institute

The primary goals of American National Standards Institute (ANSI) are to enhance global competitiveness of US businesses and improve quality of life by promoting voluntary consensus standards and monitoring their integrity. Since its formation in 1918, the Institute has been and continues to be a private, nonprofit organization supported by a diverse body of private and public organizations.

ANSI is active both nationally and internationally, addressing itself to interests of companies, government and nongovernmental entities, academia, developers, and consumers. The Institute itself does not develop national standards, but it creates a neutral venue for parties to come together to facilitate common agreement. In accordance with the Institute's principles of consensus, due process, and openness, anyone directly or materially affected by a standard under development has access to the standards process as well as the appeals mechanism.

Within the United States, individuals, companies, agencies, and labor/consumer groups voluntarily contribute their knowledge and talents to the development of these standards. As of 2017, ANSI provided a forum for more than 240 ANSI-accredited standards developers representing approximately 200 distinct organizations in the private and public sectors [5]. Furthermore, ANSI continues to provide a source of up-to-date materials on critical issues in regional and international arenas as well as background papers with external resource and supporting documents to brief members and constituents on wider areas of interest to the general body of professionals, entities, and groups.

As of January 2018, approximately 230 standards developers were accredited by ANSI representing more than 11,500 American National Standards [6].

ANSI not only works on US interests within US borders but also plays a global role by contributing to the formation of international standards by advocating US policy and technical positions with regional and international standards organizations. ANSI is the sole US representative in the International Electrotechnical Commission, which is through the US National Committee, and ANSI is a founding member of the International Organization for Standardization (ISO) as well.

ANSI development of new standards incorporates the following:

- Consensus by representatives including representatives from interested parties or parties that are materially affected
- Public peer review
- Response to peer review through modification of standards and submitted drafts

- Right to appeal

These processes ensure high quality of standards through direct, expert scrutiny while protecting participants' rights and interests and while making products more readily available and safe for consumers [7].

2.2 The international organization for standardization

The ISO is the world's largest organization to develop voluntary, global standards. Officially constituted in 1947, the ISO is currently headquartered in Geneva, Switzerland, with 164 member countries. Since its inception, the ISO has published over 22,621 international standards [8].

Similar to ANSI, the ISO is made up of volunteer member bodies, individuals, academia, and public and private entities. Experts from across these groups form a technical committee responsible for a specific subject and then develop a draft proposal based on market need for a new standard. The first draft in committee is shared for comments and discussion with expert peers who can suggest edits, encourage further development, and require necessary modification. When consensus is finally met through a voting process, the first draft is taken forward through further technical inspection and drafting stages before a final vote and proof check whereon it becomes a published international standard. Overall, the process usually takes about 3 years and requires approval by at least 75% of the member bodies casting a vote.

The key principals in ISO's development of standards are as follow:

- A direct response to a market need
- Based on a diverse assembly of experts' opinions from around the globe
- Developed through the participation of numerous, distinct entities
- Grounded in consensus.

Every 3 years, a published standard must return before its committee within the organization to be reviewed. Through a vote, it can be revoked, modified, or sustained as is.

Because of this recertification requirement, it should be noted that a standard being old does not mean that the standard is less valid. Both ANSI and ISO must recertify a process, thus keeping it current, but are not required to insert addendums or alter the standard as long as that standard continues to be valid. Therefore, a standard that is dated 15 years ago may still be current under the system of recertification outlined by both organizations and should not be disregarded simply because of how long ago it was adopted.

For example, when a member of a standards committee was presented with an already published standard in support of new material, that member rejected the existing standard as being outdated because the initial certification was a decade old, even though the standards body had continuously upheld the contents as valid and pertinent every 3 years since the standard had been proposed and accepted. Age is not a lone arbiter of validation.

2.3 Support surface standards initiative

No matter which certification body a new set of guidelines passes through, the processes of testing and becoming a recognized standard should seek to produce the use and development of consistent terminology and consistent definitions for communicating anthropometrics and support surface parameters. The development and use of such terminology and definitions need to remain consistent because these terms and the methods they describe must become uniformly useful to those receiving, managing, and transmitting the data, whether they are researchers recertifying results or purchasers intent on matching the best support surfaces with those consumers.

In 1999/2000, the National Pressure Ulcer Advisory Panel (NPUAP) in response to a proposal by Dr. Stephen Sprigle established a Support Surface Standards Initiative (S3I) committee that has evolved into a subcommittee of the Research Committee of NPUAP. After an initial period of extremely intensive and laborious work, the committee began to produce documents with standards potential. The first of these documents was the Support Surface Terms and Definitions. This document provided a validated set of terms that standardized what might be understood by terms such as "Air Fluidized" or "Pressure Redistribution." This document also eliminated some easily misunderstood or potentially incorrect terms such as "Pressure Relief." In an effort to generate impact in the industry, Dr. David Brienza suggested that the committee consider becoming an active standards committee under the Rehabilitation Engineering and Assistive Technology Society of North America (RESNA)[a], an organization with which he had significant experience as a result of work with wheelchair cushions. Originally, this evolving NPUAP committee was seen as a 3-year project tasked with the development and standardization of terminology, test methods, and reporting standards for support surfaces. It has now become a 20-year project that has generated eight different standards housed in one volume known as RESNA SS-1. This relationship between NPUAP and RESNA has proven to be a critical step in the evolution of the committee and the standards.

Through the ANSI/RESNA affiliation, the S3I committee is now the official standards body for the United States, and it became a continuing subcommittee under the NPUAP

[a] RESNA is accredited by the American National Standards Institute (ANSI). RESNA is a standards body under ANSI.

Research Committee and further charged it to implement appropriate standards. The committee not only develops terms and definitions together with standards but also validates them in addition, and once there is approval of national standards, ANSI/RESNA submits these standards for consideration as worldwide standards through ISO. S3I updates publication of these documents regularly; its most recent version was made available in December 2018 [9]. Typically standards are "revoted" or determined to continue to be valid in the industry every 3 years. This ongoing renewal cycle reflects one of the long-term benefits of standards. They are ultimately no older than 3 years as their validity and applicability are reviewed in a voting process, ensuring their continued accuracy as they are kept up to date.

3. The need for an analog body indenter

3.1 Simple problems

The need for an analog body indenter has not always been readily apparent when testing the performance of cushioning products. Initially, tests of support surfaces such as mattresses, cushions, and positioning devices focused on the cushioning properties of the material of construction but ignored looking at other factors. After all, the question being explored was how these support surfaces mitigated or distributed load of the human subject. Therefore, simple tests, such as indentation force deflection (IFD), were sufficient because the questions, as they were then being asked, required a less complex process of examination to answer.

When results from early testing were studied and applied, what they often failed to acknowledge were the controlled conditions under which the tests had been conducted. Furthermore, the human subjects are highly variable, not only in their physiology but also in their microclimate. Testing, at its best, acknowledges and incorporates these aspects as well and acknowledges the limits of any one test being able to account for them.

When considering tests that challenge cushioning devices like support surfaces, the focus turns to what is already available in the industry. Some of the types of test indenters that first come to mind are reviewed here. Generally, these are represented in one form or another in a revised shape and application to provide current focus on standards for wound care and prevention. The compressive roller is a device that applies a compression force while rolling back and forth in an oscillating fashion over carpet and foam (see Fig. 18.1).

These rollers were designed to test the material's fatigue under repetitive use, but the need for more information from a greater variety of routine scenarios became apparent. In evolving tests and expanding the challenges for cushioning products, rollers were modified so they operated in a shearing fashion, where the mechanism was canted at an angle of 15 degrees off the axis of travel. An aggressive test created an off-axis shearing effect as the

Figure 18.1
Early compression force indenter testing of carpet.

roller oscillated back and forth on the surface of the test article tearing at the material and proving how resilient it is (see Figs. 18.2 and 18.3). Note that in Fig. 18.3, the roller was weighted and traveling parallel to the scale, and the canted angle of the roller generates shear to the right in this image.

While the compression rollers shown above focus on a larger area, a different kind of test, one for compressive and impact pressure, focused on a smaller area, such as a seat cushion or the area of a bed where one sits before "getting into bed." In this scenario, a user would not move across the support area but instead would rise off of and then descend onto the surface again and again throughout the life of the support surface. An IFD indenter (see Fig. 18.4) was developed by the foam industry. One of the most basic tests measures how

Figure 18.2
Roller shear test fixture.

Figure 18.3
Roller shear test fixture.

Figure 18.4
203 mm indentation force deflection indenter.

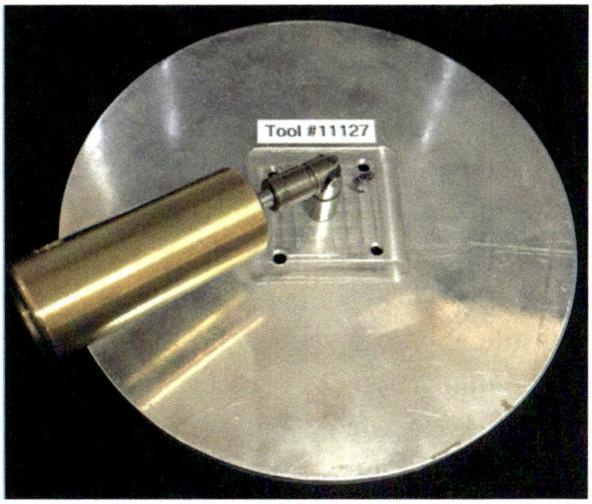

Figure 18.5
Universal joint modification of the indentation force deflection indenter.

much force is required to compress foam to set percentages of its original thickness. The indenter in Fig. 18.4 uses a 203 mm (8″) diameter indenter to press into rather than roll across the test article, and then the force to obtain specific indentation or compression values was recorded. While values were clear and results dependable, the problem of measuring compressive and impact pressure continued to evolve: what would occur when an irregular or contoured surface was tested or if the indenter took the shape of the body?

The first modification of the IFD indenter incorporated a universal swivel joint (See Fig. 18.5). This allowed the indenter plate to settle to the plane that is naturally created by the mechanical average of the contours that the indenter comes in contact with.

3.2 Complex questions

As testing begins to advance, taking into account the inherent variability of a human user of a support surface, those in the field of wound prevention and care, clinicians, academics, and engineers seek answers to the more complex problems: does the product cushion the skin, what is the true nature of the concept of cushioning, does the device add heat to the body, does the device trap heat at the body surface, is moisture allowed to escape or is it trapped, does the body's interaction with the surface compromise the skin and underlying tissues, how long can a particular cushion take a repetitive load before breaking down, at what point is the surface no longer beneficial to the occupant, and is there a safety or injury concern that is created through the use of the product? Although testing of support surfaces began with impact load and repetition, it has evolved and done so precisely because clinicians and researchers conducting tests and examining materials have left themselves open to ask more complex questions that their data do not yet begin to address.

While testing of materials often seeks to solve a specific problem or answer a direct question, the goal of creating standards, as well as the purpose for analyzing support surfaces, is to characterize the performance of the products. The best assessment techniques do not seek to establish a pass/fail dichotomy that is often the focus of regulators or purchasers who seek to justify the process or the expense. The significant danger hidden within such justification can seek a yes/no result from a test, which will in turn stifle creative thinking and innovation. The process of forming new standards should, through building on previous work, seek to expand the field for further inquiry in ways that have not yet been thought of. First, new standards should require that products meet certain minimum performance criteria to pass the tests. Second, the tests should be performed and the result reported to characterize the product. The type of data that the universal joint IFD indenter in Fig. 18.5 could gather cannot provide comprehensive answers to questions such as how does the use of varied support surfaces impact the development of pressure injuries/ulcers in human tissue, can matching a patient with the right support surface prevent tissue damage altogether, or is tissue damage a foregone conclusion even under the best of circumstances? Going even further, the question should be asked, is it possible for a support surface to replace some clinical or nursing activity thus creating an efficiency justifying the cost of the support surface?

As more complex questions are asked, more complex methods of testing support surfaces are developed, and although standards and testing cannot directly answer such questions, nor are they meant to, rather they create a reliable body of knowledge on which professionals can draw to make the best-informed decisions concerning support surface performance.

3.3 Complex tests

A study conducted in 1990 by Stephen Sprigle et al. is an example of the evolution of testing and standard creation [10]. The study sought to examine not only impacts load of a subject but also concentrates on two additional elements: first, the use of contoured cushions and second, those cushions as a means of reducing tissue distortion in a seated posture. The study focused on foam characteristics and seat contours for wheelchair cushions, building on theoretical and empirical experiments that had preceded it [11–14].

Recognizing the need for a more accurate understanding of the interface between user and cushion, the project identified and sought to study five major factors in the development of pressure injuries/ulcers:

1. The shape of the buttocks
2. The shape of the cushion

3. Biomechanical properties of the buttocks influenced by tissue thickness and stiffness
4. Mechanical properties of foam classified by indentation load deflection and density
5. Weight distribution on the cushion

To test these elements, the project employed 17 human subjects and recorded load transfer between buttocks and cushion. The results confirmed that a proper cushion should distribute pressure, minimize peak pressure at bony prominences, and encourage proper posture in the end user.

Working beyond a yes/no or pass/fail dichotomy for results, the group examined the data gathered in the study and confirmed that testing of support surfaces is complicated by multiple factors that need to be accounted and compensated for in the process of analysis, and failing a current means to do so, must be acknowledged, and their impacted stated.

For example, tests involving human subjects are less able to be generalized in a reliable way for the end user. The use of human subjects would seem to be the most obvious resource as these support surfaces are designed for human use, but the variability of subjects' physical features and state of health interferes with establishing a consistent benchmark in testing, which in turn makes test results more diverse rather than more valid.

Use of the terms variability and validity become of special importance when reporting test results and forming conclusions. The greater variability of human test subjects means that is it more difficult to measure a true mean performance. Take as an example an archery target where the bullseye represents a test's validity. A target filled with random arrows, some even hitting the bullseye, represents the variability of using human test subjects. On the other hand, bench testing, or using human analogues, generally provides better grouping of the arrows but does not guarantee that the arrows are closer to the center of the target, the measure of validity. A perfect test model would have the arrows in a tight group on or near the bullseye, meaning that variability is low and validity is high.

On using live subjects, Sprigle wrote, "The variation of humans across the world obviates the ability to define a consistent subject pool for testing." [15] Yet, the problem persists that studies often make use of human test subjects, and, with great regularity, the number of human subjects in a given test is few. Sprigle, who has published and been involved with over 90 studies concerning patients in wheelchairs, must constantly reiterate in each study in which he participates that a model who is physiologically correct is harder to position and is less representative of one human buttock, or an "average" buttock, for a relative given value of average. Therefore, such models do not easily generalize because they "lack external validity" to a larger sampling of users.

As one of the major goals of testing is to produce methods that can be repeated and verified in every lab, the physical inconsistencies of human test subjects make their use inadvisable because they directly interfere with the reproducibility of results. For example, while one patient undergoing an MRI may develop a pressure wound despite the use of a cushion, another patient may be able to sit on that same cushion for multiple hours without any negative effect.

The use of human subjects in testing becomes problematic when applying the results to a larger population. Given the inherent problem with the variability of human subjects, a better approach to creating standards is to use analog indenters that simulate the human body in a known, fixed quantity. While no indenter that perfectly equates to each human physiology can be created, a body analogue (BA) dramatically reduces the number of variables in a given study, leading to greater repeatability and to higher accuracy in test/retest scenarios across multiple test laboratories.

BA indenters have their limitations, but understanding those limitations and being able to extrapolate results helps contextualize the data being gathered and will allow a well-informed clinician to better pair a patient with a support device.

As testing progresses and evolves, better methods are developed when investigating the complexities of human interaction with support surfaces, which leads to the improvement of standards to the benefit of all parties involved, whether it is the engineer conducting experiments in the laboratory, the academic examining data and publishing the paper, or the clinician recommending a particular support surface to the end user.

When working from the prospective of characterization of the support surface, the use of an indenter does provide a tool that allows for the differentiation of products. If it is generally agreed that the nature of the indenter is at least legitimate to the intended use, then this can be acceptable and allows the comparison of products in this performance characteristic.

4. Expansion of the equipment and methods: the thermodynamic rigid cushion loading indenter

4.1 Knowledge gaps: microclimate

The 1990 Sprigle study did what it should. It built on past knowledge, asked new questions, developed new tests, gathered and analyzed the data, and the conclusions it could draw gave rise to further questions and testing which found their way into the continuing development of standards. Revealing gaps in knowledge and then working to bridge them is the natural process of evolution in any field of study.

Two of the important areas that have come under consideration in wound care and prevention since the 80s are the examination of materials used in support surfaces and the effect of human microclimate, or temperature and relative humidity observed in a specified location that is typically where the support surface and indenter interface, or the interaction between human subjects and support surfaces. Each area has drawn on experts across multiple fields such as engineers to build test mechanisms, academics to gather, analyze, and publish test results, and clinicians who receive the results of such work and in turn directly engage with patients to find best support surfaces to fit their needs.

4.2 The thermodynamic rigid cushion loading indenter

Not all standards come directly from an ANSI or ISO mandate. They can also respond directly to a market need or identify gaps in knowledge and then develop studies to begin to fill in those gaps.

Dr. Martin Ferguson-Pell and Graham Nicholson recognized such a gap in the knowledge base. It was not enough to test for sheer, load, and impact pressure alone; reliable tests needed to account for the increasing variety of materials are available in the manufacture of support surfaces.

Where basic foam had dominated common designs of wheelchair cushions and hospital beds, more recent products began to employ the use of elastic foam, viscoelastic foam, closed or open cell, elastomer, air, viscous fluid, or some combination of these [16]. Each material or combination of materials presents a type of load environment with unique characteristics, which in turn leads to different sets of variables that impact the end user in ways that can easily turn detrimental as conditions of use evolve.

While studying compression and impact load on materials was a step forward in the field of secondary complications of physical disabilities, Dr. Ferguson-Pell's group was faced with the simple problem that no one had conducted studies on the fact that human beings sweat and give off heat: "Although much attention has been directed to the reduction of ischemia-initiated soft tissue injury through mechanical means, the influence of tissue temperature and the effect of moisture have been neglected." [17].

In addition, the impact of the shape of the body and the body's mass on the performance of the product had not been considered in the literature. The sweating guarded hot plate (SGHP) test standardized in SS-1, for example, measures heat and moisture removal, but not under the conditions of a support surface being sat on.

In conjunction with the use of a support surface, the microclimate/materials interaction is of particular importance for those faced by long-term use of those surfaces, such as by individuals confined to wheelchairs, which is one of Ferguson-Pell's focuses. The interaction needed to be studied. Although many products claimed to improve heat dissipation and moisture, no reliable means of testing such claims had been developed up to that point.

Figure 18.6
Rigid Cushion Loading Indenter described by Ham Staarink.

Dr. Ferguson-Pell and Graham Nicholson reached out to Hideyuki Hirose and Evan Call. Using investigational research grants[b], the team met first in London to build a prototype load indenter for testing support surfaces, then proceeded to build a second in Salt Lake City, US, and a third in Tokorozawa, Japan. This team built a new piece of equipment, the thermodynamic rigid cushion loading indenter, or more precisely, they built three of them and proceeded with independent testing.

What made thermodynamic rigid cushion loading indenter unique was the attention to its human characteristics. While indenters had already progressed beyond devices such as simple disks that were designed to test repeated impact, the human analog indenters represented an anthropomorphic shape (see Fig 18.6), yet the focus of the new generation of indenters continued to explore the effects of repeated or persistent load. The anthropomorphic shape selected here is the indenter described by Ham Staarink and brought to our attention by Stephen Sprigle [18].

The Ferguson-Pell team created a more dynamic system, combining a buttock-shaped indenter that not only incorporated the more accurate, anthropomorphic shape, but they then built into the shape the ability to simulate a microclimate of heat and sweat while testing for load (see Figs. 18.7 and 18.8).

---

[b] This study used peer-reviewed grants from the Butterworth Foundation Japan Research Exchange, UK NHS CEP, Association for Spinal Research Rehabilitation and Education, Rehabilitation Engineering Research Society of North America, and the National Pressure Ulcer Advisory Panel (USA).

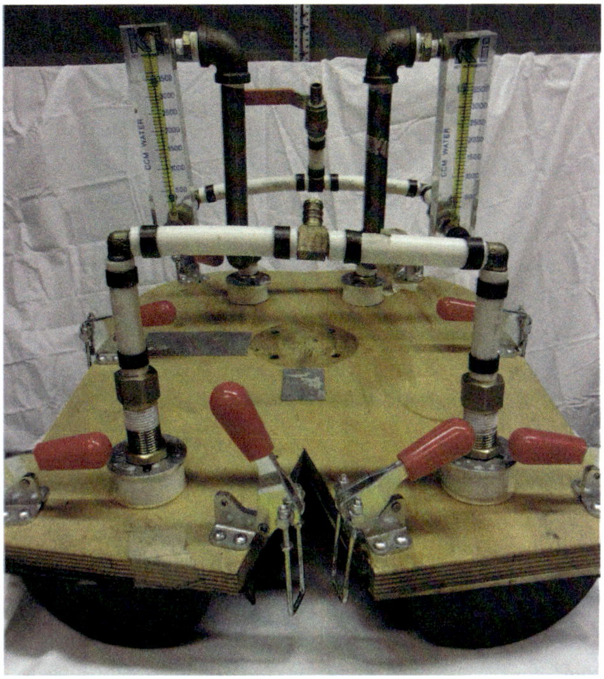

Figure 18.7
Humidity and temperature controlled indenter. See ISO 16840-7.

Figure 18.8
Humidity and temperature controlled indenter. See ISO 16840-7.

The TRCLI delivered controlled heat and water vapor (HWV) rates that allowed measurement of boundary conditions that permitted comparisons between products for measurement of the steady-state and transient conditions normally generated by a seated individual.

The tests conducted with the TRCLI took into account a cohort of 32 commercially available wheelchair cushions available at the time (2008) and addressed the possibility that elevated tissue temperature and moisture buildup from being seated over prolonged periods increased the risks of ischemia and macerative damage. Each of the members of the team tested materials in their independent labs, utilizing identical indenters and the same cohort of cushions. The results of the tests corroborated each other but were far from comprehensive.

One conclusion of the study was that the compression of material of the support surface changes the thermodynamics of the microclimate. Furthermore, that interaction becomes so increasingly complex that simple classification of cushion properties was not an adequate prediction of thermal dissipation properties. The team succeeded in establishing that "testing of individual products with the use of the TRCLI can, however, reliably differentiate wheelchair cushions by their ability to dissipate heat and water vapor." [19] This study worked toward filling in a gap in the knowledge base, yet the study of cushion/thermal dissipation performance had much room for further investigation.

The Ferguson-Pell et al. study cautioned that these tests were conducted under controlled circumstances and that microclimate of the human subject is as variable as subjects' physiologies and directly influences the interaction of user and support surface. In addition, the study cautioned that seating systems such as those found in wheelchairs have complex geometries. These geometries combined with increased variety of materials and added to heat and mass of any given subject make measuring more difficult and the results of any testing incredibly complex to interpret.

Combined with clear terms and standards set down by the S3I committee and further enlarging that knowledge base through its research and careful testing, one of the great steps forward of this study was the creation and verification of a reliable tool for measuring and testing an indenter that could continue to be developed, adapted, or specialized (see Fig. 18.9) for testing any kind of support surface and that further tests could be conducted with assurance that the results, should they proceed with the same meticulous attention to methodologies, and would continue to contribute to a knowledge base that could expand, including the study of every kind of support surfaces that would in turn lead to reliable progress in all areas of study in the fields which impact wound care and management.

The work culminating in the TRCLI study significantly contributed to the development of both the RESNA SS-1 standard for support surfaces and the ISO 16840 family of standards for seating and seat cushions, in particular, part 2, "Determination of Physical and Mechanical Characteristics of Devices Intended to Manage Tissue Integrity—Seat Cushions," part 7; "Cushion Heat and Water Vapor Testing," and part 11; "Determination of perspiration dissipation characteristics of seat cushions intended to manage tissue integrity."

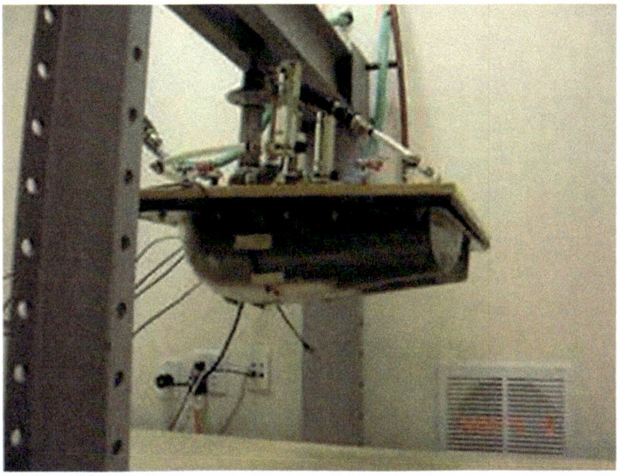

Figure 18.9
The thermodynamic rigid cushion loading indenter mounted on bed testing frame.

4.3 Knowledge gaps: pressure mapping

Describing a strictly linear narrative of cause and effect in tracing research, testing, and studies as they develop from ideas into published standards is unrealistic. Research and testing take time, and even before the process is concluded, they have already begun to drive the development of new materials and new research methods and new ideas, which in turn lead to more research, study, and the development of new guidelines. Often these processes are occurring as contemporaneous events that may influence each other immediately or may take time to permeate the various communities of academia, laboratory testing and research, standards organizations, and clinicians in their practices.

While adding the variables of microclimate to the study of impact pressure was and continues to transform the study of support surfaces, other advancements continue to reveal opportunities for further development of both testing and methodologies while highlighting the need for a greater diversity of approach even as they are maintaining a strict adherence to standards already put forward. While the technology of materials of construction of support surfaces evolves, so too does the technology of gathering data. One such area is the use of pressure mapping.

Almost as soon as it emerged, the use of pressure mapping became a benchmark for recording and studying the impact of pressure on tissue. The technology presents many advantages as it continues to evolve, yet its limitations are equally significant and must be acknowledged.

A pressure mat is a thin sheet consisting of perhaps two to three layers of sensor material introduced between an indenter or human subject and a support surface (see Fig. 18.10).

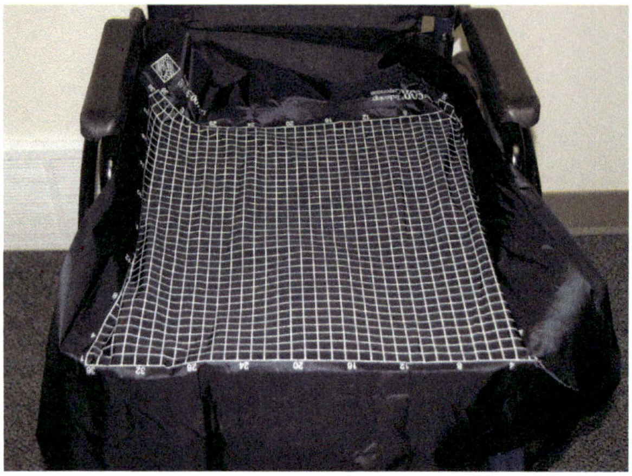

Figure 18.10
An example of a pressure mat positioned on a wheelchair.

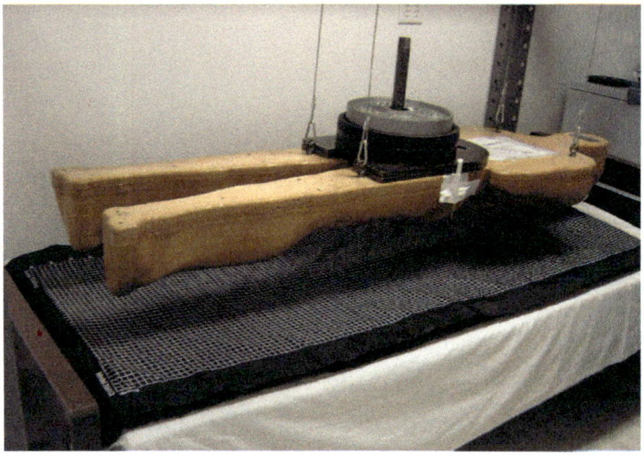

Figure 18.11
A 50th percentile male mannequin weighted for testing using a bed-size pressure mat situated on a hospital bed.

The material is fitted with sensors that can be set to record varied types of data associated with impact pressure and present them as an image or map. The mat in Fig. 18.10 positioned in a wheelchair would produce a pressure map of the buttocks. A full-body indenter (see Fig. 18.11) placed on a mat of sufficient size would produce a full body map.

The highly intuitive representation of the pressure map's results drive interest in its use in standards. Testing of the dependability of pressure mats and pressure mapping systems has been ongoing from the early 90s [20] through to the present day, but the technology has always come with severe reliability and verifiability concerns that continue to need to be acknowledged in the results of any test and should strive to be improved.

4.4 Evaluating test methods

In 2003, Stephen Sprigle, William Dunlop, and Larry Press published a study on the reliability of testing interface pressure on wheelchair cushions using pressure mats [21]. Their work was thoroughly based in previous research that had concluded that "tissue's tolerance to load varies according to the condition of the tissue, its location, age, hydration, and metabolism. All of the pressure injury/ulcer risk factors tend to influence the tissue's ability to withstand load." [22].

The Sprigle et al. 2003 study concluded that both the methods and the technology of testing impact pressure using pressure mats needed further development because they failed to reliably meet not only those standards already established before 2001 but also the new standards under consideration, standards that continue to be supported in current research from multiple laboratories worldwide, all of which demonstrate that tissue distortion is the cause of pressure injury [23–26].

During the 2003 study, the load calculated from the recorded pressure values needed to be within 10% of the error of the applied load, and even though the error allowance was generous, it was still not easy to meet. Furthermore, the use of pressure mats had not yet achieved the target of interlaboratory verifiability and repeatability.

One reason for shortcomings in the technology is the way in which the mat recorded compressive forces and load. "Pressure maps measure normal pressure, and as contouring increases, the map sensels rotate off the line of gravity. This off-axis loading creates error." [27] During the study, measurements were taken three times for each cushion within the same session, yielding a repeatability result with less than a 5% variation in most cases. Yet across multiple sessions, high variability occurred and often fell outside the 10% error allotment for the test.

The rigorousness of the 2003 study called for a high test-retest reliability measure with high verifiability so that repeatability between test sites could be ensured. While many of the variables met these goals during testing (such as dispersion index, contact area, peak pressure index, and seating pressure index), many variables did not meet these goals (such as peak pressure and seating pressure index skew), especially within the test-retest standard.

A portion of the concern for use of pressure maps was identified by Kim Davis, who showed that the calibration of pressure maps does not create a highly accurate and repeatable reading [28]. As pressure maps continue to evolve and improve and the technology becomes more robust, it is likely that the highly intuitive results of pressure mapping will drive their inclusion in standards. For now, they are not included.

Studies involving pressure mat testing were analyzed, and the underlying elements of a support surface's performance were identified. The reasoning being that if the

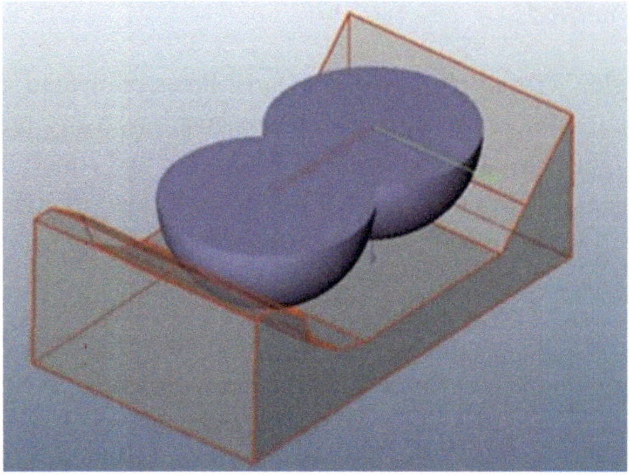

Figure 18.12
Immersion: Depth of penetration or sinking into a support surface.

characteristic that impact pressure redistribution was measured, then a more robust test may be developed. Two concepts were identified, causing two new elements for testing to be highlighted: as a load is placed on a support surface, that load may become immersed in and enveloped by the support surface.

The first element, immersion, is defined by S3I as penetration by sinking into support surface measured by depth (see Fig. 18.12) [29].

The second element, envelopment, is defined by S3I as the ability of a given support surface to conform to fit or mold around irregularities in the body (see Fig. 18.13) [30].

It should be noted that envelopment cannot occur without immersion, but immersion does not necessarily guarantee envelopment (see Fig. 18.14).

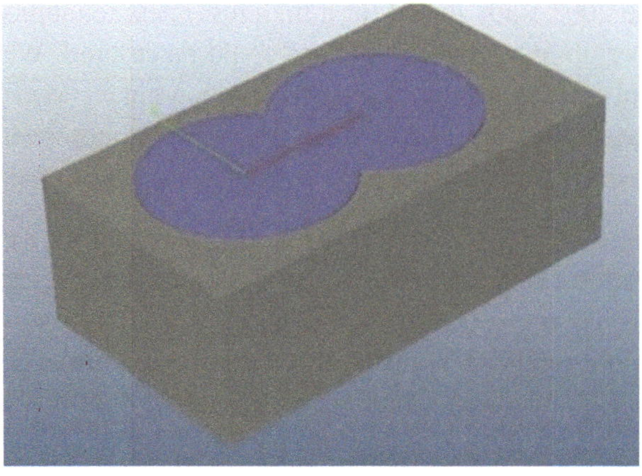

Figure 18.13
Envelopment: The ability of a support surface to conform to fit or mold around the irregularities of the test subject.

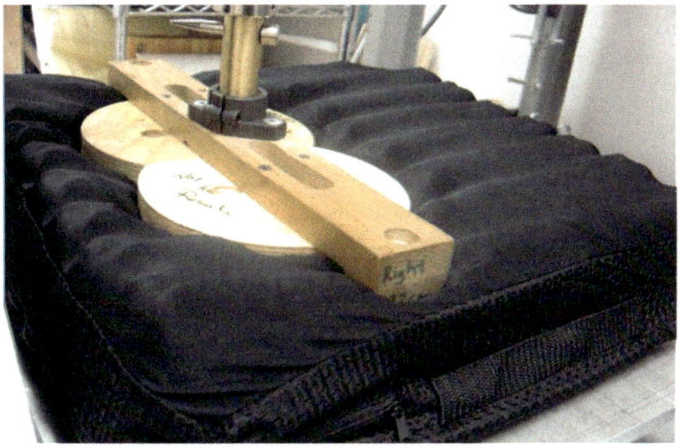

Figure 18.14
Envelopment test being conducted with a wooden dual buttock indenter.

Testing for both immersion and envelopment is conducted with specialized analog indenters for best effect, but the results need to be taken together before selecting a support surface for an end user.

Despite the development and standardization of new methods and vocabulary, heavy reliance on pressure mats as a primary means of measurement continued to be a problem, and further investigation revealed how. In 2008, a study led by Pipkin concluded that "introducing an IPM [interface pressure map] mat to the model–cushion interface alters the loading on the cushion." The modification of loading by the pressure mat is further complicated because "a significant interaction existed between cushion and mat condition and cushion and model for all variables." Put simply, the mats "bridged the contours of the model" and changed the pressure being applied. The study cautioned that "clinical and research users of pressure mats should consider the effect of mat presence, the effect of model design, and mat and buttocks interactions with cushions for successful use." [31].

A sensor-filled mat alters the environment it is meant to measure, impacting the ability to record true immersion and envelopment, which in turn confounds the test-retest reliability of readings. This presents a test environment with high variability and low validity. A highly effective alternative to pressure maps is the use of analog indenters fitted with sensors. Thus, the interference of materials that constitute the sensor mat would be completely removed from the testing environment, making the tests more reliable and guaranteeing higher test-retest accuracy.

Donati and team have pursued the use of a sensor mat technology composed of flexible matrices of optoelectronic sensors covered by a soft silicone cover [32]. The system is modular, scalable, and capable of measuring different pressure ranges. Their research directly supports the same developments that led to the development of new and better indenters such as the Sprigle indenter. While efforts along these lines directly address some of the drawbacks of using pressure maps, further testing and intralaboratory verification are necessary.

Donati's experimental tests, published in January 2013, explored human–robot interaction using a pressure-sensitive foot insole but consisted of validation on only three healthy human subjects. Yet, it must be acknowledged that these developments encourage a perspective for future application and continuing advancement of flexible sensor technology specifically and of pressure mapping technology in general.

Although the technology of pressure mats continues to develop, pressure mapping data are typically suspect in the industry. However, when it is combined with immersion data, pressure mapping can be strengthened and become a more valuable tool.

5. New standards: RESNA SS-1

5.1 SS-1:2014

The pursuit of solutions for ongoing problems in the testing of support surfaces, such as those problems presented by the high variability and low validity in the use of pressure mats, was already being addressed by NPUAP. The Support Standards committee that formed in 1999/2000 began producing documents with standards potential. The first of the documents was Support Surface Terms and Definitions, which was the initial step. The committee built on that document, including the following iterations detailed protocols and test methods for full-body support surfaces. They arrived at the document SS-1, protocols, and standards that can be readily adapted and applied to pressure testing in different ways.

The initial SS-1 document, published in 2014, began with a section dedicated to the standardization of vocabulary. Three more sections outlined testing protocols. The first protocol was for measuring immersion with full-body support surfaces. Two additional sections outlined separate methods for measuring HWV dissipation. In 2019,[c] SS-1 has expanded the total number of standard protocols to eight with additions made to the vocabulary section [33]. The three protocol sections from 2014 were reviewed and recertified for the 2019 version.

Despite having been expanded and updated, the 2014 SS-1 document should not be dismissed. In its four sections, SS-1:2014 set down a sure framework for all subsequent additions because it acted to standardize a suite of protocols for the testing of support surfaces which would be "capable of repeatedly, reliably, and accurately reporting upon characteristics common to all support surfaces believed to be related to the extrinsic risk factors associated with skin breakdown, as indicated by the literature to date." [34].

[c] Approved by RESNA on February 11, 2019 and by ANSI on February 15, 2019.

In keeping with practices of other standards bodies, such as ISO [35] and the European Pressure Ulcer Advisory Panel [36], the RESNA SS committee arrived at the development of SS-1 to provide requirements, specifications, guidelines, and characteristics that, if used correctly, would consistently ensure that materials, processes, and products would fit their proposed purpose. The committee's four guiding principles were clear:

1. Only validated test methods would be published.
2. Results would not be reduced to "good" or "bad" or "pass/fail."
3. No individual test would be sufficient to evaluate a surface.
4. Not all tests are required of all surfaces, based on applicability.

SS-1 is not a document for theory, but for practice; a publication that delineates tried and tested methods already in place, providing clear information with detailed schematics illustrating on how the methods are to be executed so that they can be applied repeatably and, if necessary, adapted to special test conditions.

Test methods are validated with multiple trials in at least three different labs, which are either independent or manufacturer labs. The validation process includes ensuring that the results from each lab are statistically consistent with the other test labs. Once lab results are shown to be consistent with other labs they are then considered eligible to be included in the standards document.

Through careful, broad research within the areas of wound care and support surface development and testing, the SS committee considered the following performance factors for standardization before releasing SS-1: terms and definitions, pressure distribution, immersion, envelopment, friction, shear forces, microclimate of temperature and humidity, life span, and safety.

5.2 Immersion

The first test protocol in is an immersion test that employs a rigid full-body mannequin pressed into the support surface where immersion occurs. The mannequin incorporates a series of three string potentiometer sensors to measure how far into the surface the mannequin immerses (see Figs. 18.15 and 18.16).

The immersion test has proved accurate to 1/10th of 1% and produced results that are valid across multiple testing facilities.

5.3 Microclimate

Given the increasing abundance of materials to choose from in the manufacture of a support surface, such as hundreds of different variations of foam based on combinations of elastic polyurethane foam, viscoelastic foam or highly elastic elastomers, open or closed

Figure 18.15
A 50th percentile male, full-body rigid mannequin indenter loaded for an immersion test. The pelvic area is weighted and laid on the support surface directly from above.

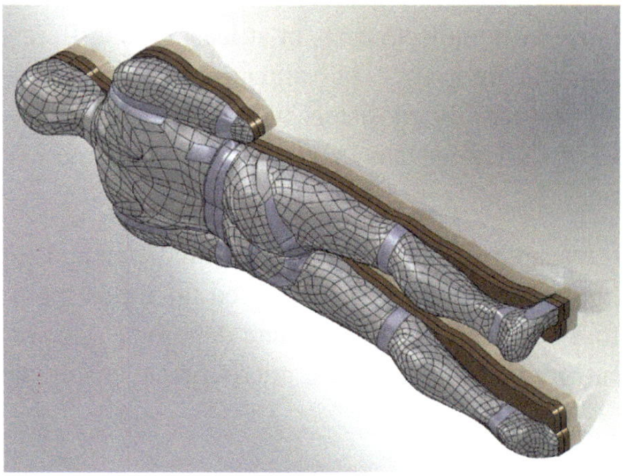

Figure 18.16
A digital representation of a male full-body rigid mannequin indenter as seen from below.

cell, small or large pore size, stiffness, various low or high flow air surfaces, viscous fluid, or some combination of these, the effects of variables involved with and caused by load on different support surfaces continue to evolve. As microclimate is a significant contributor in the development of pressure injuries/ulcers, and given that full-body support surfaces impede the natural flow of heat and humidity from skin to the environment, the interaction of microclimate with support surface needs closer scrutiny.

The significance of heat in stressing the skin and contributing to superficial pressure injury/ulcers is an ongoing concern and as such continues to be examined [37]. The interaction of moisture with skin is highly significant [38]. When skin is kept moist or in a high humidity environment, it loses its cohesive strength and increases in elasticity.

Dermal excoriation and abrasion occur in response to friction and shear when the skin is in this moistened state.

As the human body perspires, moisture is present wherever human and support surface meet. A full-body support surface, such as a mattress in a hospital, could be made of various kinds of materials that in turn are typically covered with a water-proof or water-resistant cover. Both cover and mattress affect the movement, or lack of movement, of air and the degree of evaporation or transmission of moisture from patient to support surface. In most cases, air does not flow over the patient and therefore most movement of air and moisture takes place under the mattress cover [39].

Given the variables of patient, support surface, cover, and microclimate, the complexity of such circumstances need careful testing. SS-1:2019 presents three protocols for testing microclimate, or HWV, dissipation characteristics in full-body support surfaces: (a) the BA method, (b) the sweating guarded hot plate method, and (c) the heated bladder method. The results that come from individual tests should not be taken as an overall result of surface performance but should be placed in fuller context with additional tests to create a more informed understanding of the support surfaces in question.

5.3.1 Body analog method

The first protocol is for the BA method, which was included in 2014 and recertified for 2019. This test measures the percent of relative humidity at the body/support surface interface and reports the temperature and relative humidity over time between the test indenter and the support surface (see Fig. 18.17).

The BA protocol sets down a means to construct a TRCLI and to reliably and repeatedly measure the HWV transfer characteristics of support surfaces by examining (a) material properties at the interface with skin, referred to as localized heat and localized moisture characteristics and (b) the bulk capacity of the support surface to absorb and retain HWV, referred to as bulk heat capacity and the bulk moisture capacity.

Because the characteristics of HWV dissipation are measured in a steady-state condition, the BA method is perfect for discovering and maintaining a specific relative microclimate at the body/support surface location (see Fig. 18.18).

The HWV test is unique because it applies a BA indenter to the surface being tested. This is the only test available, which challenges a support surface with the body-like displacement and compression as that seen when loaded by a human. This provides the closest realistic conditions and properly represents the insulation value and the thermal conduction characteristics of the surface when in use by a patient. This test is also the only test that properly loads and compresses viscoelastic products in the test environment or compresses the air structures of inflated and powered support surfaces.

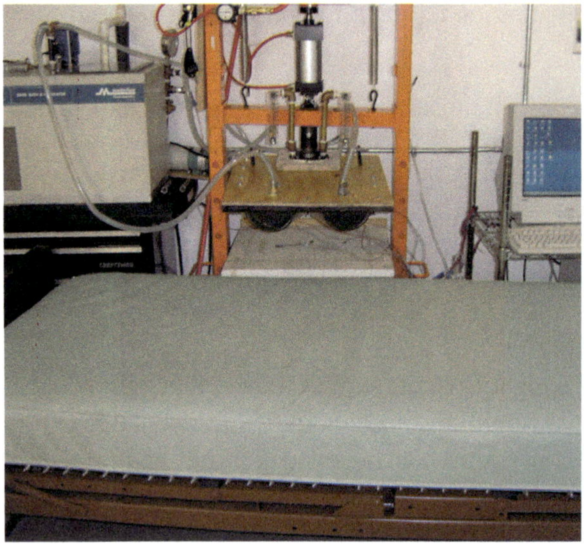

Figure 18.17
The thermodynamic rigid cushion loading indenter (TRCLI) ready for testing on a hospital bed.

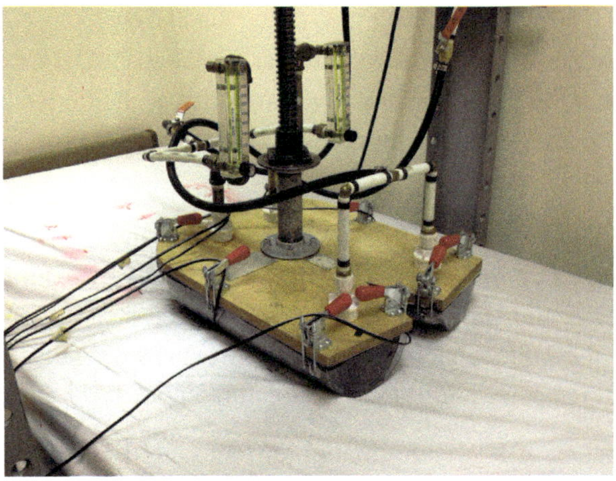

Figure 18.18
The thermodynamic rigid cushion loading indenter (TRCLI) under simulated load being used to test heat and water vapor dissipation in a full support surface.

5.3.2 Sweating guarded hotplate

A second HWV dissipation protocol, which was included in SS-1:2014 and recertified for 2019, uses a SGHP protocol. The driving force for the development of this method was the need for engineering values for heat and moisture transfer in the surfaces tested, representing the heat that the surface can carry away from the body and the moisture that can be drawn away from the same space. This method, modified by Dr. Charlie Lachenbruch and Chris Wyel, uses materials from sporting goods and safety clothing such as protective fire gear. It inverts the clothing test model and alters the moisture delivery process. The design was specifically undertaken to resolve reliability issues that had existed in previous test processes and includes multiple redundant temperature and heat flux sensors as well as a logging system to record the performance of the surface over the course of the test.

SGHP measures HWV dissipating properties of full-body support surfaces under conditions that simulate load with flat and articulated profiles. The test has two major components: (a) dry heat flux and (b) evaporative heat flux, also known as wet flux.

Dry heat flux tests how much heat is removed by the support surface and cover when moisture is not present. Evaporative heat flux can be used to calculate the evaporative capacity of the surface. Because the indenter for these tests is small, it can be readily moved to different locations on the support surface (see Figs. 18.19 and 18.20).

Most importantly, this test has the ability to measure heat removed in a dry condition, and therefore this measures moisture removal over time.

The SGHP test is particularly beneficial because it can be readily used across various support surfaces with multiple types of material.

Figure 18.19
Hot plate used in a sweating guarded hot plate protocol.

Figure 18.20
Hot plate with weight.

5.3.3 Heated bladder method

Testing for temperature and relative humidity at a specified location had begun to fill in an information gap [40], but as often occurs, the process of study and research revealed the need for further nuance within the field. A third microclimate protocol added in 2019 to SS-1 is a test for measuring water vapor dissipation characteristics through the use of a heated water bladder (HWB) (see Fig. 18.21).

Developed first by a team led by Stephen Reger [41], and modified by Richard Figliola [42], and then further modified for SS-1 [43], the HWB protocol test studies how much moisture a

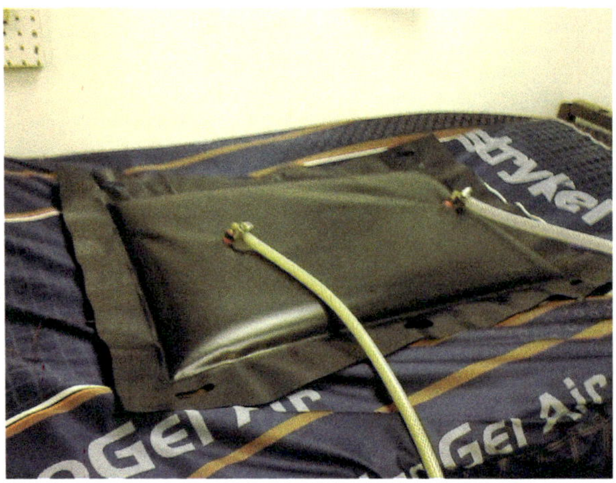

Figure 18.21
A heated water bladder being readied for a test. The bladder is not yet weighted.

support surface can remove over time, which is of special importance given the resistance to the natural flow of humidity around the skin when a patient is seated or lying down.

5.4 Shear force distribution: sliding resistance/horizontal stiffness

Alongside protocols incorporating microclimate is the issue of shear force distribution. A primary cause of stress for soft tissue is the force of material on the surface of a support surface pushing back against the soft tissue of a body. As the body shifts laterally and in parallel across the surface, such movement causes the support surface materials to deform and exert force in return, much like a spring.

The support surface horizontal stiffness test (HS) is used to determine the measurement of HS or the "bulk modulus," i.e., how resistant to compression while sliding the support surface is. The protocol for HS in SS-1:2019 specifies a BA pelvic indenter that is weighted (see Figs. 18.22A and B).

This method focuses on the pushback force against an indenter being dragged across the support surface (see Fig. 18.23).

A very stiff surface exerts a high resisting force on the indenter in response to a given amount of displacement while, as would be expected, a less stiff surface exerts less force.

(A) (B)

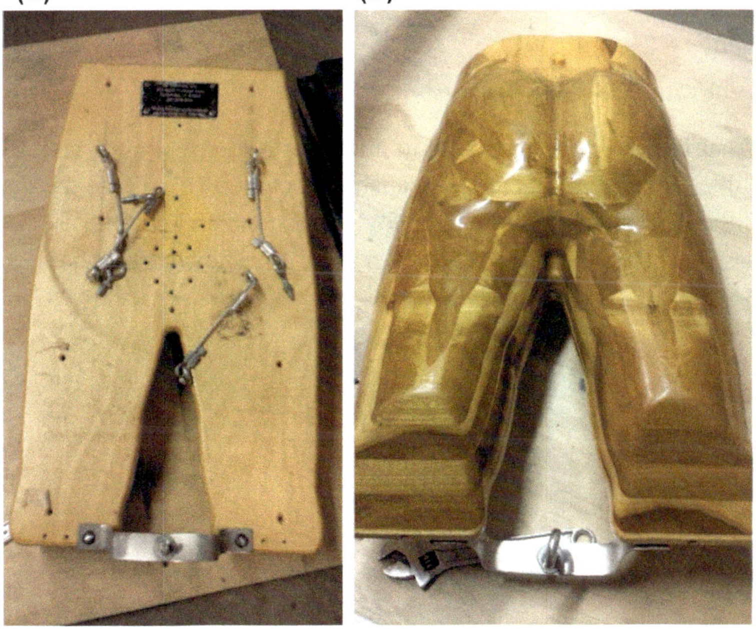

Figure 18.22
(A) Front of a body analog pelvic indenter, without weights, used to test bulk modulus, horizontal stiffness. (B): The back of the same indenter.

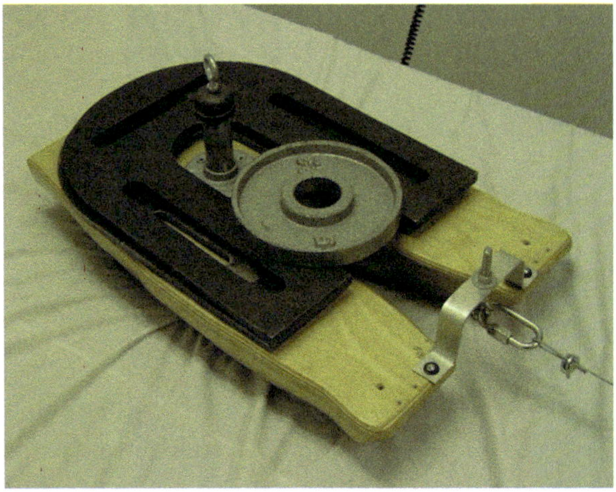

Figure 18.23
A weighted pelvic indenter being pulled across a support surface.

If the HS is high, a person would be less likely to slide toward the foot of the bed when the head of the bed is raised.

This test does not measure shear forces related to the resistance of the support surface, but it does measure the degree to which surface materials will resist or apply sustained force to the tissue of a patient moving laterally across the surface.

5.5 Envelopment

5.5.1 Envelopment with dual hemispheres

When a support surface is reactive enough to allow for a more intimate contact with the surface, the double hemisphere indenter is capable of gathering data for the distribution of pressure across a dual buttock, including analog trochanters and the crevasse between the two hemispheres (see Fig. 18.24). This allows a more significant challenge in

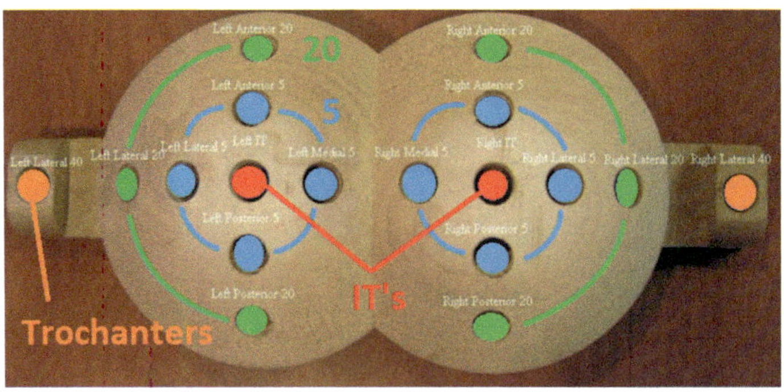

Figure 18.24
Wooden double hemisphere instrumented indenter pressure distribution rig.

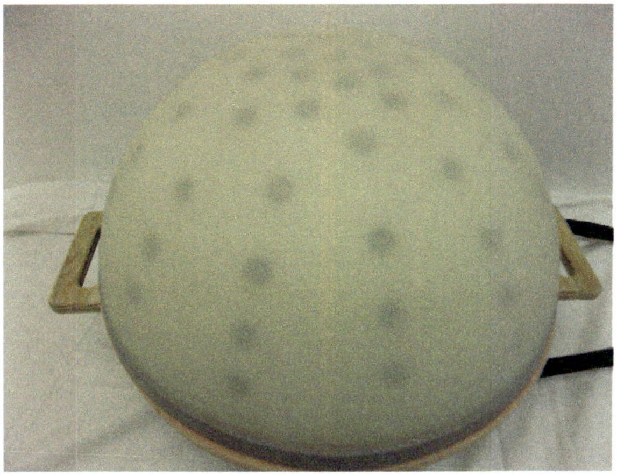

Figure 18.25

Hemispherical indenter with sensors and gel covering, bottom view. Reflects the curvature of the 50th percentile male buttocks.

characterization of the performance of a given support surface in relation to the human analogue. This challenge is based in the surface's ability to apply load to the sensors that are in the crevasse or adjacent to the trochanters.

This protocol does not test for microclimate and as such needs to take into account tests that are done to establish a fuller understanding of a support surface/human interaction.

5.5.2 Envelopment with 50.8 cm-diameter indenter

The envelopment protocol incorporates pressure sensors in a hemispherical indenter to measure the ability of a support surface to envelop or cradle bony prominences and other irregularities of the body as the support surface distributes pressure (see Figs. 18.25 and 18.26). This protocol does not incorporate testing for microclimate.

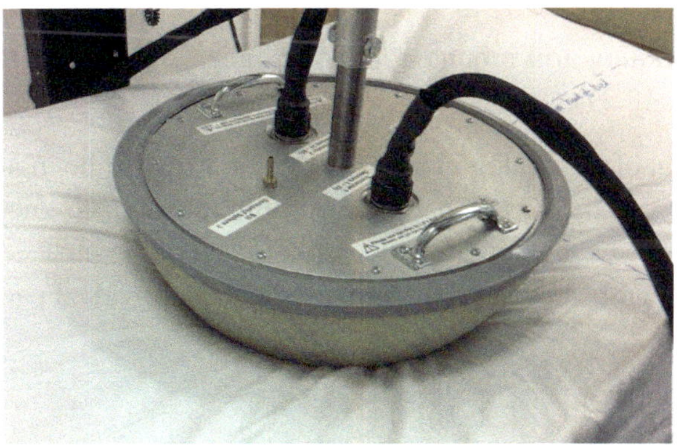

Figure 18.26

Hemispherical indenter with sensors and gel covering, top view, attached to Instron for load application.

The test begins by first immersing the indenter into the support surface and then recording envelopment.

6. Future protocols for SS-1

As SS-1 was expanded from 2014 to 2019, so will it continue to expand in the future. The following are tests and studies being conducted, which have the potential of becoming new protocols for future SS-1 updates.

6.1 Bed edge ingress/egress

An important study being considered for future SS-1 inclusion is the bed edge ingress/egress test (BEIE). Currently, there is an urgent need for trials of hospital fall prevention. "Patient falls are the most common adverse events reported in hospitals. Each year, roughly 700,000 to 1 million patient falls occur in the US hospitals, resulting in around 250,000 injuries and up to 11,000 deaths." [44] Falls that do not result in severe injury often result in minor injury, reduced quality of life, and liability to the facility.

One element of reducing incidents of bedside falls lies in identifying which beds and mattresses provide the lowest risk during ingress/egress. Determining such information allows for greater patient safety and for improved facility-based risk management.

Patients entering and exiting a bed may be exposed to either ingress (entering the bed) or egress (exiting the bed) forces due to compression of the mattress edge or bolster. Some beds/mattress combinations predispose a person to undesirable forces during egress, effectively pushing them forward out of the bed, which can readily contribute to falls. Yet during ingress, some bed/mattress combinations provide stability in the bed edge and thus provide neutral or positive ingress forces, in some cases even predisposing the occupant to fall into the bed rather than out on to the floor, thus providing an obvious safety advantage.

The BEIE test uses a falling arm mounted indenter under load to quantify both ingress and egress forces (see Fig. 18.27).

The floor-to-knee height of the analog indenter is kept constant while ingress/egress forces are recorded at various support surface heights. This allows a floor-to-knee height to bed height ratio to be generated, which will provide an estimate of the ingress/egress forces. From these data, a safe bed height can be established based on an individual patient's height.

6.2 Bed edge fall off

Directly related to BEIE is the bed edge fall off study (BEFO). BEFO occurs when patients lying in bed repositions to the bed edge, either in preparation for egress or as a

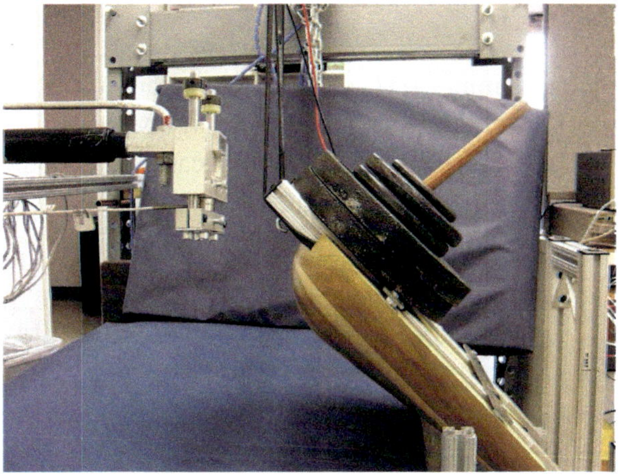

Figure 18.27
A falling arm mounted indenter under load set at bedside for testing ingress/egress.

result of sleep- or pain-related movement. Such repositioning of the patient may collapse the bed edge and subject the patient to bed ejection or entrapment.

The BEFO test characterizes a mattress's ability to support a patient by positioning a mannequin at bed edge and pushing the mannequin's legs off the edge as would be observed in preparation for sitting up on the edge of the bed (see Fig. 18.28). The force required to push the mannequin from the bed and the distance traveled before a fall are recorded and used to generate a safety profile for the test mattress.

Given the high number of ingress/egress-related falls in hospitals, the BEIE and the BEFO tests are of paramount importance to both hospitals and patients.

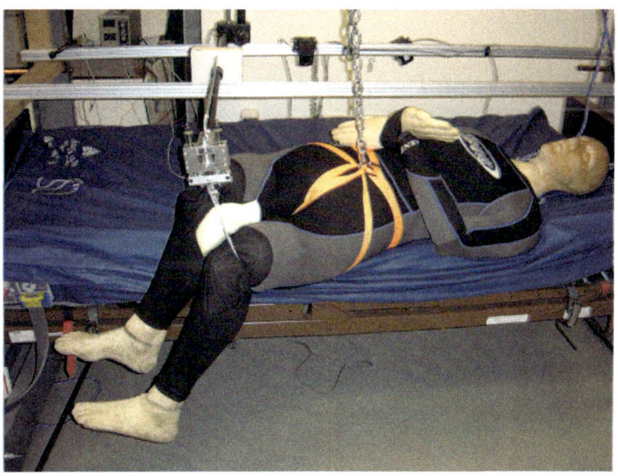

Figure 18.28
Bed edge fall off test performed using a mannequin.

6.3 Deep tissue injury

While the study of deep tissue injury (DTI) has a long history, its relationship to pressure injury in connection with support surfaces is only recently begun to be thoroughly explored [45–48]. "DTI can be difficult to diagnose because many other skin and wound problems can also appear as purple skin or rapidly appearing eschar" [49].

Pressure wounds occurring in the outermost layers of skin are affected by the support surface first and further complicated by the presence of microclimate, friction, and shear. DTI, on the other hand, is caused by prolonged load inducing deep tissue deformation, which interferes with flow of blood and clearance of metabolic by-products in vascularized soft tissues resulting in a pressure-related tissue necrosis, which can be further complicated by bony prominences [50].

DTI can develop under intact skin and often shows no signs of serious injury in the early stages. External causes are pressure and tissue deformation over bony prominence such as ischial tuberosities. Internal problems are exacerbated by immobility because patients are not able to relieve mechanical loads on soft tissue.

Skin color itself is a problematic indicator of pressure ulcer/injuries, especially DTI, given the variability of a human subject's natural skin color. A differential diagnosis with a medical history is needed to properly diagnose DTI, which makes the creation of protocols to study the effects of support surfaces on the development of DTI even more complicated.

Like pressure injuries/ulcers on the external skin, DTI may be addressed by turning the patient off the at-risk tissues and redistribution of contact pressures through use of support surfaces [51]. However, predicting risk for DTI using body-support pressure measurements alone remains problematic [52].

6.4 Linens, incontinence pads, and wound dressings

Pressure mats drew attention to the fact that additional materials present between a human subject and support surface, such as hospital gowns and night clothing, could adversely affect patient health. To date, the effects of linens, wound dressings, and similar materials on the creation of pressure injuries/ulcers need to be more thoroughly studied to create new standards in wound prevention and care.

Management of wound dressings is itself complicated. "The wound bed must remain moist to promote healing, but care must be taken to remove excess fluid to avoid maceration and subsequent breakdown of the periwound site." [53] While the moisture permeability of dressings has been studied, the wound treatment environment as a whole, i.e., support surface–microclimate–patient interface, needs further research and testing so that standards and protocols can be developed, which extend beyond support surfaces to include wound dressings.

Warp and weft of fabric add further variables and are hard to predict, which generally makes the presence of linens on bed surface adversely affect the friction and pressure-redistribution capabilities of the surface on which they are placed. Fundamentally, less material is better for the patient, but further testing is necessary because the introduction of any new variables, especially materials as complex as fabrics, makes an already involved patient–microclimate–support surface interface that is much more complicated to study.

For example, microclimate alone grows increasingly complicated based just on the type of material being used. Even a relatively thin, breathable incontinence pad can increase interface pressure over a fitted sheet as much as 56% with each additional layer providing additional damping of the pressure-redistribution properties of the surface [54].

6.5 Support surfaces

New support surfaces are being developed, which actively regulate the microclimate of the skin. Engineers throughout the industry have currently produced a line of surfaces that precisely regulates airflow and air temperature allowing the surface to maintain a more optimum microclimate [55]. While initial tests are promising, such surfaces need further testing across multiple labs to verify results and enhance methodology.

6.6 And beyond

Studies and research into areas such as BEIE, BEFO, deep tissue studies, and wound dressings are only a few examples from among many currently being explored in the field of wound care and the prevention of pressure injuries/ulcers.

SS-1 continues to develop new protocols and standards, working in such areas as product life, disinfection, and alternating pressure versus low air loss support surfaces. Additional studies currently under development include

- Bottoming out
- Weight capacity
- Bulk modulus or sliding resistance with head of bed elevated
- Influence of bed frame on load-bearing characteristics
- Air-powered and air-fluidized beds

Standardization needs a performance-based set of criteria for protocols and testing. "With respect to the three purposes identified: comfort, postural control and pressure management, only parameters that can be measured noninvasively, have been considered.

The nine parameters are life expectancy of the surface, skin moisture control, skin temperature control, redistribution of pressure, product service requirements, fail safety, infection, flammability, and patient/product friction." [56].

But for information to be better understood and more readily absorbed as it is disseminated, studies and protocols need clear, consistent language. A continuing goal of the SS committee is to make the SS-1 a living document with updated reference lists online and ongoing incorporation of new and relevant terms and definitions that continue to evolve as the study and standards of wound care and management evolve. Such a document would serve as a ready lexicon to deliver clarity and understanding in the evaluation and development of testing protocols and the use of support surfaces.

One of the pressing areas that remain unaddressed by Standards is the area of Prevention Dressings. It is now common for clinicians to apply wound dressings to persons at risk for pressure injury [59]. Bench [60] and Modeling Research [61] are beginning to identify the protective mechanisms that this practice provides to the bed-bound patient. There are a number of test methods that have been employed [60] in this characterization, and these would be the obvious first considerations for potential standards. Tests that characterize the dressings ability to prevent tissue distortion, absorb shear, when the dressing is worn and no longer providing prophylactic protection and match the anisotropic nature of human tissues will be particularly valuable in providing data to clinicians, which are relevant to purchase decisions that ultimately protect the bed-bound patient.

7. Conclusion

The road to creating standards takes time as the questions evolve, reflecting an increase in understanding as the problems within the fields of wound care and prevention continue to themselves evolve. Progress is slow but steady. It took 20 years since S3I was found to reach SS-1:2019, but the research used to create SS-1 goes back decades further.

Despite the investment of time, resources, and extensive training, the occurrence of pressure injuries/ulcers has not declined as much as has been hoped [57]. More research into understanding the causation of pressure injuries/ulcers—pressure distribution, shear, and microclimate—needs to continue.

An ongoing problem in the prevention of pressure injuries/ulcers is the fact that the degree of pressure that leads to the development of a pressure injury/ulcer is hard to determine. Even as research is constantly evolving, the identification of a specific threshold where pressure and load become harmful is impossible to generalize because any given tissue's tolerance under load incorporates a great many variables even with a single subject, such as the location of tissue, its age, degree of hydration, and metabolism. Nor do these variables remain consistent from human subject to human subject.

We cannot use interface pressures alone to make definitive decisions on support surface effectiveness, but we can make informed, relative comparisons.

Discovering what constitutes a good support surface or a bad support surface for an individual to use is not a distinction that can easily be made, nor should it be thought of in those terms because of the relativity of such a conclusion. Even for informed, educated professionals in the fields of health and well-being, there are too many factors involved to be so definitive.

People themselves vary too much. Using a greater variety in the physiology of human test subjects and the scenarios in which they use a support surface may help improve the reliability and validity of the data set results, but such complexity of factors in testing inherently leads to a greater complication and complexity in the results, expanding choices for professionals rather than refining and narrowing them down to a manageable set that would more readily aid in the choice of a support surface for a given patient.

However, "the goal of standards development is not to create tests that can be used to prescribe and support surfaces; rather, its aim is to define test methods that reflect cushion performance." [58] Therefore, standards development should not be undertaken to provide a definitive end user choice but rather to create a more reliable, more valid understanding of support surface performance to aid the clinician and patient make better choices.

While the use of guidelines from any of the standards bodies, such as ISO, NPUAP, ANSI, and RESNA, is completely voluntary, the use of standardized methods, protocols, and vocabulary will be a benefit for all parties in every stage, from researcher to patients. The continuing development and use of standards will help narrow the information gap even as it opens the door to innovation.

Guidelines and standards mean that manufacturers will be better able to develop new products or redesign existing ones. Vendors will be better able to describe and compare products from a multitude of manufacturers. Clinicians will have a more accurate, more readily understandable body of knowledge to draw on when matching patients with support surfaces, which can in turn better fulfill a patient's requirements. Active participation across all these areas is essential. High test-retest reliability and intralaboratory repeatability are essential. Tests should not be generalized beyond the exact conditions in which they were conducted.

References

[1] Pieper B. In: Bryant RA, Nix DP, editors. "Mechanical forces: pressure, shear, and friction." Acute & chronic wounds: current management concepts. 3rd ed. St. Louis (MO): Mosby, Inc.; 2007. p218.
[2] Pieper B. In: Bryant RA, Nix DP, editors. "Mechanical forces: pressure, shear, and friction." Acute & chronic wounds: current management concepts. 3rd ed. St. Louis (MO): Mosby, Inc.; 2007. p209.

[3] Ceelen KK, Stekelenburg A, Loerakker S, Strijkers GJ, Bader DL, Nicolay K, Baaijens FPT, Oomens CWJ. Compression-induced damage and internal tissue strains are related. J Biomech 2008;41(16):3399–404.

[4] Support Surface Standards Initiative (S3I). Available from: https://www.npuap.org/resources/educational-and-clinical-resources/support-surface-standards-initiative-s3i/.

[5] National Standardization. ANSI. Available from: https://www.ansi.org/standards_activities/overview/overview?menuid=3.

[6] National Standarization. ANSI. Available from: https://www.ansi.org/about_ansi/introduction/introduction?menuid=1.

[7] For a fuller description of ANSI programs and involvement with international bodies, see www.ansi.org.

[8] Our Story. ISO. Available from: https://www.iso.org/about-us.html.

[9] Terms and Definitions Related to Support Surfaces. NPUAP. Available from: https://www.npuap.org/wp-content/uploads/2012/01/Terms-and-Defs-Feb-5-2019-FINAL-3.12-2.pdf.

[10] Sprigle S, Chung K, Brubaker C. Factors affecting seat contour characteristics. J Rehabil Res Dev 1990;27(2):127–34.

[11] Crenshaw RP, Vistnes LMA. A decade of pressure sore research. J Rehabil Res Dev 1989;26:63–74.

[12] Bennett L, Kanner D, Lee BK, Trainor FA. Shear vs pressure as causative factors in skin blood flow occlusion. Arch Phys Med Rehabil 1979;60:309–14.

[13] Chow WW. Mechanical porperties of gels and other materials with respect to their use in pads transmitting forces to the human body. University of Michigan; 1974. Ph.D. Diss.

[14] Kadaba MD, Ferguson-Pell MW, Palmieri V, Cochran GVD. Ultrasound mapping of the buttock-cushion interface contour. Arch Phys Med Rehabil 1984;65:467–8.

[15] Sprigle S, Dunlop W, Press L. Reliability of bench tests of interface pressure. Asstist Technol 2003;15(1):49–57. RESNA. P51.

[16] Update Terms and definitions related to support surfaces. NPUAP; December 27, 2018. https://www.npuap.org/wp-content/uploads/2012/01/Terms-and-Defs-Feb-5-2019-FINAL-3.12-2.pdf.

[17] Ferguson-Pell M, Hideyuki H, Nicholson G, Call E. Thermodynamic rigid cushion loading indenter: a buttock-shaped temperature and humidity measurement system for cushioning surfaces under anatomical compression conditions. J Rehabil Res Dev 2009;46(7):945–56. p945.

[18] Staarink HA. Sitting posture, comfort and pressure: assessing the quality of wheelchair cushions. Delft (the Netherlands): Delft University Press; 1995.

[19] Ferguson-Pell M, Hideyuki H, Nicholson G, Call E. Thermodynamic rigid cushion loading indenter: a buttock-shaped temperature and humidity measurement system for cushioning surfaces under anatomical compression conditions. J Rehabil Res Dev 2009;46(7):945–56. p945.

[20] Ferguson-Pell M, Cardi MD. Prototype development and comparative evaluation of wheelchair pressure mapping system. Assist Technol 1993;5(2):78–91. RESNA.

[21] Sprigle S, Dunlop W, Press L. Reliability of bench tests of interface pressure. Assist Technol Summer 2003;15(1):49–57.

[22] Sprigle S, Dunlop W, Press L. Reliability of bench tests of interface pressure. Assist Technol Summer 2003;15(1):49–57. p50.

[23] Bhattacharya S, Mishra R. Pressure ulcers: current understanding and newer modalities of treatment. Indian J Plast Surg January–April 2015;48(1):4–16.

[24] Gefen A. Reswick and Rogers pressure-time curve for pressure ulcer risk. Part 2. Nurs Stand July 22–28, 2009;23(46):40–4.

[25] Hsu MY, Lin JP, Lyu JY. Medical device related pressure injury prevention strategies. Hu Li Za Zhi June 2019;66(3):106–11.

[26] Edsberg LE, Black JM, Goldberg M, McNichol L, Moore L, Sieggreen M. Revised national pressure ulcer advisory Panel pressure injury staging system: revised pressure injury staging system. J Wound Ostomy Cont Nurs November/December 2016;43(6):585–97.

[27] Sprigle S, Dunlop W, Press L. Reliability of bench tests of interface pressure. Assist Technol Summer 2003;15(1):49–57. p55.
[28] Davis K, Call E. Evidence-based guidelines for interface pressure mapping: final research findings. In: PowerPoint presentation. 2013 International seating symposium. Nashville, TN; March 7, 2013.
[29] "Support Surface Standards Initiative (S3I)." NPUAP. Available from: https://www.npuap.org/resources/educational-and-clinical-resources/support-surface-standards-initiative-s3i/.
[30] "Support Surface Standards Initiative (S3I)." NPUAP. Available from: https://www.npuap.org/resources/educational-and-clinical-resources/support-surface-standards-initiative-s3i/.
[31] Pipkin L, Sprigle S. Effect of model design, cushion construction, and interface pressure mats on interface pressure and immersion. J Rehabil Res Dev 2008;45(6):875–82. p875.
[32] Donati M, et al. A modular sensorized mat for monitoring infant posture. Sensors (Basel). January 13, 2014;(1):1021–45.
[33] "RESNA Standards Committee on Support Surfaces (SS)." RESNA: Rehabilitation Engineering and Assistive Technology Society of North America. Available from https://www.resna.org/resna-standards-committee-support-surfaces-ss.
[34] Stone A, Brienza D, Call E, Fontaine R, Goldberg M, Hong K, Jordan R, Lachenbruch C, LaFleche P, Sylvia C. Standardizing support surface testing and reporting: a national pressure ulcer advisory panel executive summary. J Wound Ostomy Cont Nurs September 2015;42(5):445–9. NPUAP.
[35] "Standards." International Standards Organization. Available from: http://www.iso.org/iso/home/standards.htm.
[36] NPUAP and European Pressure Ulcer Advisory Panel. Prevention and treatment of pressure ulcers: clinical practice guideline. Washington, DC: NPUAP; 2009.
[37] Kokate J, Leland K, Held A, Hansen G, Graig K, Johnson B, Wilke M, Sparrow E, Iaizzo P. Temperature-modulated pressure injurie/ulcers: a porcine model. Arch Phys Med Rehabil July 1995;76:666–73.
[38] Phipps L, Gray M, Call E. Single exposure to clinically relevant levels of moisture contained by an incontinence pad compromises skin healthy. Hill-Rom; 2009.
[39] Jordan R, Sylvia C. What's in a name? terms, definitions & performance evaluation in S3I. Powerpoint Present. 2019 National Pressure Ulcer Advisory Panel.
[40] Lachenbruch C, Tzen Y, Brienza D, Karg P, Lachenbruch P. Relative contributions of interface pressure, shear stress, and temperature on ischemic-induced, skin-reactive hyperemia in healthy volunteers: a repeated measures laboratory study. Ostomy Wound Manag 2015;61(2):16–25.
[41] Reger S, Adams T, Maklebust J, Sahgal V. Validation test for climate control on air-loss supports. Arch Phys Med Rehabil 2001;82(5):597–605.
[42] Figliola R. A proposed method for quantifying low-air-loss mattress performance by moisture transport. Ostomy Wound Manag January 2003;49(1):32–42.
[43] Reger S, Adams T, Maklebust J, Sahgal V. Validation test for climate control on air-loss supports. Arch Phys Med Rehabil May 2001;82(5). 597-203.
[44] LeLaurin J, Shorr R. Preventing falls in hospitalized patients. Glin Geriatr Med May 2019;35(2):273–83. p273.
[45] Gefen A, Gefen N, Linder-Ganz E, Margulies S. In vivo muscle stiffening under bone compression promotes deep pressure sores. J Biomech Eng 2005;127:512–24.
[46] Gefen A. Reswick and Rogers pressure-time curve for pressure ulcer risk. Part 2. Nurs Stand 2009;23(46):40–4.
[47] Oomens C, Bader D, Loerakker S, Baaijens. Pressure induced deep tissue injury explained. Ann Biomed Eng 2015;43(2):297–305.
[48] Call E, Jones R, DeMonja K, Burton J, Jellum S, Bernkopf A, Oberg C. 'The elastic limit': introducing a novel concept tin communicating excessive shear and tissue deformation. WCET J 2017;37(4): 16–20.

[49] Black J, Brindle C, Honaker J. Differential diagnosis of suspected deep tissue injury. Int Wound J 2016;13:531–9. https://doi.org/10.1111/iwj.12471.

[50] Gefen A. Plantar soft tissue loading under the medial metatarsals in the standing diabetic foot. Med Eng Phys 2003;(25):491–9. p491.

[51] Gefen A. Plantar soft tissue loading under the medial metatarsals in the standing diabetic foot. Med Eng Phys 2003;(25):491–9. p498.

[52] Gefen A, Levine J. The false premise in measuring body-support interface pressures for preventing serious pressure ulcers. J Med Eng Technol 2007;31:375–80.

[53] Lachenbruch C, VanGilder C. Estimates of evaporation rates from wounds for various dressings/support surface combinations. Adv Skin Wound Care 2012;25(1):29–34. p34.

[54] Williamson R, Lachenbruch C, VanGilder C. The effect of multiple layers of linens on surface interface pressure: results of a laboratory study. Ostomy Wound Manag 2013;59(6):38–47. p47.

[55] Lachenbruch C. Microclimate management: so much more than just airflow.... Hill-Rom; 2008.

[56] Krouskop T, Van Rijswijk L. Standardizing performance-based criteria for support surfaces. Ostomy Wound Manag 1995;41(1):34–6. 38, 40-5.

[57] Dealy C, Brindle CT, Black J, Alves P, Santamaria N, Call E, Clark M. Challenges in pressure ulcer prevention. Int Wound J 2013. https://doi.org/10.1111/iwj.12107.

[58] Sprigle S, Dunlop W, Press L. Reliability of bench tests of interface pressure. RESNA Assist Technol 2003;15(1):49–57. p51.

[59] Santamaria N, Gerdtz M, Sage S, McCann J, Freeman A, Vassilou T, DeVincentis S, Wei A, Manias E, Liu W, Knott J. A randomised controlled trial of the effectiveness of soft silicone foam multi-layer dressings in the prevention of sacral and heel pressure ulcers in trauma and critically ill patients: the border trial. Int Wound J 2013. https://doi.org/10.1111/iwj.12101.

[60] Call E, Pedersen J, Bill B, Black J, Alves P, Brindle CT, Dealey C, Santamaria N, Clark M. Enhancing pressure ulcer prevention using wound dressings: what are the modes of action? Int Wound J 2013. https://doi.org/10.1111/iwj.12123.

[61] Levy A, Gefen A. Assessment of the biomechanical effects of prophylactic sacral dressings on tissue loads: a computational modeling analysis. Ostomy Wound Manag 2017;63(10):48–55.

Index

Note: 'Page numbers followed by "*f*" indicate figures and "*t*" indicate tables'.

T

9780128150283